AF696320

Osseointegration in Skeletal Reconstruction and Joint Replacement

Library of Congress Cataloging-In-Publication Data

International Workshop on Osseointegration in Skeletal Reconstruction and Joint Replacement (2nd : 1994 : Rancho Santa Fe, Calif.)
Osseointegration in skeletal reconstruction and joint replacement: Second International Workshop on Osseointegration in Skeletal Reconstruction and Joint Replacement, Rancho Santa Fe, California, October 27–29, 1994 / editors, Per-Ingvar Brånemark, Björn L. Rydevik, Richard Skalak.
p. cm.
Includes bibliographical references.
ISBN 0-86715-325-3
1. Osseointegration—Congresses. 2. Tissue-integrated prostheses—Congresses. I. Brånemark, Per-Ingvar. II. Rydevik, Björn. III. Skalak, Richard. IV. Title.
[DNLM: 1. Osseointegration—congresses. 2. Bone and Bones—surgery—congresses. 3. Arthroplasty—congresses. WE 200 I64o 1997].
RD684.I56 1994
617.4'710592—dc21
DNLM/DLC
for Library of Congress 96-51486
CIP

quintessence books

Published by Quintessence Publishing Co, Inc
551 N. Kimberly Dr.
Carol Stream, IL 60188

Editor: Patricia Bereck Weikersheimer
Assistant Editor: Lane Evensen
Production Manager: Timothy M. Robbins
Cover design: Lisa Ream

Printed in the USA

Osseointegration in Skeletal Reconstruction and Joint Replacement

SECOND INTERNATIONAL WORKSHOP ON OSSEOINTEGRATION IN SKELETAL RECONSTRUCTION AND JOINT REPLACEMENT

RANCHO SANTA FE, CALIFORNIA, OCTOBER 27–29, 1994

EDITORS

Per-Ingvar Brånemark
Professor and Director
The Institute for Applied Biotechnology
Gothenburg, Sweden

Björn L. Rydevik, MD, PhD
Professor and Chairman
Department of Orthopedics
Sahlgren's University Hospital
University of Gothenburg
Gothenburg, Sweden

Richard Skalak, PhD, MD (hon)
Professor
Department of Bioengineering
University of California, San Diego
La Jolla, California

quintessence books

Quintessence Publishing Co, Inc
Chicago, Berlin, London, Tokyo, São Paulo, Moscow, Prague, and Warsaw

Contents

Contributors vii

Introduction
Shu Chien ix

Definition of Osseointegration
Richard Skalak and Per-Ingvar Brånemark xi

1 Osseointegration: Biotechnological Perspective and Clinical Modality
Per-Ingvar Brånemark 1

2 History, Development, and Current Status of Osseointegration as Revealed by Experience in Craniomaxillofacial Surgery
Philip Worthington 25

3 Biomechanics of Osseointegration
Richard Skalak 45

4 Modeling of Ingrowth into Implant Cavities Along the Bone-Implant Interface
Stephen C. Cowin and Gangming Luo 57

5 Matrix Molecules in Connective Tissues with a Potential for Use in Guiding Osseointegration
Dick Heinegård, Björn Klinge, and Mikael Wendel 73

6 The Impact of Bone Marrow Biology on Skeletal Reconstruction and Osseointegration
Dennis G. Osmond and M. Dwayne Rahal 79

7 Skeletal Mechanics as Related to Bone Mineral Content and Bone Remodeling
Tommy Hansson 91

8 Osseointegration and Mechanical Capacity in Bone Tissue of Varying Quality
Rickard Brånemark 101

9 Tumor Resection and Skeletal Reconstruction in the Extremities
Björn Gunterberg 111

10 Tumors of the Spine
James N. Weinstein and Stefano Boriani 121

11 Spine Reconstruction, Stabilization, and Fusion
Steven R. Garfin 143

12 The Interfaces Between Humans and Limb Replacement Components
Dudley S. Childress 157

13 Osseointegrated Metacarpophalangeal Joint Prostheses in the Hand
Göran Lundborg 167

14 Amputation Prostheses and Osseoperception in the Lower and Upper Extremity
Björn L. Rydevik 175

15 Soft-Tissue Management in Extremity Surgery and Amputation Prostheses
Elof Eriksson, Dennis P. Orgill, and Julian J. Pribaz 183

16 Managing the Failed Total Hip Arthroplasty
Richard D. Coutts 195

17 The Implant-Bone Interface and the Effects of Particulate Debris
William J. Maloney and Charles A. Engh 203

18 General Discussion
Göran Lundborg and Richard Skalak 215

19 Concluding Remarks
Richard Skalak 227

The editors would like to thank all workshop participants and authors for their important contributions to this book. A special appreciation is extended to Professor Shu Chien, who introduced this interdisciplinary and cross-fertilizing workshop.

Per-Ingvar Brånemark, Björn Rydevik, and Richard Skalak

Contributors

Stefano Boriani, MD
Head
Department of Orthopedics
and Traumatology
Ospedale Maggiore
Bologna, Italy

Per-Ingvar Brånemark, MD, PhD
Professor and Director
The Institute for Applied Biotechnology
Gothenburg, Sweden

Rickard Brånemark, MD, PhD
Assistant Professor
Department of Orthopedics
Sahlgren's University Hospital
University of Gothenburg

The Institute for Applied Biotechnology
Gothenburg, Sweden

Shu Chien, MD, PhD
Professor of Bioengineering and Medicine
Chair, Department of Bioengineering
Director, Institute for Biomedical Engineering
University of California, San Diego
La Jolla, California

Dudley S. Childress, PhD
Professor of Biomedical Engineering and
Orthopedic Surgery
Director, Prosthetic Research Laboratory
Northwestern University Medical School
Chicago, Illinois

McCormick School of Engineering and
Applied Science
Northwestern University
Evanston, Illinois

Richard D. Coutts, MD
Orthopedic Medical Group
San Diego, California

Stephen C. Cowin, PhD
Distinguished Professor
Department of Mechanical Engineering
City College
City University of New York
New York, New York

Charles A. Engh, MD
Anderson Orthopedic Research Institute
Arlington, Virginia

Elof Eriksson, MD
Professor and Chief of Plastic Surgery
Brigham and Women's Hospital
Harvard Medical School
Boston, Massachusetts

Steven R. Garfin, MD
Professor of Orthopedic Surgery
Department of Orthopedics
Medical Center
University of California, San Diego

Björn Gunterberg, MD, PhD
Associate Professor
Department of Orthopedics
Sahlgren's University Hospital
University of Gothenburg
Gothenburg, Sweden

Tommy Hansson, MD, PhD
Professor and Head
Unified Department of Orthopedics,
Rheumatology, and Hand Surgery
Sahlgren's University Hospital
University of Gothenburg
Gothenburg, Sweden

Dick Heinegård, MD, PhD
Professor
Department of Cell and Molecular Biology
Lund University
Lund, Sweden

Björn Klinge, LDS, Odont Dr
Professor
Department of Periodontology, Clinical Oral Sciences

Dean
Faculty of Odontology
Karolinska Institute
Stockholm, Sweden

Göran Lundborg, MD, PhD
Professor
Department of Hand Surgery
Malmö University Hospital
Lund University
Malmö, Sweden

Gangming Luo, PhD
Research Associate
Department of Rehabilitation Medicine
New York University
School of Medicine
New York, New York

William J. Maloney, MD
Associate Professor
Department of Orthopedic Surgery
Washington University School of Medicine

Chief of Service and Head of Joint Replacement Surgery
Department of Orthopedic Surgery
Barnes-Jewish Hospital
St. Louis, Missouri

Dennis P. Orgill, MD, PhD
Assistant Professor
Associate Chief of Plastic Surgery
Brigham and Women's Hospital
Harvard Medical School
Boston, Massachusetts

Dennis G. Osmond, MB, ChB, DSc, FRCS
Robert Reford Professor of Anatomy
Department of Anatomy and Cell Biology
McGill University
Montreal, Canada

Julian J. Pribaz, MD
Associate Professor
Director of Hand and Microsurgery Unit
Brigham and Women's Hospital
Harvard Medical School
Boston, Massachusetts

M. Dwayne Rahal, DDS, PhD
Assistant Professor
Faculty of Dentistry
McGill University
Montreal, Canada

Björn L. Rydevik, MD, PhD
Professor and Chairman
Department of Orthopedics
Sahlgren's University Hospital
University of Gothenburg
Gothenburg, Sweden

Richard Skalak, PhD, MD (hon)
Professor
Department of Bioengineering
University of California, San Diego
La Jolla, California

James N. Weinstein, DO, MS
Director
Surgical Outcomes Assessment Program
Center for Evaluative Clinical Sciences
Dartmouth Medical School
Hanover, New Hampshire

Mikael Wendel, Odont Dr
Research Assistant
Department of Cell and Molecular Biology
Lund University
Lund, Sweden

Centre for Oral Health Sciences
Malmö, Sweden

Philip Worthington, MD, DDS, BSc, FDSRCS
Professor and Chairman
Department of Oral and Maxillofacial Surgery
School of Dentistry
University of Washington
Seattle, Washington

Introduction

Shu Chien

It is indeed an honor and pleasure to be here with this distinguished group of scientists to discuss osseointegration. Customarily, when I go to a meeting I thank the host and say, "I enjoy being here in this lovely city," but here I am a local person. Therefore, according to Chinese custom I should say, "Welcome to this underdeveloped little town, and I hope you all enjoy staying here." Although I have been in San Diego for seven years, this is actually the first time I have been at this resort, which is very lovely. I hope we all have a good time here discussing osseointegration, not only in this meeting room, but also outside of it.

The First International Workshop on Osseointegration in Skeletal Reconstruction and Joint Replacement was held in Aruba in 1990. P-I Brånemark told me that this is indeed a follow-up of the first workshop to encourage further discussion among participants who are experts in various fields working on problems related to skeletal reconstruction and joint replacement. The two aims are: first, to review and evaluate the progress made to date in dental and craniofacial areas by the application of the osseointegration method, and second, to assess the possibilities and problems that may arise in further applications of osseointegration in other areas of skeletal reconstruction and joint replacement. As we will hear from the presentations at this workshop, there are certain areas that have already been developed and other areas that remain to be further explored.

Several consensus statements emerged from the first workshop. One is about promising areas for future development and another is about the applications of osseointegration that are already clinically reliable in craniomaxillofacial surgery to other fields. It seems possible to apply osseointegration to all areas of the body, including hand surgery and orthopedic surgery. There is a logical progression of applications from the smaller, more lightly loaded joints, like the joints in the fingers and wrists, to the more heavily loaded joints, such as the hip and the knee.

Osseointegration of an implant was defined at the previous workshop as an implant in which there is no progressive relative motion of the living bone and the implant. Some footnotes are given in which it is implied that only the bone tissue is involved and that the relative motion should be no larger than the elastic deformation of the materials. There is always some motion due to the elastic deformation, and this modality of motion should not be progressive. The interface is assumed to be bonded without slip. As we move from the smaller to the more heavily loaded joints, it will be important to have osseointegration more clearly defined.

I knew Dr Brånemark when he was working on microcirculation research, showing us how the red blood cells and white blood cells flow in the microcirculation. His pioneering work on the intravital microscopy of the microcirculation in the bone led him to develop the concept of osseointegration. A titanium chamber was used in his long-term intravital observation of the bone and marrow tissues. With time, he has successfully used this titanium implant with a great variety of tissues.

I know many who have titanium implants in their jaws and elsewhere. They tell me that, as patients, they know Dr Brånemark's name and appreciate having benefited from this technology. In Dr Brånemark's laboratory are studies on not only the clinical application, but also basic science. For example, in his work on the interactions of macrophages and polymorphonuclear granulocytes with the implant, he has elucidated the roles of ions and proteins and applied cell biology and molecular biology to this type of study. Research on osseointegration involves not only clinical investigation, but also engineering basic biology from the cellular to the molecular level, including the study of effects of mechanical forces and chemical factors acting on the various tissues. In the future, we may aim to modulate gene expression to affect the interactions between the cells and the implant at the interface. These new frontiers will be covered in this workshop.

The workshop is organized to proceed from the more fundamental aspects of osseointegration to applications in reconstruction and prosthetic devices in skeletal reconstruction and joint replacement. There are scientists from different disciplines, including engineering, biology, and medicine. Scientists from these different disciplines work together with the physicians and all other professionals involved in health care, with the ultimate aim of helping the patients, who are our top priority (Figure). With the diversity of participants at this workshop, there is an excellent environment and opportunity for us to interact. What better way is there to achieve this than bringing together experts from many disciplines, including basic science and clinical application, at this workshop?

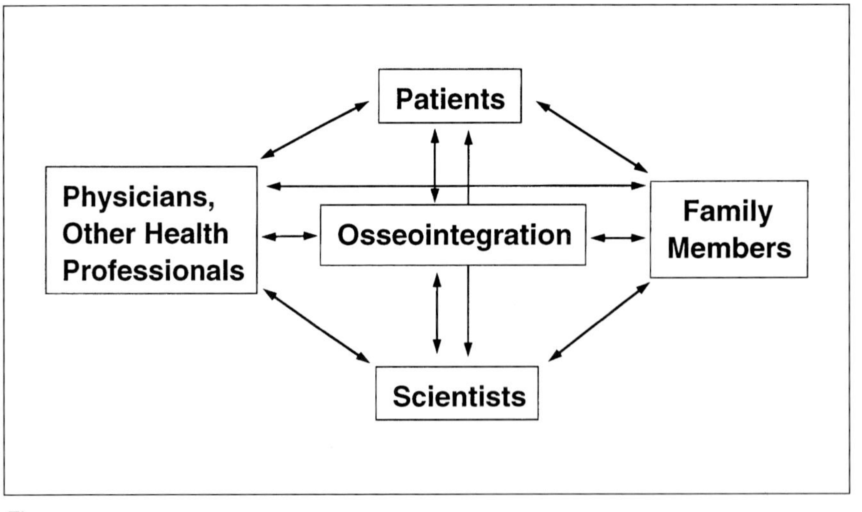

Figure

Definition of Osseointegration

The term *osseointegration* may be defined from various viewpoints and with respect to several scientific scales of interest. Only with the sum total of all of these viewpoints is a complete definition of osseointegration provided.

From the Viewpoint of the Patient

A fixture is osseointegrated if it provides a stable and apparently immobile support of a prosthesis under functional loads, without pain, inflammation, or loosening.

From the Viewpoint of Macroscopic and Microscopic Biology and Medicine

Osseointegration of a fixture in bone is defined as the close apposition of new- and reformed bone in congruency with the fixtures, including surface irregularities, so that at the light microscopic level there is no interpositioned connective or fibrous tissue and that a direct structural and functional connection is established, capable of carrying normal physiologic loads without excessive deformation and without initiating rejection mechanisms.

From the Macroscopic, Biomechanical Point of View

A fixture is osseointegrated if there is no progressive relative motion between the fixture and surrounding living bone and marrow under functional levels and types of loading for the entire life of the patient. Also required is that deformations are of the same order of magnitude as when the same loads are applied directly to the bone.

From the Microscopic, Biophysical Point of View

Osseointegration implies that, at the light microscopic and electronmicroscopic levels, the identifiable components of tissue within a thin zone of a fixture surface are identified as normal bone and marrow constituents that continuously grade into a normal bone structure surrounding the fixture. This implies that mineralized tissue is found to be in contact with the fixture surface over most of the surface within nanometers so that no functionally significant intervening material exists at the interface.

—Richard Skalak and Per-Ingvar Brånemark

Chapter 1

Osseointegration: Biotechnological Perspective and Clinical Modality

Per-Ingvar Brånemark

Osseointegration means permanent incorporation of a nonbiologic component in the human skeleton with a capacity to carry unlimited functional load in endoprosthetic and exoprosthetic replacement of structure and function. Successful osseointegration requires that the bone tissue accepts the presence of the prosthetic anchorage and adapts to the transfer of functional load but does not—even in the long term—react to the presence of the nonbiologic component.

There are two basic aims of this workshop: to explore osseointegration as a biotechnologic phenomenon; and to consider its clinical applicability today and its potential for the future. Accordingly, it is essential to cover topics ranging from design and dimensions of the anchoring elements and prostheses to interaction at the molecular level as well as clinical practicalities, such as selection of patients, understanding of the healing process, and load carrying and remodeling capacity of bone tissue.

A key issue is how bone tissue can be made to accept the intrusion of a piece of titanium without rejection phenomena, even over decades. For an exoprosthesis, it is crucial to identify the limits for penetration of the integumentum without initiating untoward soft-tissue reactions, which would require unrealistic or otherwise inappropriate maintenance efforts.

The continuing turnover mechanisms in the exterior surface of the anchoring material as well as in the nearby and distant biologic tissues and organs require careful recognition, identification, and description. The dynamic events at the interface should be considered in the overall design of a therapeutic recipe for unlimited function, minimizing the risk of fatigue and other untoward phenomena occurring or being initiated in synthetic or biologic components and compartments (Figs 1-1 to 1-9).[1-4]

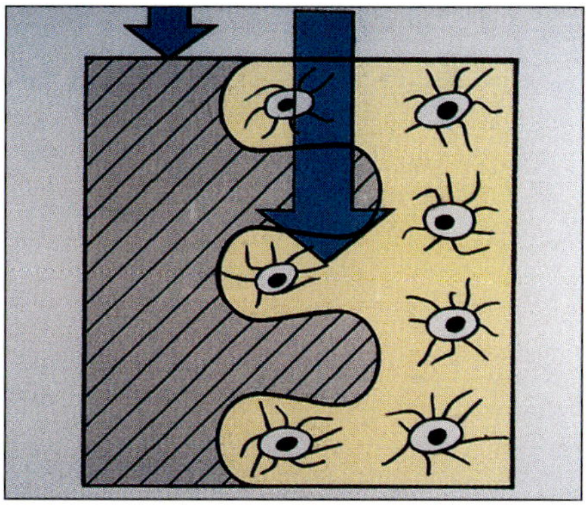

Fig 1-1 Direct connection between anchoring element and living bone. Experimentally and clinically, this type of anchorage has proven to be reliable and the only one in which the long-term outcome in the individual case is predictable. (Figs 1-1, 1-3, 1-27, and 1-28 from Brånemark P-I. Introduction to osseointegration. In Brånemark P-I, Zarb G, Albrektsson T [eds]: Tissue-Integrated Prostheses: Osseointegration in Clinical Dentistry. Chicago: Quintessence; 1985:11–76. Used with permission.)

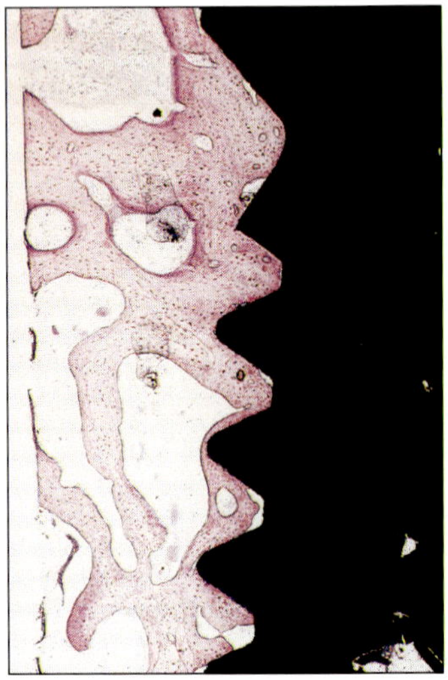

Fig 1-2 Ground section of a mandibular bone-anchored fixture supporting a bridge construction after 2 years in function. (From Brånemark P-I, Worthington P. Osseointegration in maxillofacial reconstruction. In Williams J [ed]: Maxillofacial Injuries, ed 2, vol II. Edinburgh: Churchill Livingstone; 1994:1027–1052. Used with permission.)

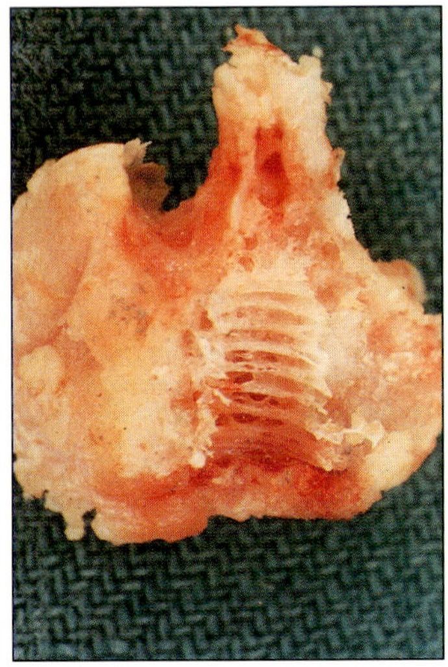

Fig 1-3 Fixture in upper jaw of dog removed from bone to illustrate remodeling of bone with formation of dense capsule in the trabecular bone; 32 months.

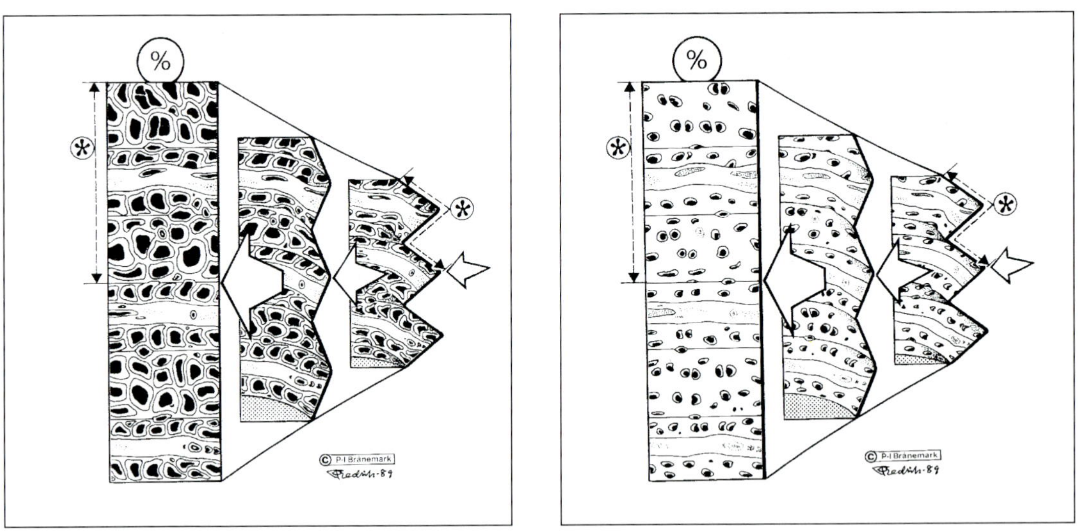

Fig 1-4 Expanded view of bone tissue facing screw-shaped anchorage device (fixture) at time of installation (left) and after functional load has been applied for some time (right). (From Brånemark P-I, Worthington P. Osseointegration in maxillofacial reconstruction. In Williams J [ed]: Maxillofacial Injuries, ed 2, vol II. Edinburgh: Churchill Livingstone; 1994:1027–1052. Used with permission.)

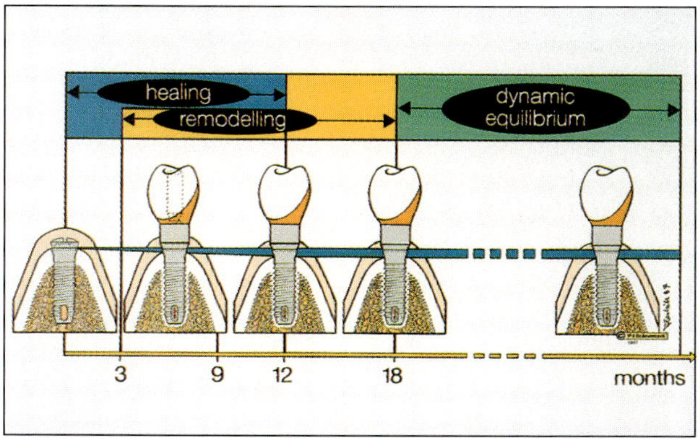

Fig 1-5 Diagram summarizing the sequence of events in bone tissue incorporating and integrating the titanium fixture (From Brånemark P-I, Worthington P. Osseointegration in maxillofacial reconstruction. In Williams J [ed]: Maxillofacial Injuries, ed 2, vol II. Edinburgh: Churchill Livingstone; 1994:1027–1052. Used with permission.)

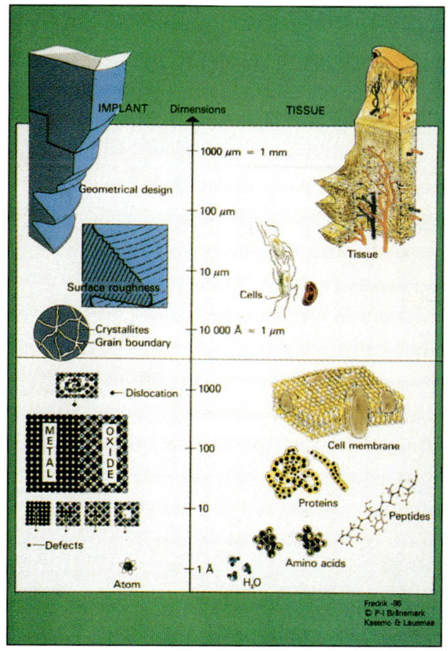

Fig 1-6 Schematic illustration of some relevant structures at the biomaterial-biosystem interface and their approximate linear dimensions. Note that many of the relevant structures are not resolvable at the light microscopy level. (From Kasemo B, Lausmaa J. Surface science aspects of inorganic biomaterials. CRC Crit Rev Biocompatibility 1986;2:4, 335–380. Used with permission.)

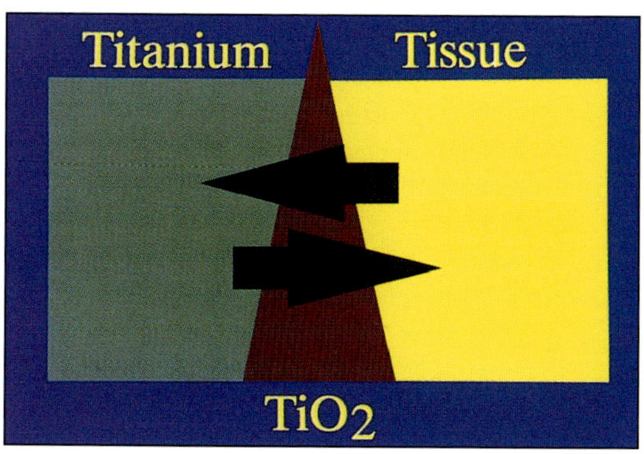

Fig 1-7 Diagrammatic simplification of decisive components in osseointegration, also indicating dimensional adaptation over time.

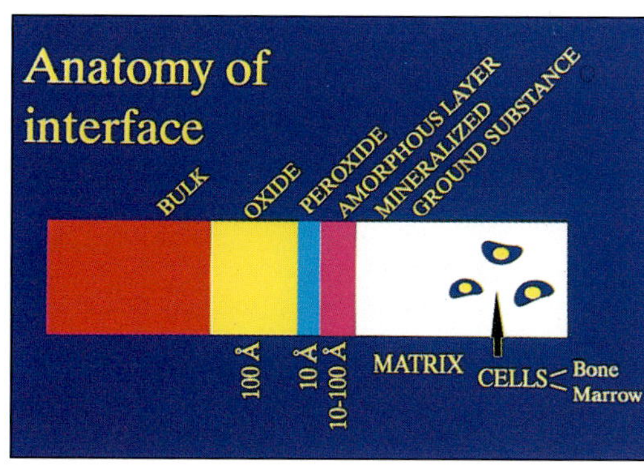

Fig 1-8 Diagrammatic anatomy of participating components at the interface indicating dimensional proportions.

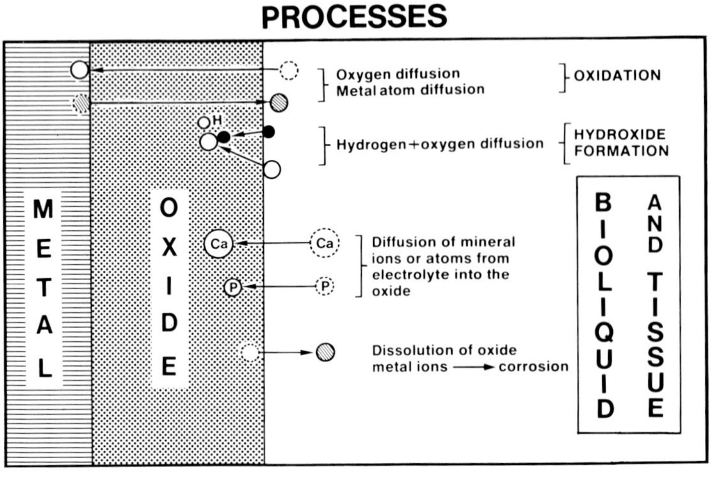

Fig 1-9 Chemical processes that can take place at the interface at a molecular scale. (From Kasemo B. Biocompatibility of titianium implants: Surface science aspects. J Prosthet Dent 1983;49:832. Used with permission.)

The fact that the equilibrium obtained is dynamic requires careful clinical follow-up based on objective criteria for success or failure over long periods in a large number of patients before a reliable treatment modality can be considered as established.

In the historical perspective, there have been a number of statements in the literature suggesting that lasting bone incorporation of a synthetic component would *not* be possible:

> "Although histologically inert, an implanted object never becomes incorporated into the bone. It remains a foreign body sequestrated from the inner structure of the bone by the wall of tissue, first fibrous, and later both fibrous and bony, that is built around it."[5]
>
> "Implants always suffer from one formidable disadvantage—they can never become part of the body structure because actual fusion is impossible."[6]
>
> "Long-term success of implants depends upon a stable fixation which will resist the forces applied to it. It is only by obtaining a truly biological fixation of new bone around the implant that a permanent result can be expected."[7]
>
> "When any metallic appliance is implanted in bone, a layer of fibrous tissue will always develop around the appliance which subsequently will never be as secure in the bone as it was at the time it was implanted."[8]

However, in 1967, Palmer[9] wrote:

> A material was created for skeletal substitution and bone connection that did not rust and did not hurt the tissue—vitallium. But it did not heal into the bone but was positioned in a site with reduced mineralization during healing. What we were missing was a material that healed into the bone tissue, was accepted, and became incorporated. I am convinced that this material will soon be available. Skeletal surgery will then take a big leap forward.

Over the years, our definitions of osseointegration have primarily been focused on its clinical capacity:

> ". . . . bone established direct contact with screw-shaped implants of commercially pure titanium."[10]
>
> "The re- and new-formed bone tissue enclose the implant with perfect congruency to the implant form and surface irregularities, thus establishing a true osseointegration of the implant without any interpositioned connective tissue."[11]

A modified version of this definition is included in Dorland's Illustrated Medical Dictionary[12]: "Osseointegration—direct anchorage of an implant by the formation of bony tissue around the implant without the growth of fibrous tissue at the bone-implant interface."

> "Osseointegration is defined as a direct structural and functional connection between ordered, living bone and the surface of a load-carrying implant. Creation and maintenance of osseointegration, therefore, depends on the understanding of the tissue's healing, repair, and remodeling capacities."[1]
>
> Osseointegration in theory and practice is defined as continuing structural and functional coexistence, possibly in a symbiotic manner, between differentiated, adequately remodeling biologic tissues and strictly defined and controlled synthetic components, providing lasting, specific clinical functions without initiating rejection mechanisms.[13]

Since the ultimate aim of osseointegration is to provide stability for a prosthesis, a more mechanically oriented definition has been suggested by Skalak (personal communication, 1991): "An implant is said to be mechanically osseointegrated if there is no progressive relative motion of living bone and implant under functional levels and types of loading for the entire life of the patients."

It is suggested that this workshop could consider the following proposed definitions of osseointegration based on different evaluation criteria (1994):

1. **From the viewpoint of the patient:** A fixture is osseointegrated if it provides a stable and apparently immobile support of a prosthesis under functional loads, without pain, inflammation, or loosening over the lifetime of the patient.
2. **From the viewpoint of macroscopic and microscopic biology and medicine**: Osseointegration of a fixture in bone is defined as the apposition of new and reformed bone in congruency with the fixtures, including surface irregularities, so that at the light microscopic level there is no interposed connective or fibrous tissue and a direct structural and functional connection is established, capable of carrying normal physiologic loads without excessive deformation and without initiating rejection mechanisms.
3. **From the macroscopic, biomechanical point of view:** A fixture is osseointegrated if there is no progressive relative motion between the fixture and surrounding living bone and marrow under functional levels and types of loading for the entire life of the patient. The fixture should undergo movements of the same order of magnitude as when the same loads are applied directly to the bone.
4. **From the microscopic biophysical point of view:** Osseointegration implies that, at the light microscopic and electron microscopic levels, the identifiable components of tissue within a thin zone of a fixture surface are identified as normal bone and marrow constituents that continuously grade into a normal bone structure surrounding the fixture; this implies that mineralized tissue is in contact with the fixture surface over most of the surface within nanometers so that no functionally significant intervening material exists at the interface.

Basic Biology of Osseointegration

The structural and functional interaction between bone and marrow, vasculature, and cytology became obvious at a very early stage of the experimental exploration of bone and marrow healing and regeneration.[14] After formation of a medullary hematoma, there is a continuing organization originating from the endosteal compact bone or from endosteal marrow remnants. The process of organiza-

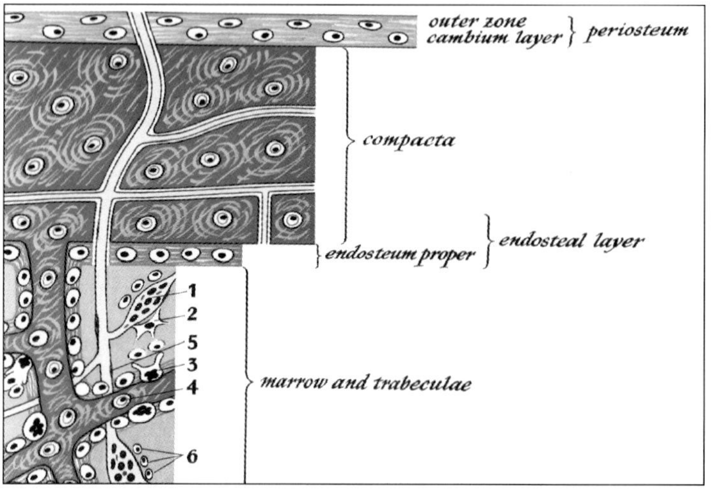

Fig 1-10 Schematic representation of relationship between structures possibly concerned in regeneration of hemopoietic tissue. 1, Sinusoid; 2, reticulum cell; 3, osteoblast with processes in an intertrabecular space; 4, osteocyte in trabeculum; 5, vascular endothelium; 6, hemopoietic cells lining sinusoid. Note osteoblasts lining trabeculae. (From Brånemark P-I, Breine U, Johanson B, Roylance PJ, Röckert H, Joffey JM. Regeneration of bone marrow: A clinical and experimental study following removal of bone marrow by curettage. Acta Anat 1964;59:1–46. Used with permission.)

tion commences with formation of a delicate network that soon develops into a trabecular system.

The regeneration of endosteal bone and marrow was observed in an experimental study after reaming the marrow of the rabbit's tibia (50 rabbits, 6 to 10 months old, followed for 150 days). In a parallel study on cleft palate repair in 15 children (7 to 8 months old), marrow and trabecular bone grafts were harvested from the upper tibial metaphysis and placed in the maxillary defect. The regeneration of tissue at the donor site was followed up with trephine biopsies at intervals from 7 to 410 days to ascertain the safety of the procedure (Figs 1-10 and 1-11).[14-16]

Because of the decisive importance of the coexistence of bone and marrow at the interface (Figs 1-12 to 1-14), we include a suggestion by Osmond (personal communication, 1994):

> "The term *myelointegration* may be defined from various viewpoints and with respect to several scientific scales of interest.
>
> 1. **From the viewpoint of the patient:** A fixture is myelointegrated if it provides a prosthesis without pain, inflammation, or other injurious effects on health over the lifetime of the patient.
> 2. **From the viewpoint of macroscopic and microscopic biology and medicine:** Myelointegration of a fixture in bone marrow is defined as the apposition of new- and re-formed bone marrow with no perturbation, either locally or systemically, in immune or hemopoietic mechanisms. Myelointegration implies that at the light microscopic and electron microscopic levels, the identifiable components of tissue adjacent to a fixture surface are identified as *normal* adjacent bone marrow constituents: this implies lymphohemopoietic tissue is in contact with either the fixture surface itself or with intervening newly reconstituted bone and bone-lining cells, re-creating a normal lymphohemopoietic microenvironment."

Our clinical experience of osseointegrated reconstructions hitherto concerns major craniomaxillofacial defects, as well as severe joint pathology in advanced rheumatoid arthritis and amputees (Figs 1-15 to 1-22).[2,17-25] Tissue-anchored prostheses have proven useful in the physical and social rehabilitation of these patients.

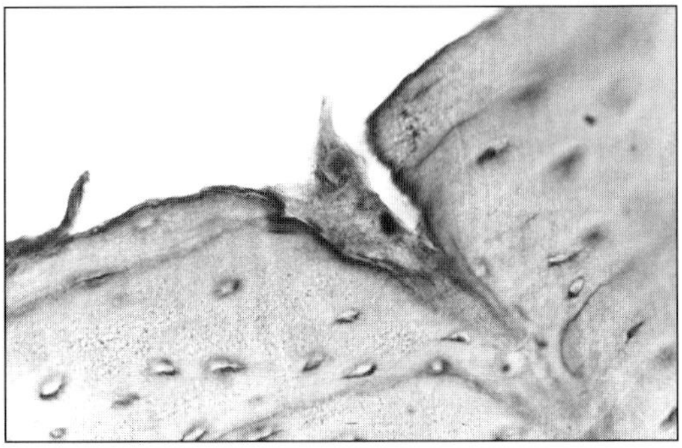

Fig 1-11a Demonstrates a haversian canal opening into the empty marrow cavity 3 days postoperatively. Rabbit. (Figs 1-11 a to i from Brånemark P-I, Breine U, Johanson B, Roylance PJ, Röckert H, Joffey JM. Regeneration of bone marrow: A clinical and experimental study following removal of bone marrow by curettage. Acta Anat 1964;59:1–46. Used with permission.)

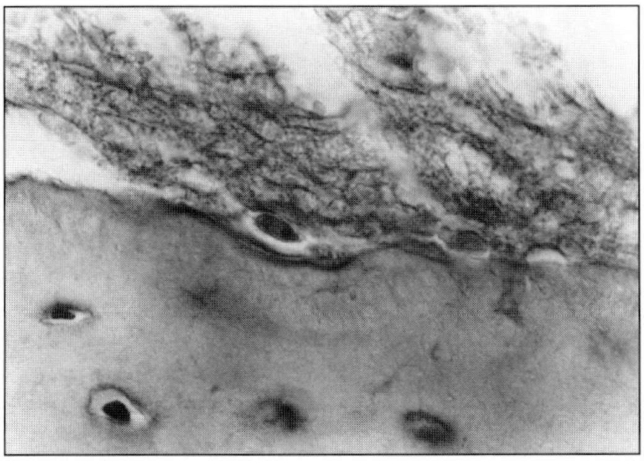

Fig 1-11b Different stages of regeneration in endosteal layer (6 days postoperatively).

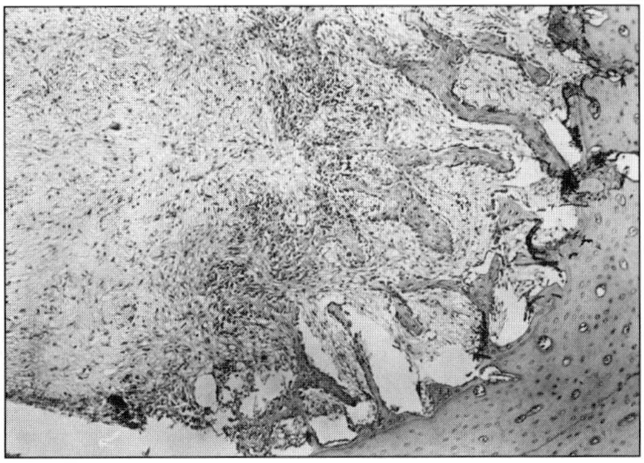

Fig 1-11c Typical appearance of endosteal trabeculae 12 days postoperatively. Rabbit.

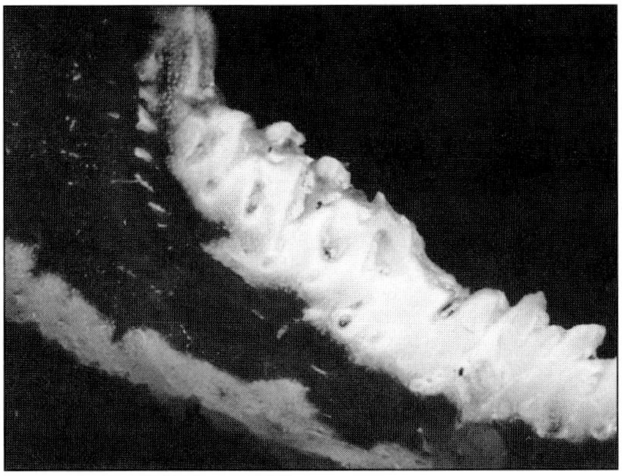

Fig 1-11d Photomicrograph in ultraviolet fluorescence after tetracycline administration 2 weeks after the operation showing the endosteal fluorescence as an indication of the metabolic activity in calcification. Rabbit.

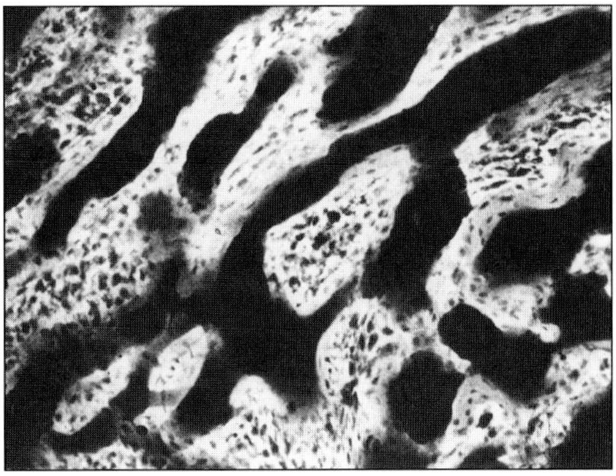

Fig 1-11e Endosteal trabeculae 3 weeks after the operation. On the microradiographs the number of osteocyte lacunae is clearly seen. Rabbit.

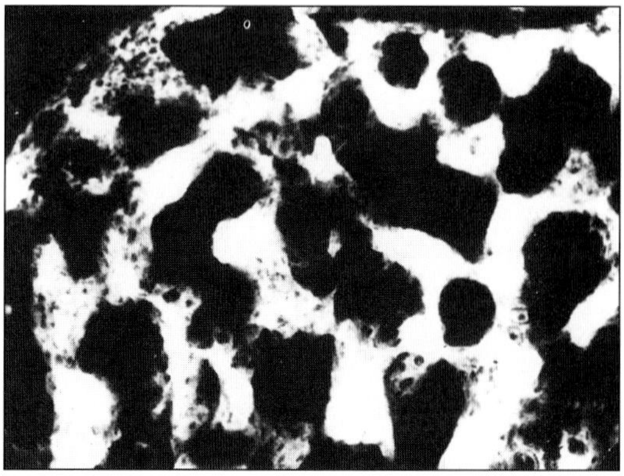

Fig 1-11f Microradiograph of endosteal trabeculae 3 weeks after the operation with an appearance reminiscent of the tibial trabeculae on the unoperated side. Rabbit.

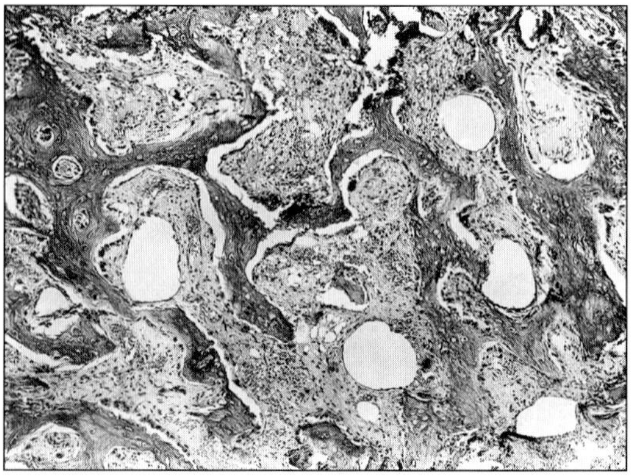

Fig 1-11g At 2 weeks, trabeculae appear. Human.

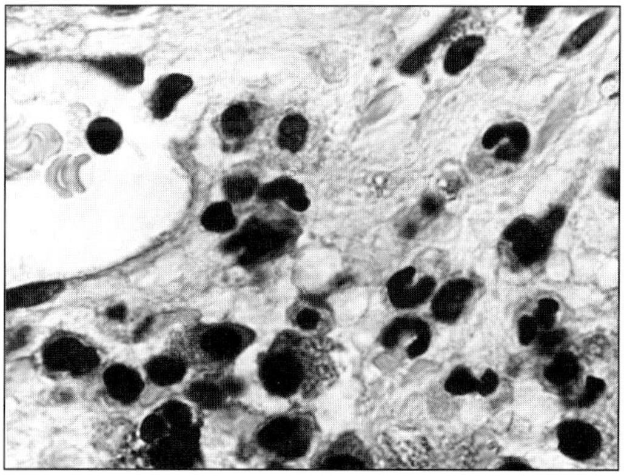

Fig 1-11h From same stage as Fig 1-11g illustrating the intertrabecular cells and the large sinusoids. Human.

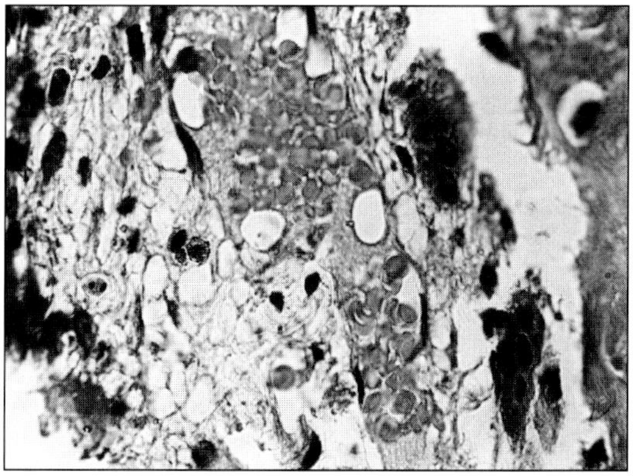

Fig 1-11i At 5 weeks, maturing intertrabecular tissue with thin endothelial lining. Sinusoids and osteoclasts are prominent features. Human.

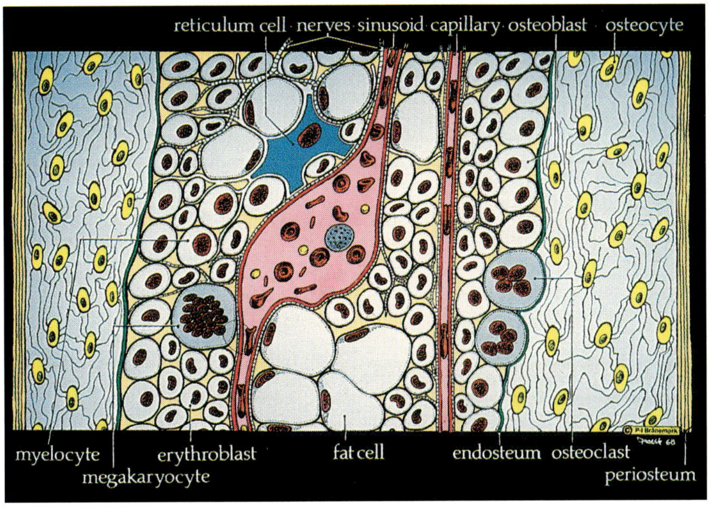

Fig 1-12 Tissue components of bone marrow relevant for osseointegrated reconstructions. (From Brånemark P-I. Bone marrow microvascular structure and function. Adv Microcirc 1968; 1:1–65. Used with permission.)

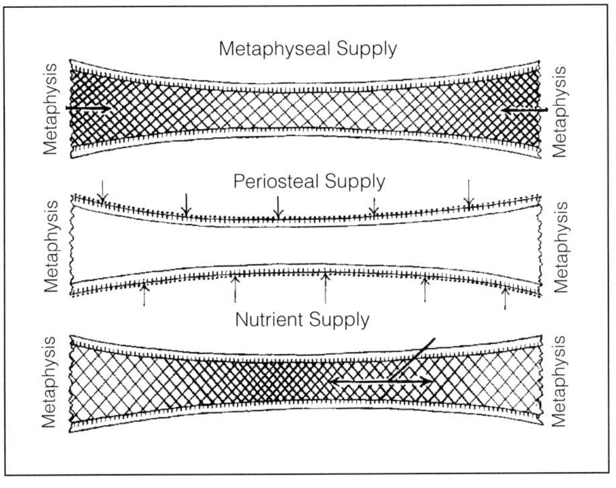

Fig 1-13a Vascular supply of bone and marrow from three main sources: nutrient, periosteal, and metaphyseal vessels.[26] (From Brånemark P-I. Bone marrow microvascular structure and function. Adv Microcirc 1968;1:1–65. Used with permission.)

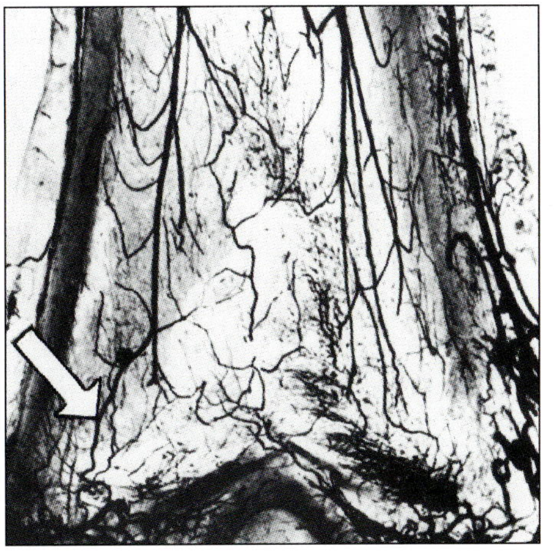

Fig 1-13b Microangiogram of rabbit tibia, showing anastomoses (arrow) between metaphyseal and medullary arteries.[27] (From Brånemark P-I. Bone marrow microvascular structure and function. Adv Microcirc 1968;1: 1–65. Used with permission.)

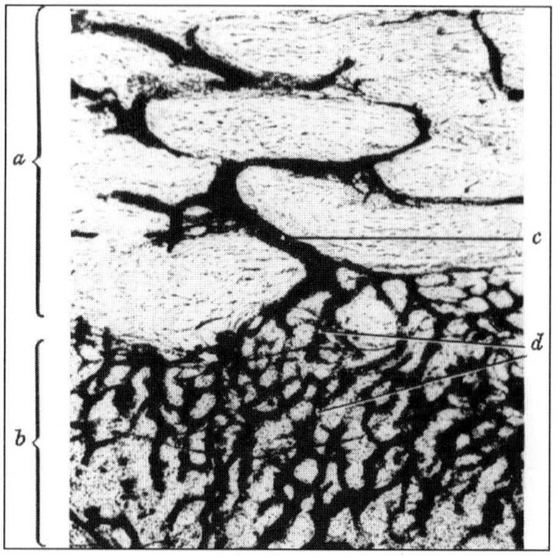

Fig 1-13c Relations between compactum veins and marrow sinusoids. a, Compactum; b, marrow; c, compactum vein; d, marrow sinusoids.[28] (From Brånemark P-I. Bone marrow microvascular structure and function. Adv Microcirc 1968;1:1–65. Used with permission.)

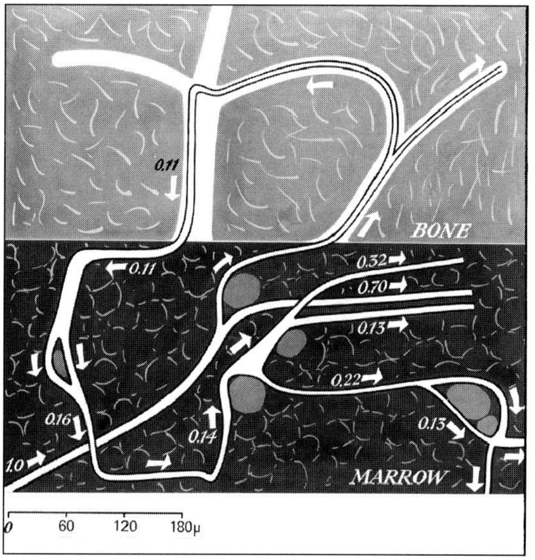

Fig 1-13d Marrow capillary dipping into osseous tissue and swinging back into marrow sinusoids. (Actual flow velocities are indicated in millimeter per second rabbit's fibula.) (From Brånemark P-I. Vital microscopy of bone marrow in rabbit. Scand J Lab Invest 1959; suppl 38:11;1–82. Used with permission.) (From Brånemark P-I. Bone marrow microvascular structure and function. Adv Microcirc 1968; 1:1–65. Used with permission.)

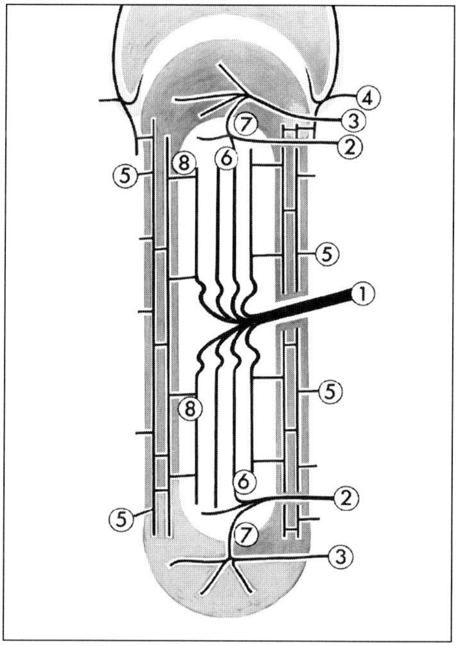

Fig 1-13e Interconnected vascular system in bone marrow and joint tissues. (From Brånemark P-I. Bone marrow microvascular structure and function. Adv Microcirc 1968; 1:1–65. Used with permission.)

Fig 1-14a Intramedullary factors influencing marrow pressure. (From Brånemark P-I. Bone marrow microvascular structure and function. Adv Microcirc 1968;1:1–65. Used with permission.)

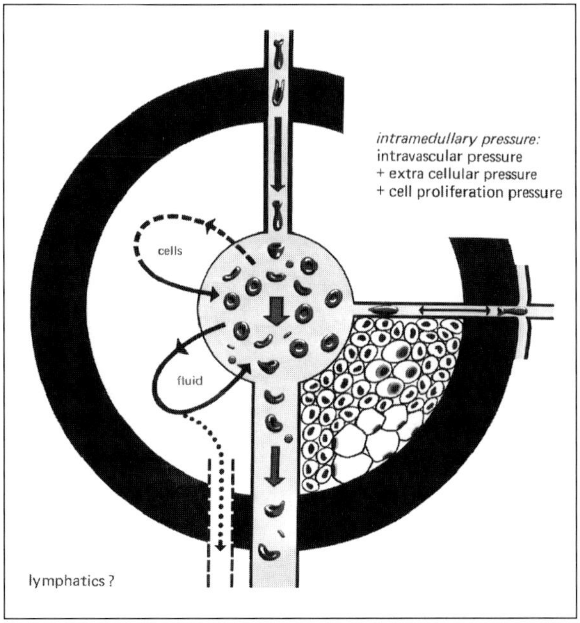

Fig 1-14b Diagrammatic representation of intramedullary and extramedullary factors influencing the pressure situation in the marrow cavity. 1 and 2, Nutrient vessels; 3, extraskeletal arteriovenous connections; 4, coiling nutrient artery along the central vein; 5, venae emissariae that connect the marrow veins with muscular and other extraskeletal venous systems (6); 7; denotes arterial pathways from surrounding tissues to the compact bone and intra-osseous connections with the nutrient arteries of the marrow; 8, possible connections between joint compartment and marrow cavities; 9, cell proliferation pressure; 10, extracellular fluid.

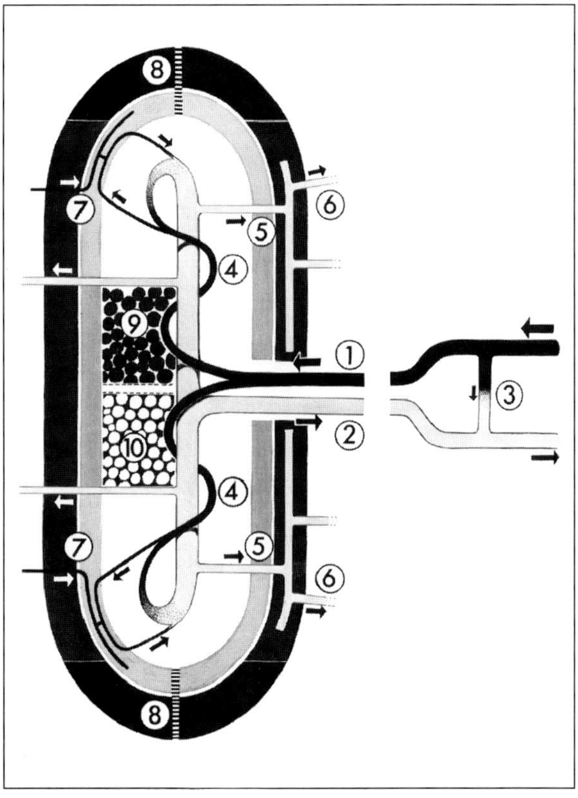

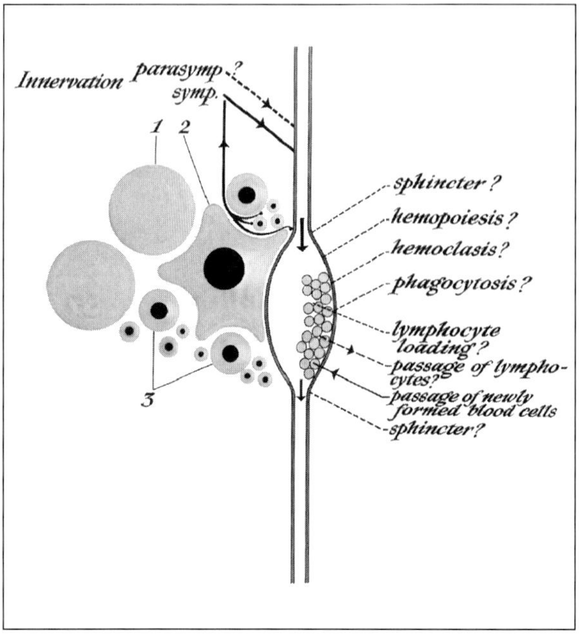

Fig 1-14c Diagrammatic representation of possible connections between the marrow sinusoid and the different hematologic activities of the marrow.

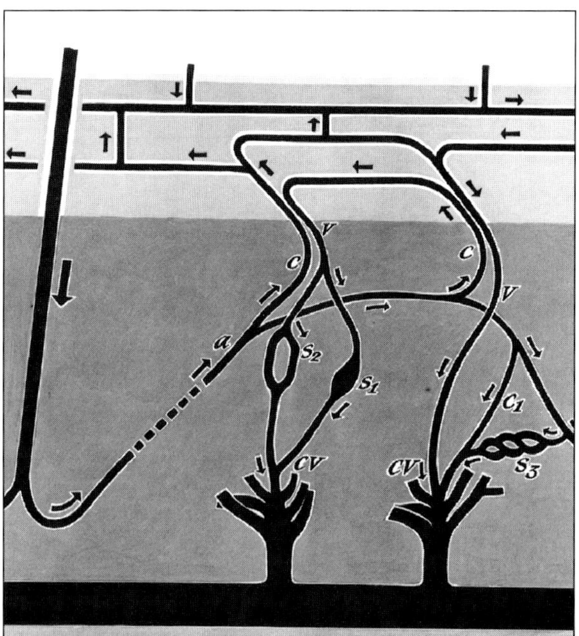

Fig 1-14d Vascular anatomy and flow pattern in bone and marrow as observed by vital microscopy in the rabbit's fibula. The bone marrow arteriole (a) divides dichotomously into capillaries (c). These capillaries run to sinusoids, which are sometimes hexagonal (S_2) and sometimes spindle-shaped (S_1). Sinusoids unite to form sinusoidal systems (S_a). The sinusoids are drained by venules into collecting venules (CF). The sinusoids vary rhythmically in degree of dilation with accompanying variation in the velocity of flow within them. Some blood cells bypass a sinusoid by flowing from a capillary via a shunting capillary (C_1) directly into a venule. Capillaries stemming from marrow arterioles enter the haversian canals to supply endosteal parts of the diaphyseal bone. The capillaries then swing back into the marrow to empty via venules (V) into sinusoids or directly into collecting venules. Sometimes two vessels with flow in opposite directions are seen in one and the same haversian canal. The blood flow in the bone capillaries is fairly steady and the velocity is higher than in the marrow capillaries. (From Brånemark P-I. Bone marrow microvascular structure and function. Adv Microcirc 1968;1:1–65. Used with permission.)

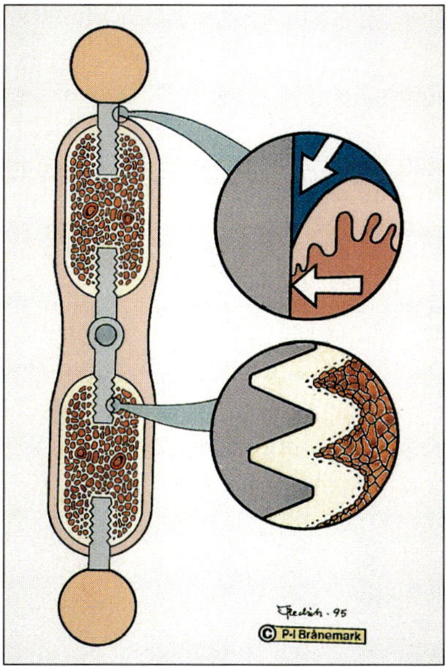

Fig 1-15 Basic biotechnologic principles for a bone-anchored structural and functional prosthesis according to the osseointegration principle. The diagram indicates anchoring into original bone or grafted bone tissue for exoprostheses and endoprostheses. It also emphasizes the importance of establishing and maintaining a soft-tissue barrier at the connection of the exoprosthesis.

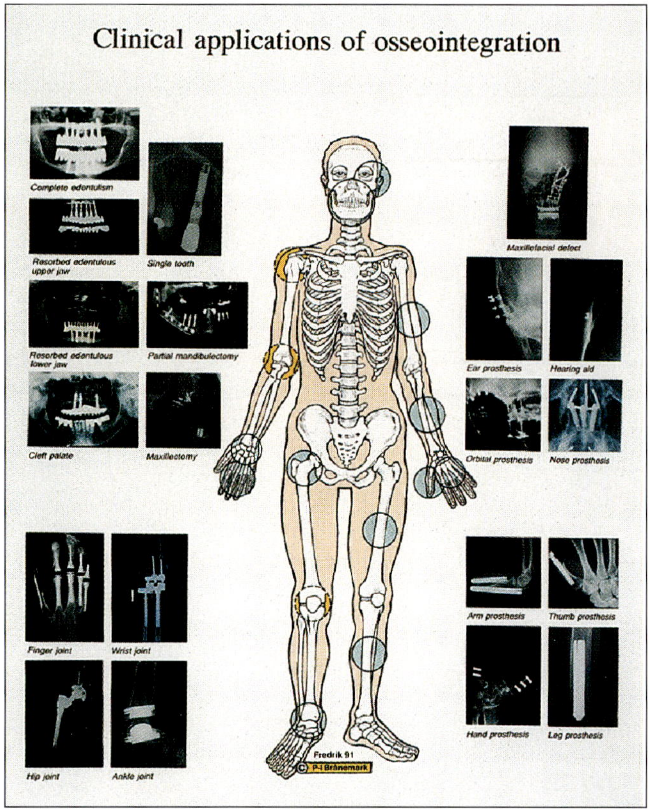

Fig 1-16 Radiographic illustration of osseointegrated exoprostheses and endoprostheses.

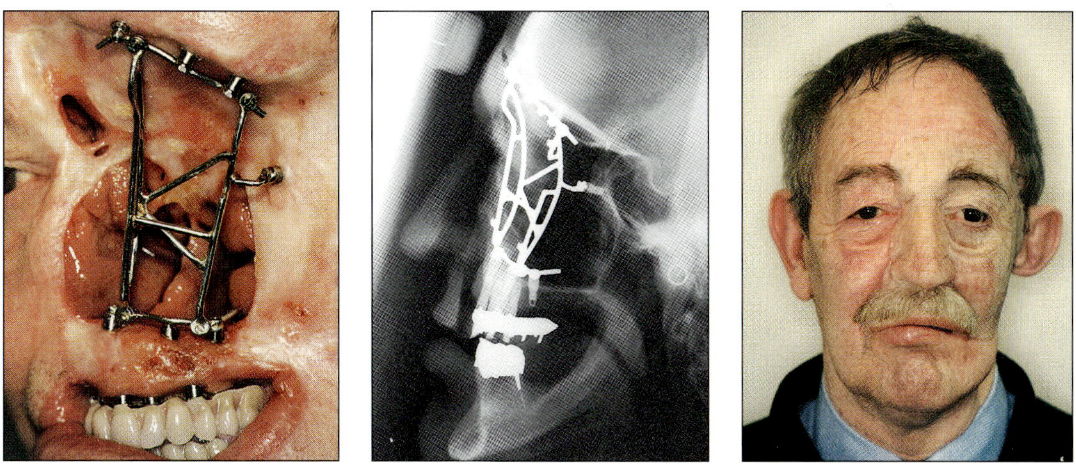

Fig 1-17 a–c Major maxillofacial defect after tumor resection and radiation therapy. Reconstruction with autologous bone graft and osseointegration were performed after preoperative hyperbaric oxygen treatment. A bone-anchored maxillary bridge and a fixture-retained maxillofacial prosthesis restores adequate anatomy and function. (From Tjellström A, Jansson K, Brånemark P-I. Craniofacial defects. In Worthington P, Brånemark P-I [eds]. Advanced Osseointegration Surgery: Applications in the Maxillofacial Region. Chicago: Quintessence; 1992:293–312. Used with permission.)

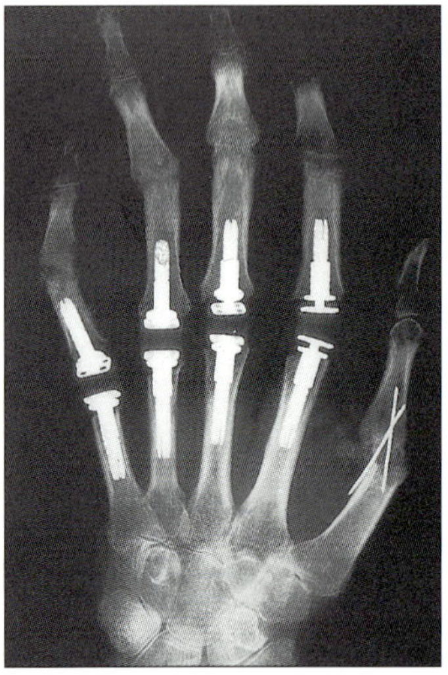

Fig 1-18 Osseointegrated metacarpophalangeal joint prostheses in a case of rheumatoid arthritis.

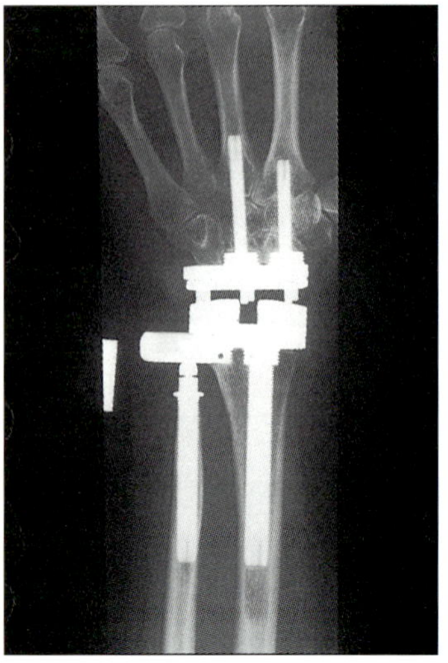

Fig 1-19 Osseointegrated wrist joint prosthesis in a case of rheumatoid arthritis.

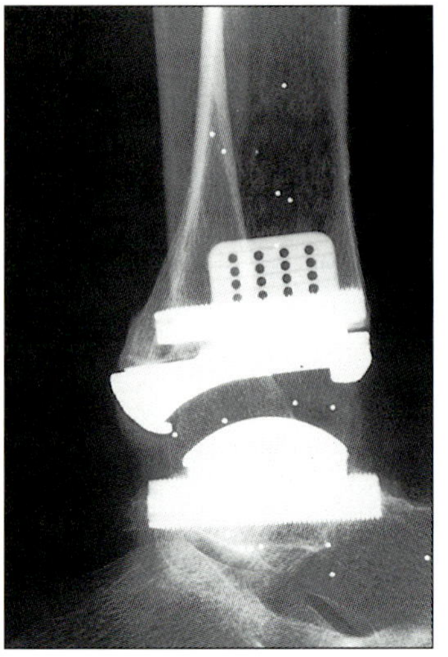

Fig 1-20 Osseointegrated ankle joint prosthesis in a case of rheumatoid arthritis at 6-year follow-up examination.

Fig 1-21 a–d Bilateral arm amputee provided with myoelectric hand prostheses with stable fixation via osseointegrated fixtures in the radius and the ulna.

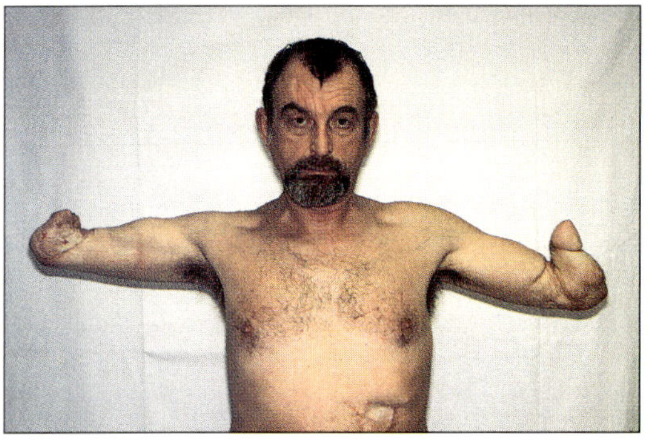

Fig 1-21a

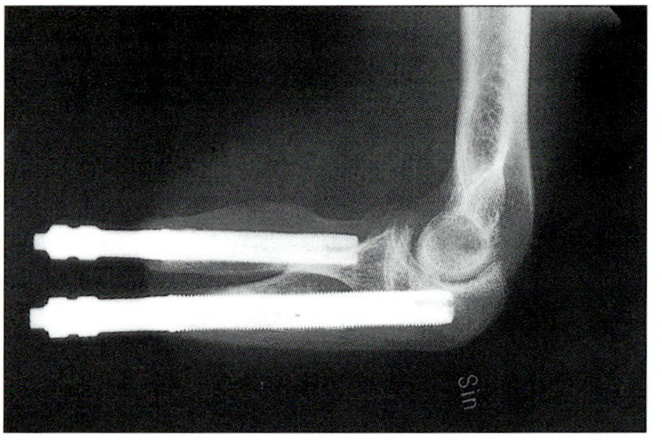

Fig 1-21b

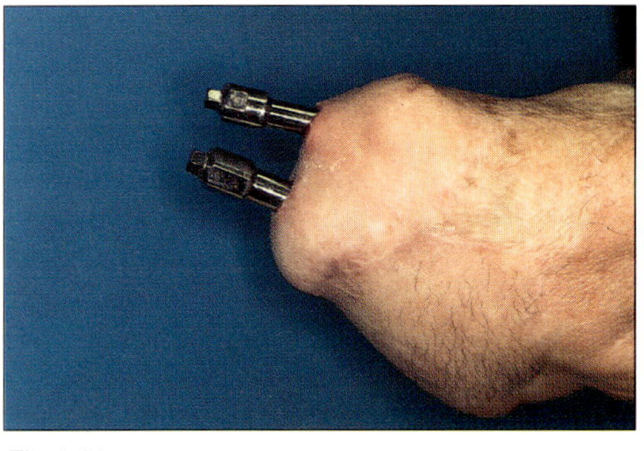

Fig 1-21c

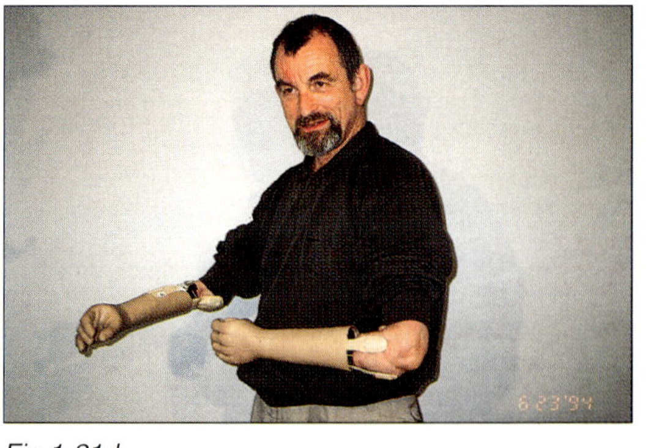

Fig 1-21d

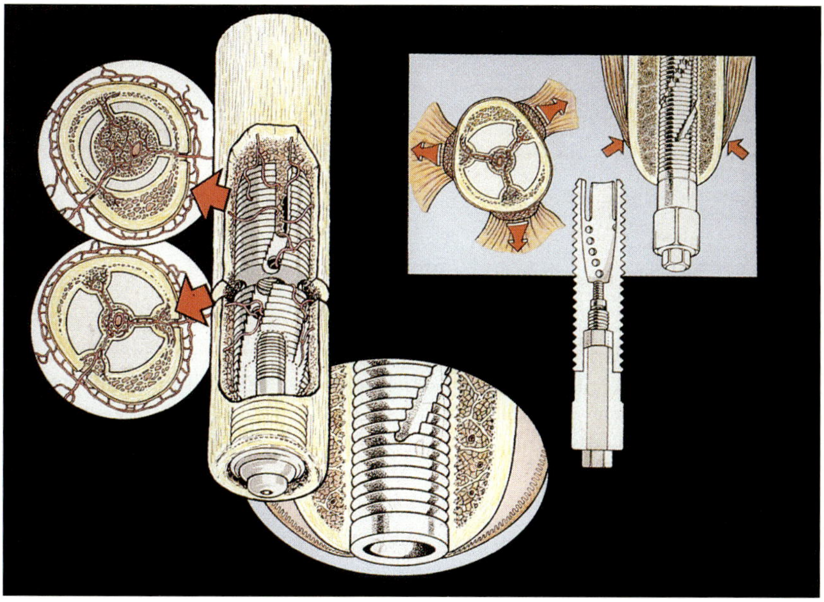

Fig 1-22a Diagrammatic representation of basic principles for bone-anchored limb prostheses.

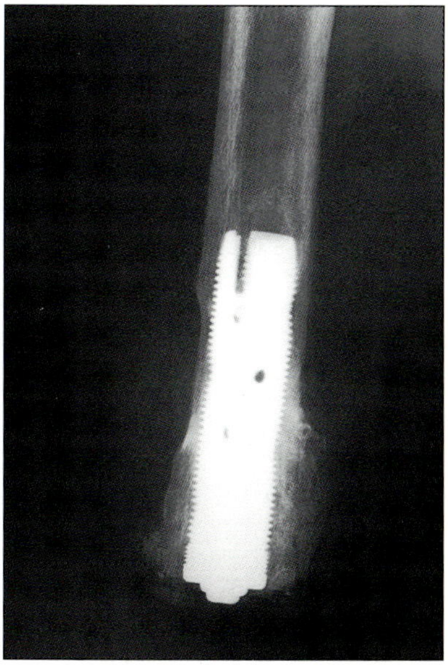

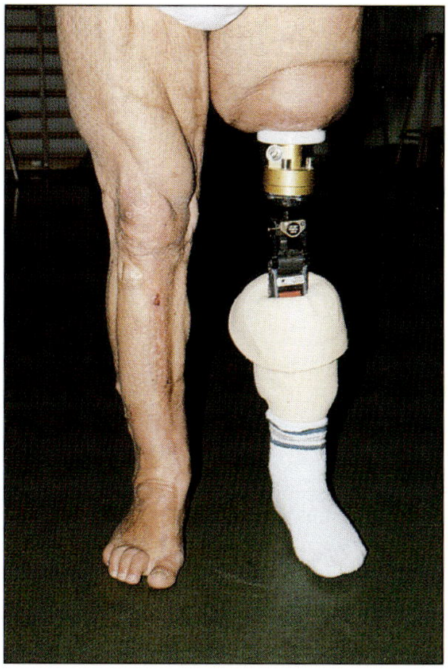

Fig 1-22 b and c Femur amputee with osseointegrated limb prosthesis.

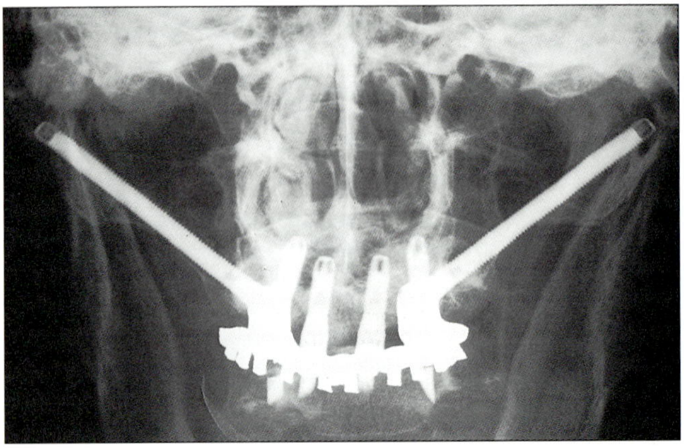

Fig 1-23 Maxillary fixed prosthesis anchored to the zygomatic bone.

Fig 1-24 a–c Edentulous maxilla and mandible provided with osseointegrated prosthetic constructions.

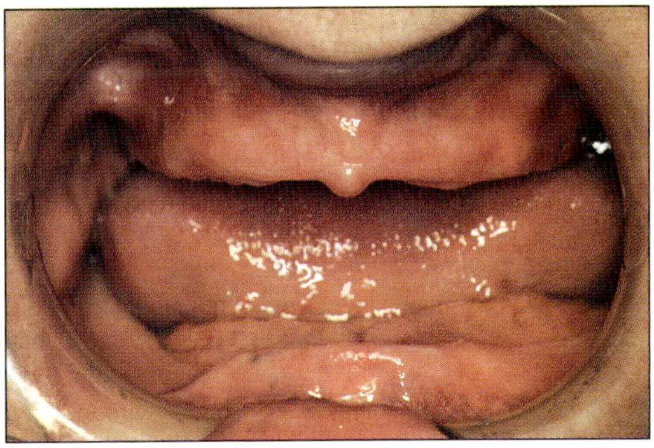

Fig 1-24a

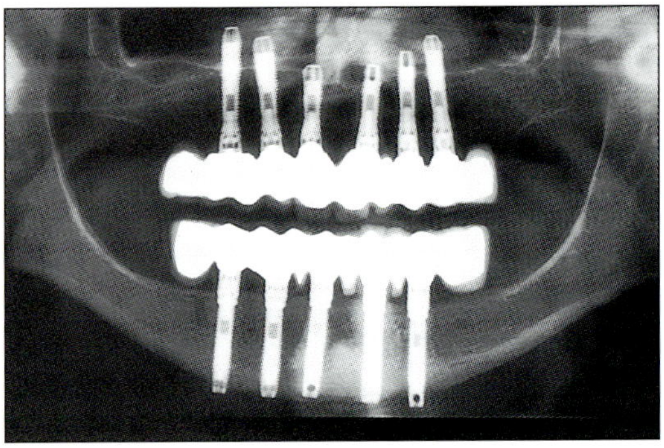

Fig 1-24b

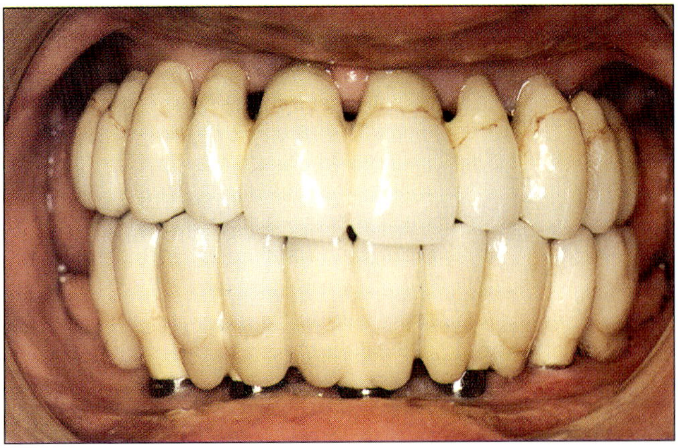

Fig 1-24c

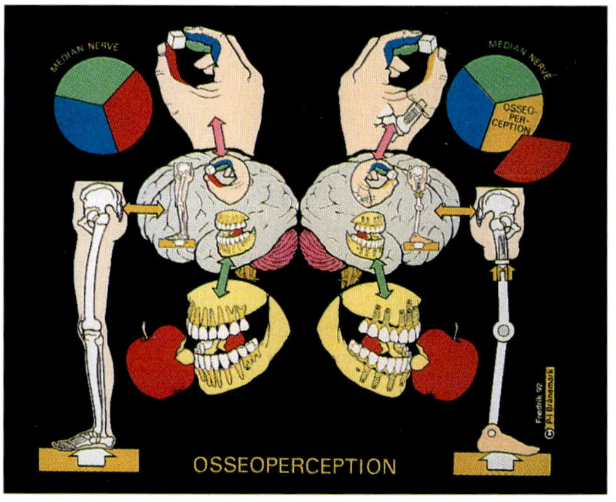

Fig 1-25 Diagrammatic representation of the phenomenon of osseoperception.

Material	Minimal volume & surface	Design	Surface architecture
Surface energy	Minimal contamination	Bone growth factors	No relative motion
Soft tissue condition	?	?	?
Controlled healing and remodeling			
Minimal tissue injury			
Precision		Experience	

Fig 1-26 Experimentally and empirically identified decisive fixture factors for predictable osseointegration.

In addition, a large number of edentulous patients have been provided with a third dentition (Figs 1-23 and 1-24).[29–31] In large, multicenter clinical studies, the long-term results indicate that for intraoral reconstructions and for extraoral craniomaxillofacial reconstructions continuing stability of the prosthesis can be expected in 95% to 99% and 95% of cases, respectively.

When therapeutic irradiation was applied, there were initially fewer favorable results concerning anchoring and soft-tissue behavior in craniomaxillofacial anchoring of epistheses. The introduction of modified procedures in the healing of craniomaxillofacial fixtures, with hyperbaric oxygen treatment included in the treatment plan, has established success rates equal to those of nonirradiated tissue.[32]

Direct bone anchorage of prosthetic teeth provided the patients with an increased sensory perception capacity. Similar discriminatory capacity has also been observed in limb prostheses directly connected to the bone tissue. This phenomenon has been called *osseoperception* (Fig 1-25).[33,34]

Ever since the first clinical application of osseointegration, careful observations of success, complications, and failure have provided a *reconstruction system* based on meticulously controlled *hardware* and *software*, where the common denominator of precision provides predictable clinical results (Fig 1-26).

The requirements for achieving and maintaining osseointegration were originally identified as a consequence of in vivo and in situ visualization of blood as a mobile tissue in microvascular physiology and pathology using vital microscopy and chamber techniques with particular emphasis on tissue injury and repair (Figs 1-27 to 1-29).[35] Because of the complexity of the problem and the diversity of the experimental approaches, the development until now has taken more than 40 years (Fig 1-30).[36,37]

We are convinced that this interactive workshop based on interdisciplinary exchange of information and interpretation together with explanation, comments, criticism, advice, and suggestions in an open atmosphere will not only increase our understanding of osseointegration as a biotechnologic occurrence but also simplify and expand its clinical applications in the future.

Fig 1-27 (opposite page) Microvascular pathophysiology after different types, level, and duration of injury was analyzed with respect to immediate reactions as well as long-term consequences and recirculation or revascularization. Shown is a diagrammatic representation of microvascular pathophysiology in tissue injury including intravascular, endothelial, and perivascular changes. Arteriolar constriction is a common phenomenon (**1**). Sometimes arteriolar dilatation occurs, or even an alternation of constricted and dilated arteriolar segments. In the true capillaries, the endothelial cells may swell (**2**), thereby reducing the capillary lumen and impairing nutritive flow of blood. This phenomenon is structurally and functionally connected with changes in the periendothelial granular cells and pericytes. Even adjacent tissue mast cells may play a role in this mechanism by disrupting and liberating the content of their granules into the ground substance and the extracellular and perivascular space. Dilatation of venules and venular stasis (**3**) is an early and important phenomenon in tissue injury. Microvascular architecture differs in various tissues. If capillary shunts or arteriolar-venular shunts exist (**4**), blood may be short-circuited and thereby bypass the nutritive part of the capillary bed. Granulocytes are often rigid and may block the nutritive capillaries temporarily or even permanently (**5**). Wall-adhering erythrocytes (**6**) and granulocytes and platelets (**7**) are characteristic findings in tissue injury. Increased numbers of granulocytes that adhere to and roll along the endothelium (**8**) are a constant finding. Microthrombi are formed and may consist of platelets (**9**), fibrin (**10**), or a combination of fibrin and platelets (**11**) and may even include red blood cells and white blood cells, thereby forming a classic microthrombus (**12**). These wall-adhering thrombi often allow single blood cells to pass (**9–12**). They may, however, also completely block the lumen (**13**), temporarily or permanently. The erythrocytes (**14**) may change in shape to crenated cells, namely, acanthocytes (**15**), which often maintain a great deal of the plasticity of the normal red blood cell (**16**). Sometimes the red blood cells turn into spherocytes (**17**), which may disrupt, causing hemolysis. When spherocytes are compressed in a vessel, they change into characteristic hexagonal bodies (**18**). The white blood cells may exhibit different degrees of plasticity (**19**), but may also turn into rigid corpuscles (**20**). Disruption of granulocytes with liberation of their granules into the lumen has been established as an in vivo phenomenon (**20**), which appears to be significant in the development of tissue injury. An aggregation of red blood cells may, to some extent, influence nutritive flow of blood (**21**). Often, however, the aggregates are plastic and deformable enough to pass even narrow nutritive capillaries, and when the flow rate is increased, the cells move as solitary bodies. The noncorpuscular plasma constituents of blood may change appreciably in composition, accompanied by disturbances in microvascular rheology. Thus significant abnormalities occur, for example, in plasma proteins, lipoproteins, and lipids. The endothelial wall disrupts functionally with leakage of plasma, resulting in edema and the passage of single blood cells (**22**), or structurally, resulting in bleeding into the tissue and more or less complete breakdown of the microcirculation (**23**). This is then followed by derangement and necrosis of the tissue.

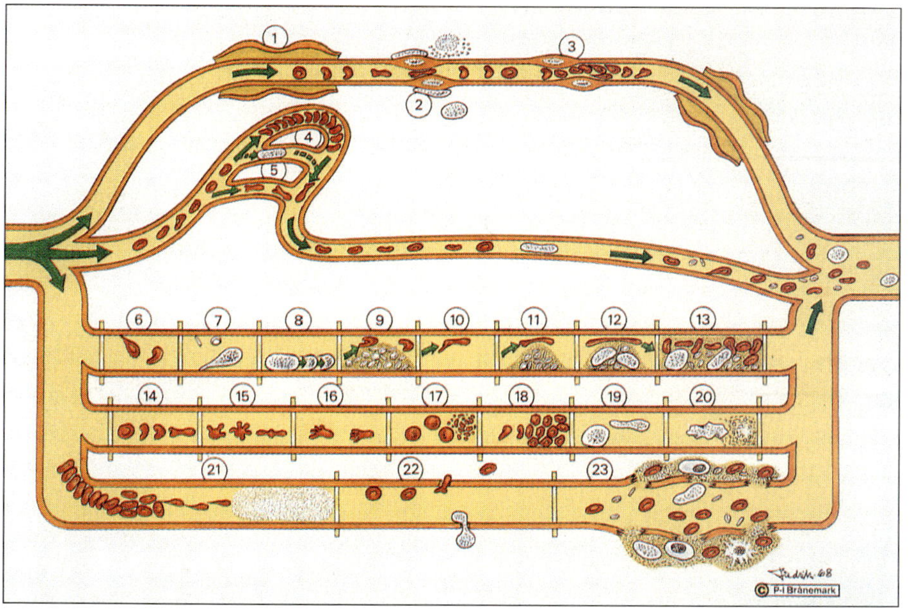

Fig 1-27

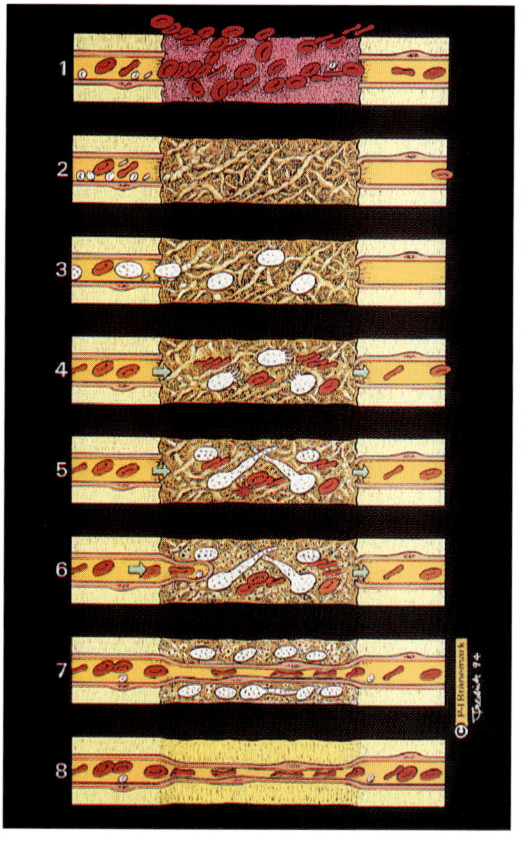

Fig 1-28 Tissue-healing phenomena were studied with different techniques including vital microscopic chamber procedures. This figure shows diagrammatically the events in the revascularization of a tissue defect, illustrating different steps in wound healing and formation of granulation tissue with respect to the microvascular system. A tissue defect, that is, a wound (1), is first filled with a fibrin reticulum formed from plasma, leaking from damaged vessels at the edge of the wound (2). After 6 to 10 hours, granular cells have already invaded the wound (3). These cells then extend cytoplasmatic processes, but are still moving around in the fibrin reticulum (4). Three to five days after the trauma, erythrocytes are perfusing the healing wound, thereby constituting an open circulation. At the next stage (5), granular cells stop moving and their processes connect with each other to form a cellular network that is still being perfused by red blood cells from vessels at the edge of the original defect. At the same time capillary sprouts are growing into the granulation tissue (6), and, finally, at 5 to 7 days after the injury the wound is penetrated by large numbers of wide, thin-walled, newly formed capillaries (7), which are then, within 3 to 4 weeks, reduced in number to give the vascular pattern of connective tissue.

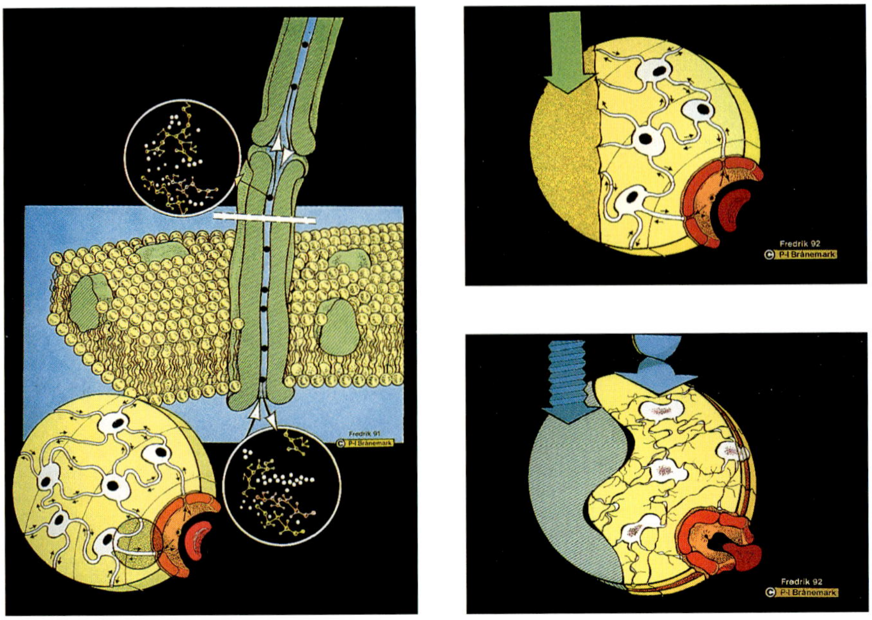

Fig 1-29 a–c Diagrammatic illustration of the possible effects of preparation trauma to the nutrient system in bone tissue at installation of fixtures. (From Brånemark P-I, Lausmaa J, Ericson L, Thomsen P, Brånemark R, Skalak R. Anatomy of osseointegration and the transfer of load. In Fonseca RJ, Davis WH [eds]: Reconstructive Preprosthetic Oral and Maxillofacial Surgery, ed 2. Philadelphia: WB Saunders 1995:165–224. Used with permission.)

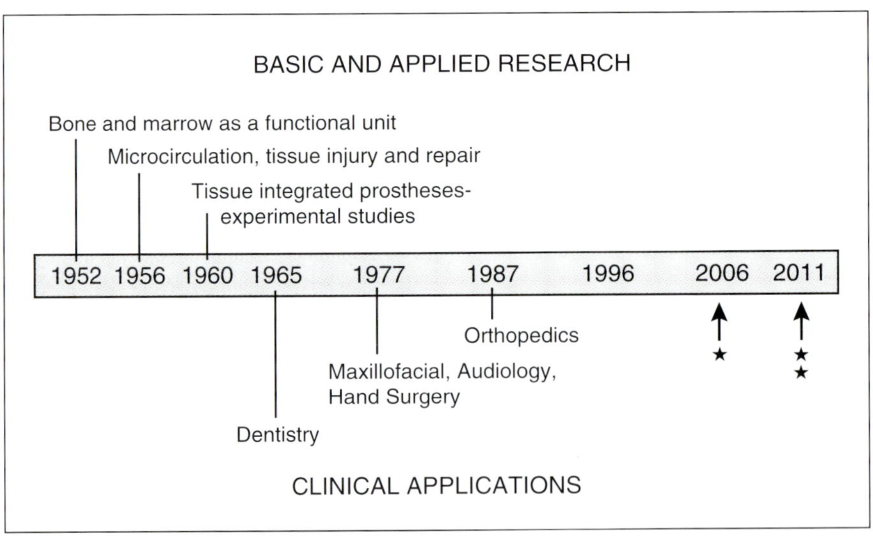

Fig 1-30 Development of the osseointegration concept from basic research to clinical application in long-term follow-up has covered four decades. Experience from endo-prosthetic reconstructions in orthopedics emphasizes the need for 10 to 15 years of follow-up of new or modified procedures before evaluation of clinical predictability can be established.

DISCUSSION

Björn Rydevik Thank you very much for this elegant and very interesting account of the development from the basic science level to the clinical applications and also the insight into possible future developments of osseointegration in various regions of the body. I would like to invite questions and comments to this presentation by Professor Brånemark. First, I would like to ask if there will be any comments that you may want to give in further detail on the possibility of extrapolating from the experience of the oral reconstructions, particularly for the completely edentulous patient vs a patient who has lost parts of an upper or lower extremity, ie, the amputee's problems.

Per-Ingvar Brånemark This is a very interesting question because I have been lecturing to dentists over the past 30 years and I thought in the beginning that it was straightforward. Then I understood that even dentists don't understand the impact of total edentulism. The dentureless patient loses teeth, loses jawbone, and loses self-confidence. So in many cases it could be worse for the patient to lose all teeth and be considered as socially inferior than to lose an arm. You might find this peculiar but that's a fact because if you lose an arm, people pity you, but they don't pity you for edentulism because you should not lose your teeth if it is not an accident or a result of trauma.

I now have tried to introduce the idea that the completely edentulous patient is equivalent to an amputee. He is an amputee; there is a part departed. It is not the volume or the consequence of removing that is important. People who are in their seventies have said that each tooth is as valuable as a diamond, and even if the dentist should not charge according to that, I think it is still a fact.

Then comes the important question, which I don't have an answer to—does it matter whether it's above or below the collar bone? In the beginning it was said, of course, from the orthopedic quarters, that of course it works in the mouth but everything works in the mouth for reasons that we don't understand. But outside the mouth below the collar bone it is entirely possible that it wouldn't work. But it seems to work there, as well.

It took us 10 years to understand how to provide the routine recipe for the treatment of straightforward total edentulism. Dr Worthington will tell you more about that. I think we have to accept that it will take another 10 years to understand how bone other than jaw bone behaves differently in relation to the osseointegrated prosthesis. But as far as I am concerned the question is: how can the osteoblast, osteocytes, and so forth know that they are not sitting here, but they are sitting there, or vice versa? Of course, you might argue that this is not true, but at least from a practical point of view, clinically, it seems to me that it works in both locations.

It is very interesting that there has been quite a heavy debate when it comes to dimensions, that the bigger and the more anchorage elements, the better. I believe, on the contrary, fewer. A small surface is more congenial. It is remarkable that if you have six teeth and you make a full bridge extended two or three teeth distally, it doesn't work predictably because there is too much load. If we take four fixtures, tiny ones, 3.72 mm in diameter and 10 mm long, 10-year follow-up studies show that they can take the full bridge with no problem. I am not suggesting that one should substitute all teeth with fixtures, but I am suggesting that we should challenge some prevailing ideas or possibly that there should be some change of prosthetic techniques in dentistry.

I think this implies that, in the other parts of the body, we need to follow that concept if it works. We are not sure yet, but we have seen, for instance, in ankle joints with Dr Carlsson that sometimes we overdid it. So there was too much anchorage, and we had fatigue because of the differences in bone capacity. Dr Skalak will possibly comment on that also and stress shielding. That is a current topic. I believe that if the body likes it, if mother nature enjoys it, then we need not worry much if we have the recipe. The sum of it all is that this could turn out as a cross-fertilization and help explain the limits for what is acceptable to the patient. Then I think we would have achieved something in a real beginning.

Richard Coutts I enjoyed your comments tremendously because they reminded me of the time when I was training and I was taught, as you had said, that nothing attaches to bone permanently. Then I went to Birmingham around 1970 and first visited Dr Charnley. That was the first time that this concept had been challenged in my own mind. Then we started seeing failures of those prostheses and then I said, yes, the basic principle still applies that bone will not take anything permanently. Then your work has come along with the long-term follow-up studies that suggested maybe there is something unique about the material and that bone and titanium have a certain affinity for each other. I am intrigued by the ability to put these titanium rods into bone and have them project through the skin and not get infected. Does the skin have an affinity for the titanium, as the bone does? Do fibroblasts like titanium surfaces, as well as osteoblasts?

PIB I don't know. What we know is that we had tremendous problems in the craniofacial region. I think there are by now 5,000 patients treated. We discovered that if you thin the skin to and bring it down to surface of the bone there is very little relative motion. Then there is virtually no problem even with the skin-penetrating prosthesis. We take a split skin graft or we trim the skin and we roughen the bone. Then you have bleeding bone and split skin on top of it and it is remarkable how it works over many years. So it seems to me that no relative motion is required, not only for the bone interface but also for the soft-tissue interface. Certainly it is easier in the skull than in a *high amputee femur*. One problem is, of course, how much can you remove ethically when you don't know how it will look after 20 to 30 years. We have tried to modify procedures and it seems that it is working quite well. I think that possibly Dr Rydevik will show you an interesting case of a colleague of ours, where the skin penetration is very obvious. We are still working on it. The problem is now, since we started some collaboration with Lebanon, that they have a large number of amputees. Can we really do it and let the patients return out there? We can do this for the thumb, but whether we can do it for the lower limb we don't know yet. It is still in the infancy.

RC Do any of them get infected?

PIB Yes, we get some skin infection, but not deep infection and then we can very easily handle it. Basically those cases in which we didn't do a radical resection of the subcutaneous tissue may have trouble. Now with the group of Dr Gunterberg, we tried to see to it also that we secure the muscles to the bone. Then the bone provides a very rigid soft tissue or connective tissue platform. We have placed bone grafts on top so that we could get the skin facing bone.

RC Then when you put the prosthesis on, all of the load is going through the prosthetic stem and so you are not getting any pressure on the skin.

PIB Yes, but we keep a space there for ventilation.

Göran Lundborg I have two questions. The first has to do with dental implants. We are having a problem with loosening in total hip/knee arthroplasty even in situations where we considered the components to be osseointegrated or rigidly fixed to the skeleton. We have attributed that to the biologic reaction of wear debris. I am wondering, have you seen pathologic bone resorption around the dental titanium implants, or defects in the bone, and if so, what do you attribute that to?

PIB We saw that in the early days before we understood that we should minimize injury. Bone can only take 42°C and still integrate. Deep down in the site it is very difficult to control the temperature. So we believe that the injury to the bone tissue is the crucial aspect. So we try to minimize injury by not using high-speed instruments. We use very slow or even manually driven things and I believe that avoiding damage to the bone is the key to it. Even Dr Charnley knew that there was a limit because he said, "Well, I killed the bone so it doesn't react." He took a patient who was about 75 years of age and then it worked, of course. I think this is very important, and even in dentistry people have not always understood that there is a limit to what the bone can take. I believe that the injury to the bone is an important part of this story.

GL That brings me to my second question. I noticed on your time line that you had orthopedics starting in 1987. Does that mean that you believe that cemented cases cannot become osseointegrated? Certainly Charnley has been doing these procedures since the 1960s. Do you feel that those are not osseointegrated? Maybe, if you do, we will have some discussion later about that.

PIB That is a very interesting question. I have to be a little bit careful now because one of my coworkers, Lars Linder, who worked with Robin Ling in England, has a slightly different opinion than I do. I recently read two reports on the 20-year follow-up studies of the Charnley prosthesis with very good results. I was impressed, but they also said that those patients had applied carefully controlled low loading, so they were not athletes. I believe that it is possible to make the bone accept even bone cement. Then it's a matter of very subtle balance between the functional level of the debris and the remodeling capacity of bone. So if we could pick the patients who were to have this range of activity, then I think it would work, but it is difficult to know in which patients you could do it or if you could do it with an individual case.

However, when Robin Ling and Lars Linder took bone cement and squeezed it around a prosthesis, it seems that here is a case in which there seemed to be living bone. This is crucial because in cases where there is a long-standing amputation, the bone may be in very bad condition. How can we improve the condition for retention? What can we add there? It is a common biologic problem and I don't think we, at this time, can predict the results in every case. We have to observe each case individually to document what happens.

GL Speaking about the concept of osseointegration, this can be defined in many ways. We have defined it earlier as a process where there is no intervening tissue between implant and bone. It is also defined as no relative motion. Even if the implant never becomes incorporated into the bone, you quoted somebody as saying, "At least the bone does not regard the implant as a foreign body." So I wonder, if you put a prosthesis on such an implant, does that prosthesis or that tooth become integrated into that person in his mind? In other words, does that person prefer to take out that prosthetic tooth or teeth construction when he goes to sleep, or does he take off the limb prosthesis, or does he regard it as a part of his body? Do you see my question?

PIB For the parts of the body below the collar bone, we don't know yet but we know surely people say, "It's my own teeth: I do not consider this as a prosthesis. It's part of my body, it's a third dentition." The intriguing thing is that the only patients who have wanted a fixture to be removed are the *dysmorphophobic* patients. This is very important because if we consider the finger prosthesis or the limb prosthesis, it could be that one reason it is incorporated in their entire personality is because of the perception that it is a part of them.

BR So you are implying that the prosthesis may become psycho-integrated; the prosthesis affects the function of the mind.

Shu Chien I have really enjoyed your wonderful presentation. You have shown these very impressive clinical results with the great success of osseointegration; also the lovely pictures on the microcirculation. Could you comment on the role of the microcirculation in the success of the osseointegration?

PIB I believe that if we cannot maintain an adequate microcirculation and fluid transfer, it will not work. That is why I am so concerned about situations where people try to speed up the procedure just to save a few minutes and they loose the possibility for the blood to reach the interface. Then it will not work, but it might take several years before the nearby bone degenerates and the fixture loosens. I believe that if we do not use a surgical preparation technique that preserves the microcirculation of the bone, osseointegration will not work predictably, but it is difficult to convince everyone of this.

References

1. Brånemark P-I. Introduction to osseointegration. In Brånemark P-I, Zarb G, Albrektsson T (eds). Tissue-Integrated Prostheses: Osseointegration in Clinical Dentistry. Chicago: Quintessence; 1985:11–76.
2. Brånemark P-I, Worthington P. Osseointegration in maxillofacial reconstruction. In Williams J (ed): Maxillofacial Injuries, ed 2, vol II. Edinburgh: Churchill Livingstone; 1994:1027–1052.
3. Brånemark P-I, Worthington P. Rehabilitation after mandibular reconstruction: Osseointegrated implants. In Komisar A (ed): Mandibular Reconstruction. New York: Thieme Medical (in press).
4. Kasemo B, Lausmaa J. Surface science aspects of inorganic biomaterials. CRC Crit Rev Biocompatibility 1986;2:4, 335–380.
5. Collins DH. Tissue changes in human femurs containing plastic appliances. J Bone Joint Surg 1954; 30B:458.

6. Gillis L. Modern Trends in Surgical Materials. London: Butterworth; 1958.
7. Cook HP. Immediate reconstruction of the mandible by metallic implant following resection for neoplasm. Ann R Coll Surg Engl 1968;42:233.
8. Southam JC, Selwyn P. Structural changes around screws used in the treatment of fractured human mandibles. Br J Oral Surg 1970;8:211.
9. Palmer I. Foga Samman. Sweden: Bonniers Förlag; 1967.
10. Brånemark P-I, Breine U, Lindström J, Adell R, Hansson BO, Ohlsson Å. Intra-osseous anchorage of dental prostheses. I. Experimental studies. Scand J Plast Reconstr Surg 1969;3:81–100.
11. Brånemark P-I, Hansson BO, Adell R, Breine U, Lindström J, Öhman A. Osseointegrated implants in the treatment of the edentulous jaw: Experience from a 10-year period. Scand J Plast Reconstr Surg 1977;11: suppl 16.
12. Dorland's Illustrated Medical Dictionary, ed 28, s.v. "osseointegration."
13. Brånemark P-I. In Program book of the Fourth Annual Meeting of the Academy of Osseointegration, March 3–4, 1989, Dallas. Academy of Osseointegration, 1989.
14. Brånemark P-I. Vital microscopy of bone marrow in rabbit. Scand J Lab Invest 1959; suppl 38:11, 1–82.
15. Brånemark P-I, Breine U, Johanson B, Roylance PJ, Röckert H, Joffey JM. Regeneration of bone marrow: A clinical and experimental study following removal of bone marrow by curettage. Acta Anat 1964;59:1–46.
16. Brånemark P-I. Bone marrow microvascular structure and function. Adv Microcirc 1968;1:1–65.
17. Hallén O, Magnusson S, Jacobsson M, Marké L-Å. Bone-Anchored Implants in the Head and Neck Region. Report from the State of the Art Conference, Gothenburg, September 1, 1988. Swedish Council on Technology Assessment in Health Care.
18. Hallén O, Brånemark P-I, Lindström J, Tjellström A. Preformed autologous ossicles: Experimental studies. Acta Otolaryngol 1976;82:394–401.
19. Tjellström A. Osseointegrated systems and their applications in the head and neck. Adv Otolaryngol Head Neck Surg 1989;3:39–70.
20. Laney W, Tolman D (eds). Tissue Integration in Oral, Orthopedic & Maxillofacial Reconstruction. Proceedings of the Second International Congress on Tissue Integration in Oral, Orthopedic and Maxillofacial Reconstruction. Mayo Medical Center, Rochester, Minnesota, 1990. Chicago: Quintessence; 1992.
21. Worthington P, Brånemark P-I (eds). Advanced Osseointegration Surgery: Applications in the Maxillofacial Region. Chicago: Quintessence; 1992.
22. Rydevik B, Brånemark P-I, Skalak R (eds). International Workshop on Osseointegration in Skeletal Reconstruction and Joint Replacement. Report from Conference, Aruba, Apr 1990. Gothenburg: The Institute for Applied Biotechnology; 1991.
23. Brånemark R, Skalak R, Brånemark P-I. Osseointegration and rigid fixation. In Yaremchuk MJ, Gruss JS, Manson PN (eds): Rigid Fixation of the Craniomaxillofacial Skeleton. New York: Butterworth-Heinemann; 1992.
24. Lundborg G, Brånemark P-I, Carlsson I. Metacarpophalangeal joint arthroplasty based on the osseointegration concept. J Hand Surg 1993;118B: 693–703.
25. Eriksson E, Brånemark P-I. Osseointegration from the perspective of the plastic surgeon. Plast Reconstr Surg 1994;93:1, 628–637.
26. Johnson RW. A physiological study of blood supply of the diaphysis. J Bone Joint Surg 1927;9:153.
27. Göthman L. Vascular reactions in experimental fractures. Acta Chir Scand 1961;suppl 284.
28. Röhlich K. Über die Verbindungen zwischen den Gefässen des Knochens und des Knochenmarkes. Verh Anat Ges 1939;88:242.
29. Editorial. Osseointegrated dental implants. Lancet 1986;ii:26.
30. Editorial. Osseointegrated dental implants here to stay. Lancet 1990;335:1190–1191.
31. van Steenberghe DA. A retrospective multicenter evaluation of the survival rate of osseointegrated fixtures supporting fixed partial prostheses in the treatment of partial edentulism. J Prosthet Dent 1989;61:2, 217.
32. Granström G, Bergström K, Tjellström A, Brånemark P-I. A detailed analysis of titanium implants lost in irradiated tissues. Int J Oral Maxillofac Implants 1994;9:6.
33. Haraldson T. Functional evaluation of bridges on Osseointegrated implants in the edentulous jaw [thesis]. Gothenburg, Sweden: University of Gothenburg; 1979.
34. Lundborg G, Brånemark P-I, Rosén OT. Osseointegrated thumb prostheses: A concept for fixation of digit prosthetic devices. Am J Hand Surg 1996; 21A:216–221.
35. Brånemark P-I. Intravascular anatomy of blood cells in man. Monograph. Basel: S Karger; 1971: 1–80.
36. Brånemark P-I, Lausmaa J, Ericson L, Thomsen P, Brånemark R, Skalak R. Anatomy of osseointegration and the transfer of load. In Fonseca RJ, Davis WH (eds): Reconstructive Preprosthetic Oral and Maxillofacial Surgery, ed 2. Orlando, FL: WB Saunders (in press).
37. Williams E. A Matter of Balance. Sweden: AB Akademiförlaget; 1992.

Chapter 2

History, Development, and Current Status of Osseointegration as Revealed by Experience in Craniomaxillofacial Surgery

Philip Worthington

The purpose of this chapter is to set the scene for later contributors by describing the history, development, and, based on experience in craniomaxillofacial surgery, the current status of osseointegration. Within the given constraints, this can be only a synoptic account, and it cannot begin to do justice to the many years of hard work devoted by Professor Brånemark and his associates to the relevant research, clinical trials, and progressive refinement of osseointegration surgery.

Osseointegration has been defined as "a direct structural and functional connection between ordered living bone and the surface of a load carrying implant." At a later stage this definition was augmented by saying that "an implant is regarded as osseointegrated when there is no progressive relative movement between the implant and the bone with which it is in direct contact." What it means in effect is that in osseointegration we have an *anchorage mechanism* whereby nonvital components can be reliably and predictably incorporated into living bone and that this anchorage can persist under all normal conditions of loading.

To understand how revolutionary this concept was at the time osseointegration was discovered, it is necessary for us to return in time to the thinking of the 1950s and 1960s. For example, in 1954 Collins[1] wrote " . . . although biologically inert, an implanted object . . . never becomes incorporated into bone." As late as 1971, Southam et al[2] wrote, "Whenever any metallic appliance is implanted into bone, a layer of fibrous tissue will always develop around the appliance which subsequently will never be as secure in the bone as it was at the time it was implanted."

Those of us who for years repaired broken bones with stainless-steel screws and wires were familiar enough with the fact that screws did eventually loosen and bone did recede from metallic surfaces. The concept of osseointegration was novel, and initially it was scorned and rejected. Over the years, it has gained widespread acceptance by the scientific community. Its discovery came about in this way; in the course of investigating the microcirculation of bone and its repair processes, Brånemark used an established method of microscopic observation through an implanted observation chamber[3] (Fig 2-1). This device, set into the long bone of an experimental animal, had in it a small gap through which vital bone could grow and where it could be observed in vivo. The observation chamber technique was not new—but what *was* new was the finding that when the time came for retrieval of the observation chambers, those made of the metal titanium could not be removed from the bone. The liv-

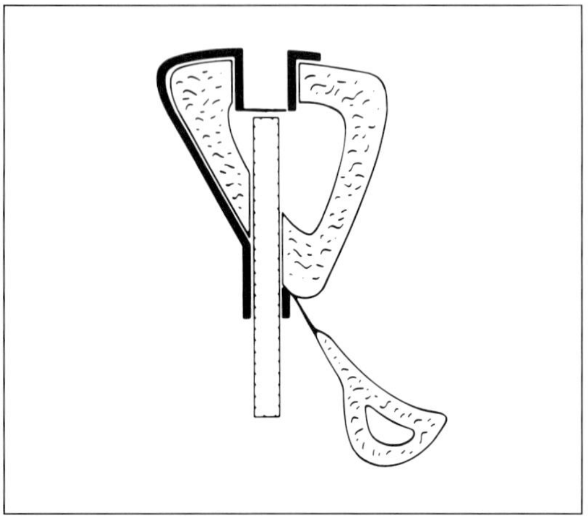

Fig 2-1 Cross-sectional diagram showing observation chamber in rabbit tibia. (Figs 2-1 to 2-3 from Brånemark P-I. Introduction to Osseointegration. In Brånemark P-I, Zarb G, Albrektsson T [eds]: Tissue-Integrated Prostheses: Osseointegration in Clinical Dentistry. Chicago: Quintessence; 1985:11–76. Used with permission.)

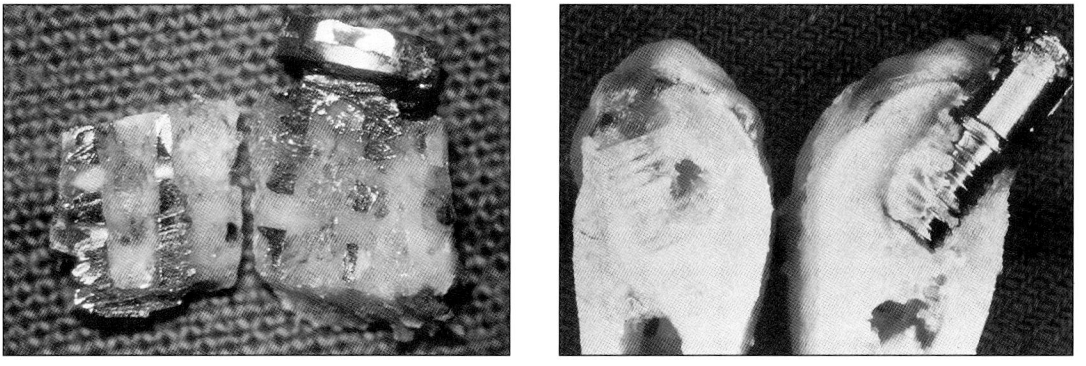

Fig 2-2 a and b Implants securely integrated into bone.

ing bone had so fused with the titanium oxide layer on the surface of the metal that the device was strongly incorporated into the bone. Brånemark had the foresight to recognize the enormous significance of this finding and subsequently spent his time and energies investigating the phenomenon of osseointegration and its potential application.

There followed years of animal experiments during which the reproducibility of osseointegration was confirmed (Fig 2-2). The strength of the implant-bone union was tested (Fig 2-3) and the factors essential for successful osseointegration were identified. It must be admitted that even now we do not know enough about the biology of osseointegration. We do know, however, that at the bone/implant interface there is ionic interchange between the body tissues and the layer of oxides that coats the bulk metal (Fig 2-4). Hydrogen, oxygen, calcium, and phosphorus ions diffuse into the oxide layer. We also know that this layer of oxides increases in thickness with the passage of time; it is essentially a ceramic layer of several different oxides (Fig 2-5), and if disturbed it immediately re-forms in any oxygen-containing atmosphere.

The bonding between implant and bone is unique to titanium. It is as if the body cannot

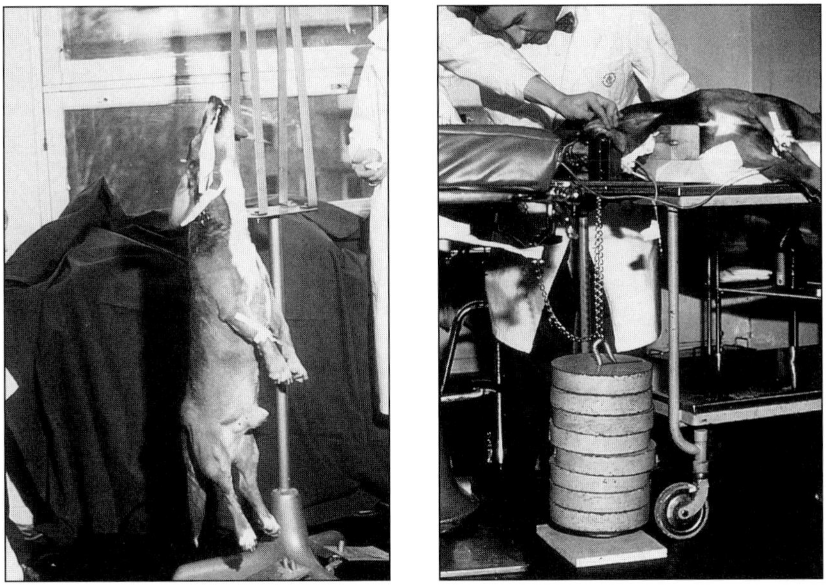

Fig 2-3 a and b Experimental testing of the strength of the bone-implant bonding.

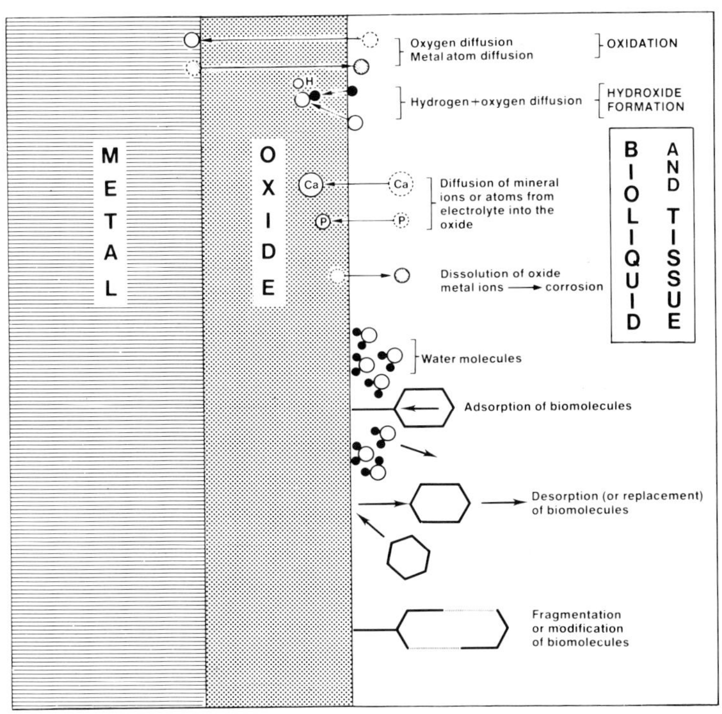

Fig 2-4 Diagrammatic representation of chemical processes that can take place at the interface at a molecular level. (From Kasemo B. Biocompatibility of titanium implants: Surface science aspects. J Prosthet Dent 1983;49:832. Used with permission.)

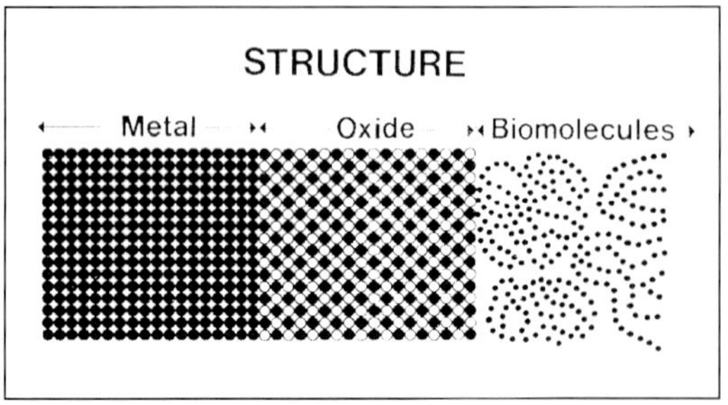

Fig 2-5 Between the biomolecules of the body and the bulk metal of the titanium implant is a layer of oxides of titanium.

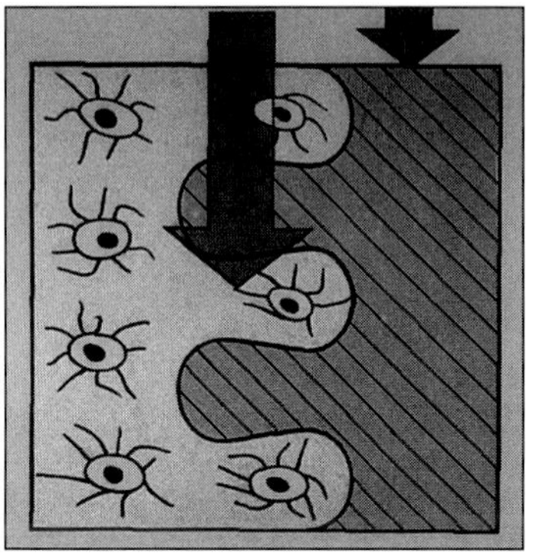

Fig 2-6a Diagrammatic depiction of the direct bone-titanium contact without any intervening layer of less differentiated tissue, thus allowing direct transfer of load from implant to bone.

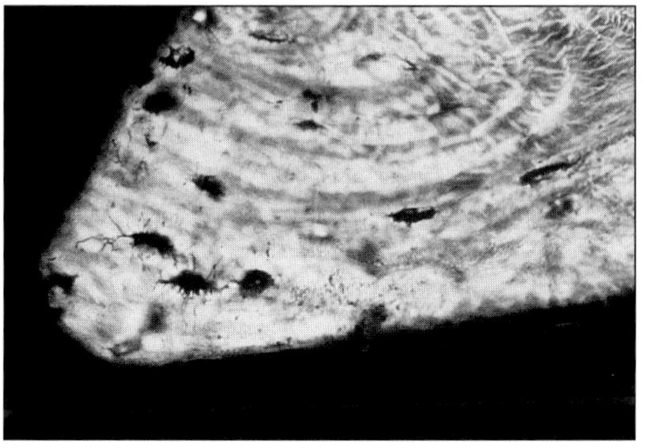

Fig 2-6b The direct bone-implant contact at the light microscopic level. Viable osteocytes are close to the implant surface.

recognize that the titanium is nonself—possibly because titanium and calcium are so close together in the periodic table of elements. The essential feature of osseointegration is that at the light microscopic level, bone is in direct apposition to the titanium with no intervening layer of fibrous tissue (Fig 2-6).

Eventually animal experimentation led to clinical trials, starting in 1965. Curiously, by choosing to investigate the dental application of osseointegrated titanium implants, Brånemark chose perhaps the most severe testing laboratory—the mouth. Teeming with bacteria, bathed in saliva, and dependent on oral hygiene in a situation where mucoperiosteal penetration was necessary for function—this was not an easy situation.

It was soon established that the completely edentulous mandible could be predictably re-equipped with an artificial dentition by means of prostheses anchored to the jaws with endosseous titanium implants (Fig 2-7).

Fundamental to success were several factors, defined by Brånemark and assuredly bound to appeal to any orthopedic surgeon. These factors were:

- Complete sterility
- Avoidance of contamination
- Minimal tissue violence
- Maximal precision

Some explanation of these items may be in order. Minimal tissue violence implies a very gentle handling of the tissues, both soft

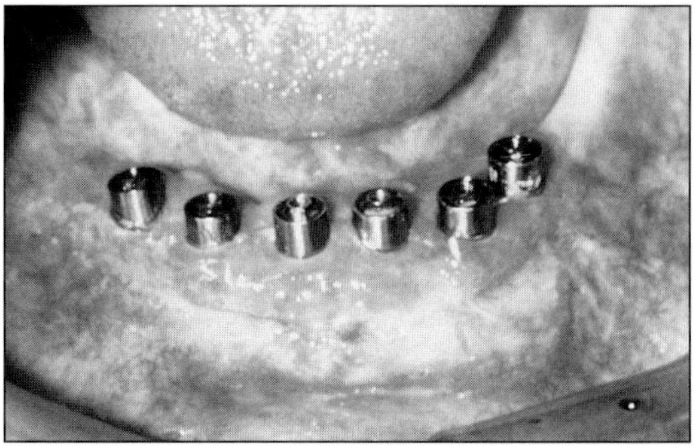

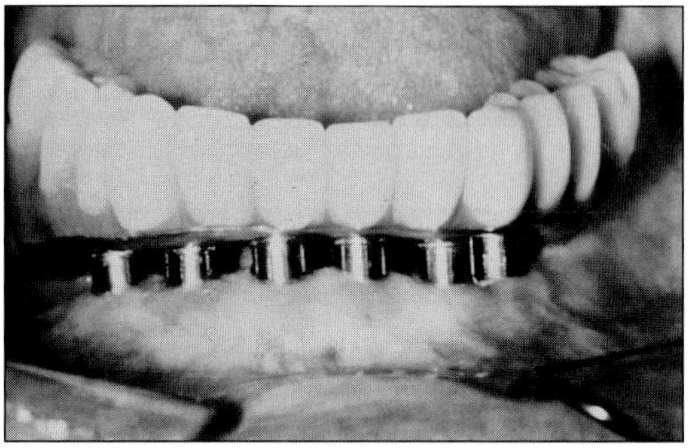

Fig 2-7 a and b Many years ago, clinical trials showed that it was feasible to provide the edentulous mandible with a complete implant-supported dental prosthesis that would withstand the physiologic loads of mastication.

and hard; the use of sharp, disposable drills with light and intermittent drilling pressure; the incremental removal of bone by using drills of increasing diameters; and above all, the use of copious coolant irrigation.

Maximal precision implies manufacture of instruments and components to fine tolerances to achieve close congruence between the implant and its bony site and an accurate fit of components.

Avoidance of contamination means that the integrating titanium surfaces, supplied in a reactive state, should not be allowed to touch other metals, rubber gloves, cotton towels, etc, because chemical elements adhering to the surface may deny the chance of osseointegration. Complete sterility implies the exercise of strict aseptic operating room technique and needs no justification.

In the limited time available the surgical technique of implant installation cannot be described in detail, except to say that it is a two-stage procedure.[4] At the first stage, implants are placed into the bone by means of a very gentle surgical technique avoiding as far as possible any thermal injury to the bone. The implants are then buried under the mucosa and excluded from the oral environment during the obligatory healing period, during which there must be no loading (Fig 2-8). During this time the bone heals and reorganizes. Permucosal attachments are then added to the implants, and the definitive fixed prosthesis is attached to these abutments, which now penetrate the mucosa (Fig 2-9).

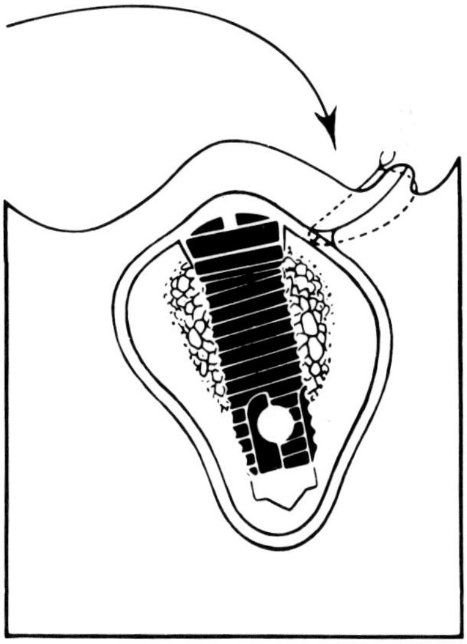

Fig 2-8 Cross section of mandibular symphysis showing buried implant in place, with mucoperiosteal flap replaced to exclude the implant from the oral environment. Note that the incision is remote from the implant head. (Figs 2-8 and 2-9 from Adell R, Lekholm U, Brånemark P-I. Surgical Procedures. In Brånemark P-I, Zarb G, Albrektsson T [eds]: Tissue-Integrated Prostheses: Osseointegration in Clinical Dentistry. Chicago: Quintessence; 1985: 211–232. Used with permission.)

It is essential that the healing period be respected both in duration and in avoidance of

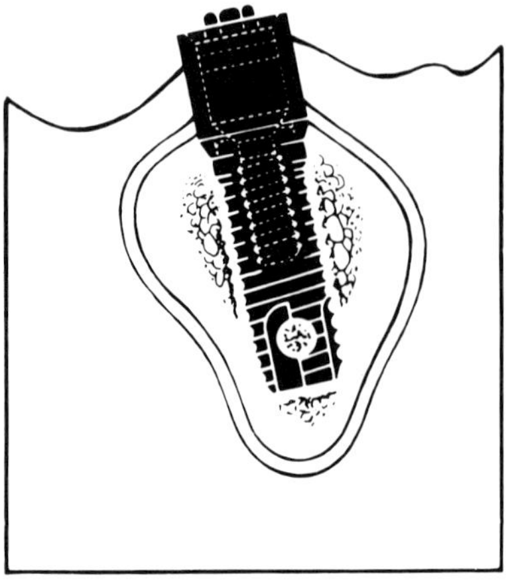

Fig 2-9 Diagram of second-stage surgery at which the abutment cylinder has been attached to the implant and penetrates the oral mucosa.

any transmitted load. Initial implant stability is also very important. When loading is eventually applied, it must be within certain physiologic limits to provide the bone with stimulation without causing overload. The natural wound healing processes are harnessed and respected.

This path of development was what one could call an ethical one. We should consider unethical vs ethical development. By unethical development we mean the premature marketing of an idea without proper testing, without animal trials, without replication studies. It implies the use of the patient as the experimental animal. If successful, the inventor retires in affluence; if unsuccessful, he retires into obscurity. In the long and not always distinguished history of oral implantology, such unethical development is not exactly unknown. By contrast, ethical development involves the development of a design prototype, animal research leading to modifications of design, further experiments, controlled clinical trials, and finally a product that can be released for general clinical use with appropriate training being provided. In oral implantology this latter path has seldom been followed.

Over the years, the development of osseointegration has progressed and at varying intervals its validity has been confirmed by replication studies in several countries and by many multicenter trials. The applications extended from the completely edentulous mandible to the edentulous maxilla, to the partially edentulous case and the single tooth replacement.

The oral applications were then studied in special patients; the patient with severe bony atrophy; the medically compromised patient; the ectodermal dysplasia patient who not only had oligodontia but also hypoplasia of the jaws. The irradiated patient can also be treated, although with great care and with preliminary prophylactic hyperbaric oxygen. Osseointegration has been of great help in head and neck cancer reconstruction. And finally it was used in extraoral sites for craniomaxillofacial ablations where it has been invaluable for the attachment of maxillofacial prostheses. Let us consider some examples.

The completely edentulous patient is typified by the patient shown in Figure 2-10. The face is prematurely aged, with sunken cheeks and heavy facial lines. Although the mandible is atrophic, the reconstruction based on osseointegrated implants provides not only a firm fixed lower partial denture, but also improved lip support. The patient enjoys better chewing ability and better esthetics, and is psychologically uplifted.

Fig 2-10 a–f Typical complete denture patient with advanced alveolar bone resorption. Note the sunken cheeks, heavily lined face, and the marked bone loss on radiographs. Function and appearance are much improved with a lower prosthesis anchored by osseointegrated implants, which provide absolute stability. (Prosthodontic treatment by Dr CL Bolender.)

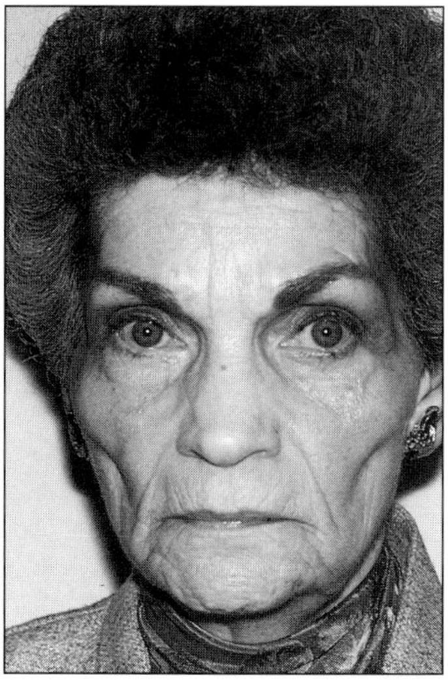
a

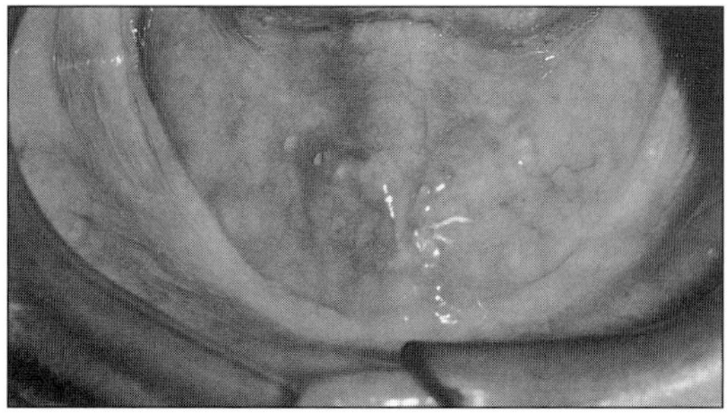
b

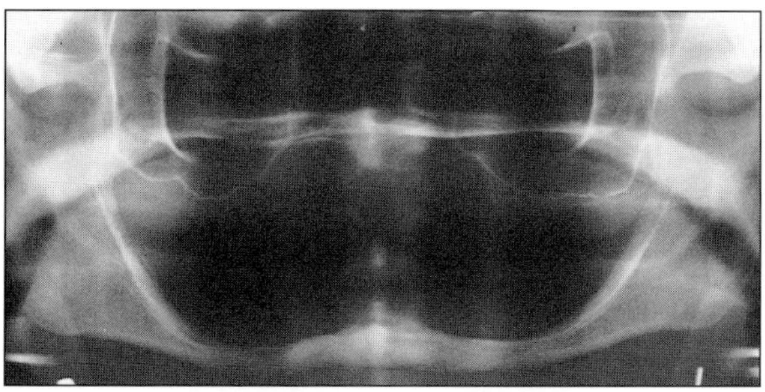
c

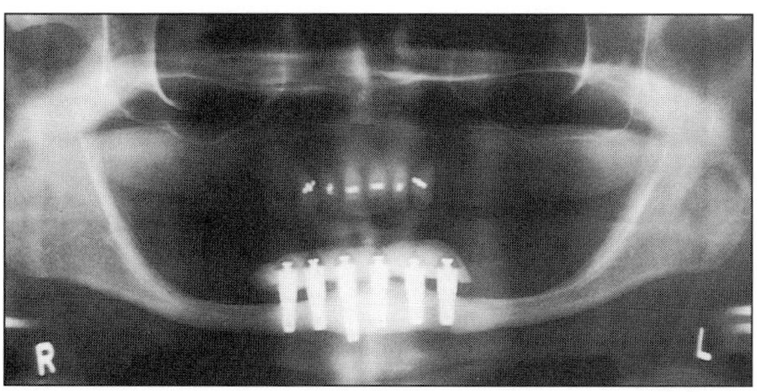

d

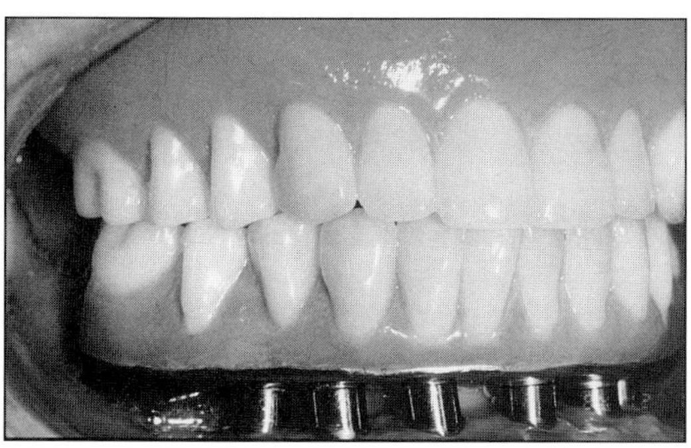
e

f

Fig 2-11 a–d This patient had had a benign but extensive lesion of the left mandible removed. Reconstruction with implants to support a segmental prosthesis avoids the need to prepare natural, healthy teeth as abutments and also avoids the disadvantages of a removable partial denture. (Prosthodontic treatment by Dr TD Taylor.)

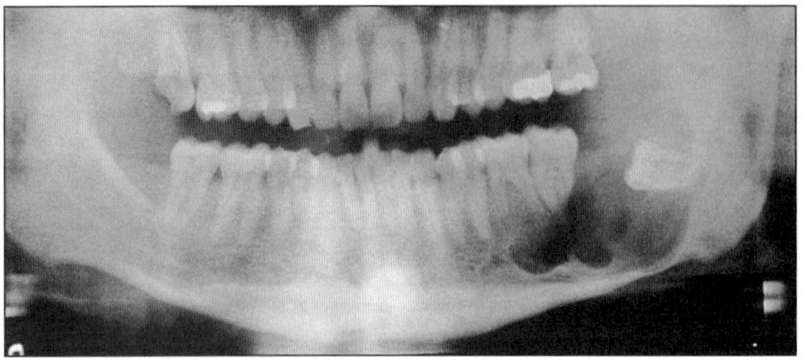

Fig 2-11a

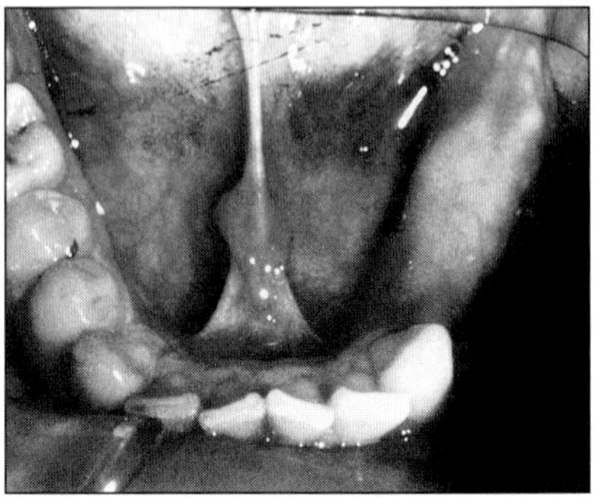

Fig 2-11b

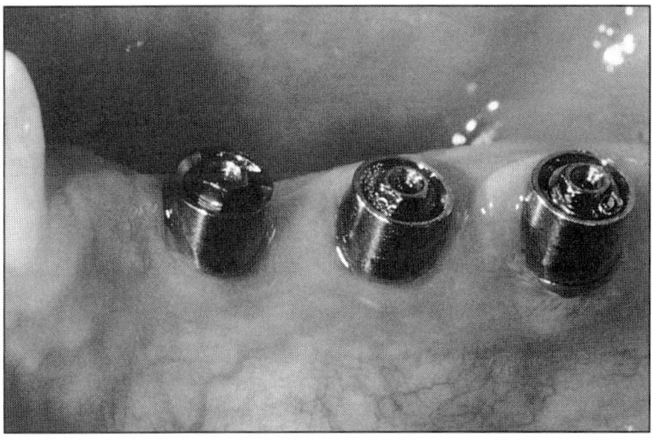

Fig 2-11c

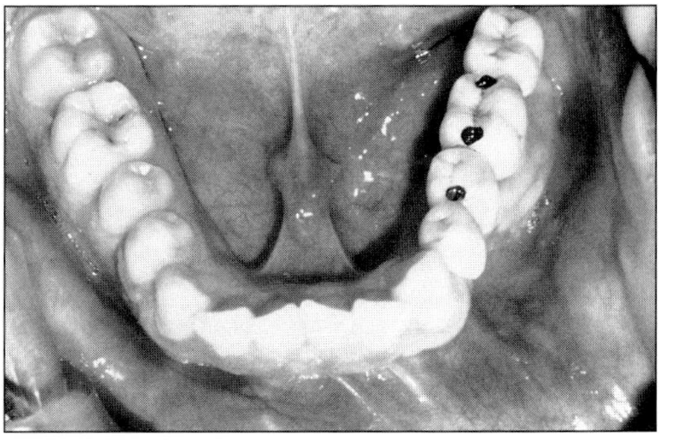

Fig 2-11d

An example of a partially edentulous patient is shown in Figure 2-11. This Native American patient had regenerated a reasonable amount of posterior mandibular bone after removal of a benign but extensive lesion. It was possible to place three implants to secure a partial fixed prosthesis. This avoids preparing the healthy natural teeth for use as abutments and also avoids the disadvantages of a removable partial denture.

The patient with ectodermal dysplasia may show significant underdevelopment of the jaw bones, which usually contain very few, if any, developing teeth (Fig 2-12). The child may then be seriously handicapped, not only in speech and mastication, but in a psychosocial sense too. Fixed anchorage by means of osseointegrated implants provides the mechanism whereby prostheses may be attached, and occlusal forces are then transmitted into the center of the mandible, which provides helpful stimulation to the bone, rather than surface pressure as from a conventional denture, which promotes bony resorption.

The young woman shown in Figure 2-13 suffered from an unusual form of jaw sarcoma, for which a full-thickness resection was necessary. Eventually reconstruction by means of a bone graft from the ilium was

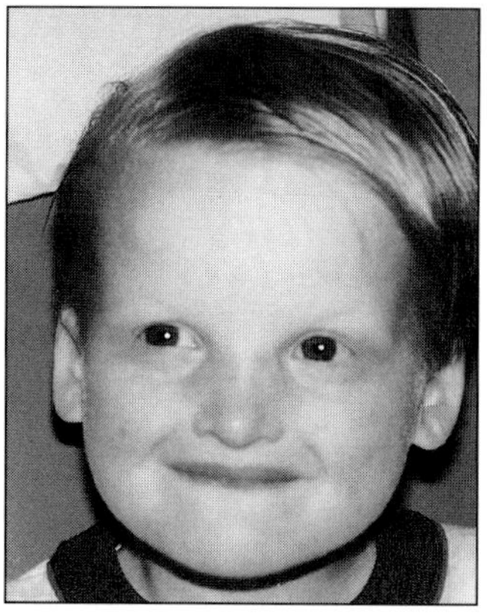

Fig 2-12a

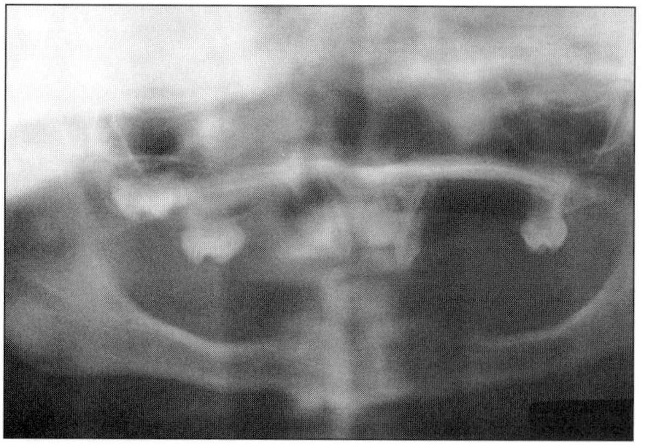

Fig 2-12b

Fig 2-12 a–d Ectodermal dysplasia patient aged 5 years with hypodontia and mandibular hypoplasia. Anchorage for a lower prosthesis was provided by osseointegrated implants, with a vast improvement in function and psychosocial state for the patient. (Prosthodontic treatment by Dr TD Taylor.) (From Worthington P, Brånemark P-I. Advanced Osseointegration Surgery: Applications in the Maxillofacial Region. Chicago: Quintessence; 1992. Used with permission.)

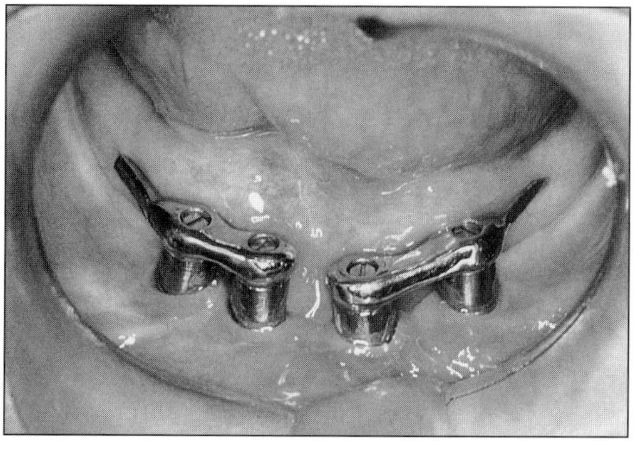

Fig 2-12c

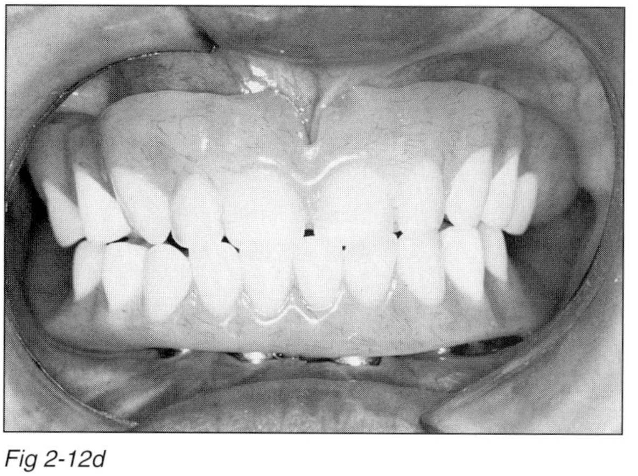

Fig 2-12d

Fig 2-13 a and b A young woman with a large, unusual form of sarcoma of the mandible.

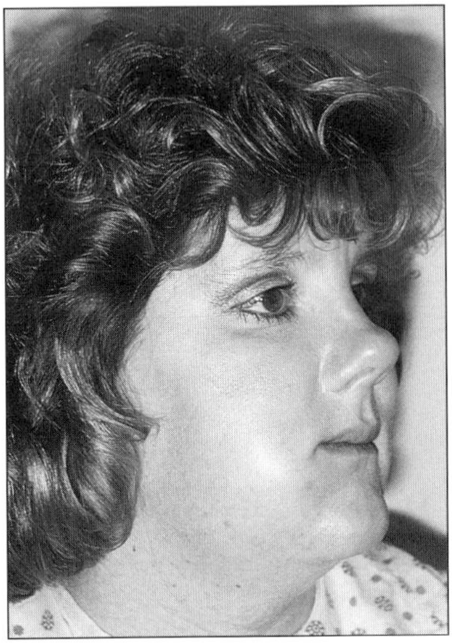

Fig 2-13a

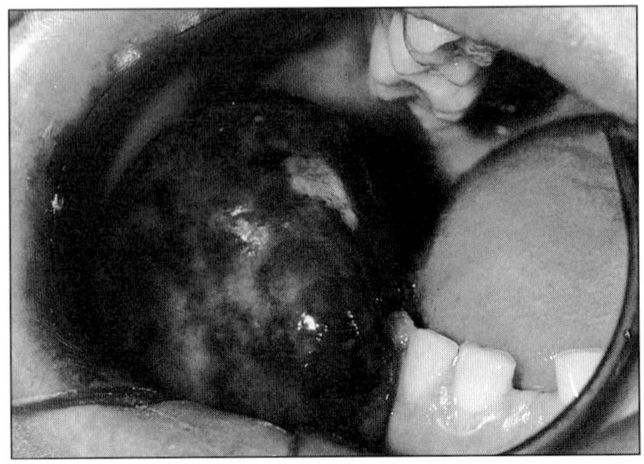

Fig 2-13b

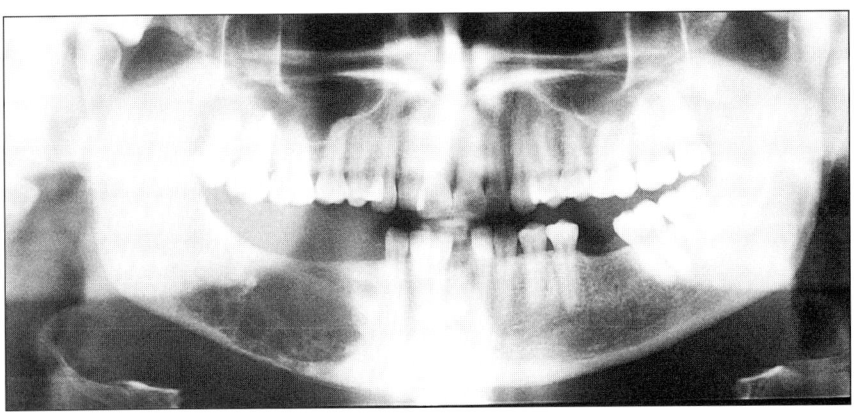

Fig 2-13c Intraosseous extent of tumor (dark area, lower left).

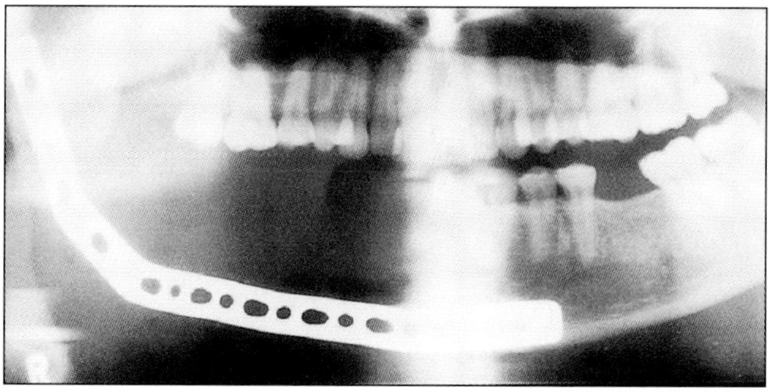

Fig 2-13d

Fig 2-13 d–f After resection a reconstruction plate was used to maintain continuity of the jaw pending bone grafting (d). Eventual rehabilitation was by means of bone grafting from the ilium and endosseous implants to support a segmental prosthesis (e). Note that the left lower first molar has been similarly replaced (f). (Temporary prosthodontic treatment by Dr J Rubenstein.)

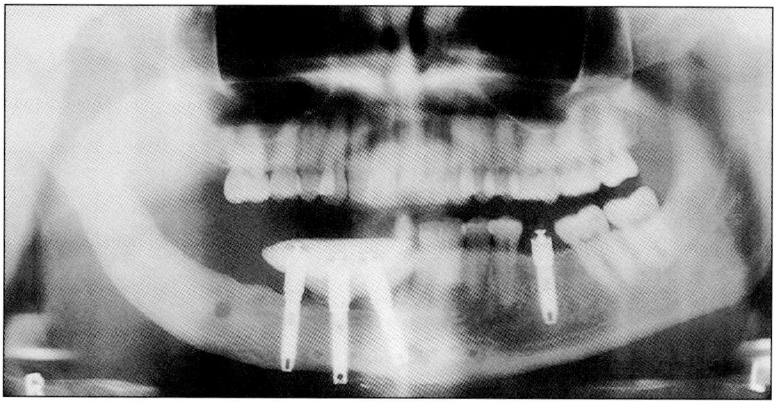

Fig 2-13e

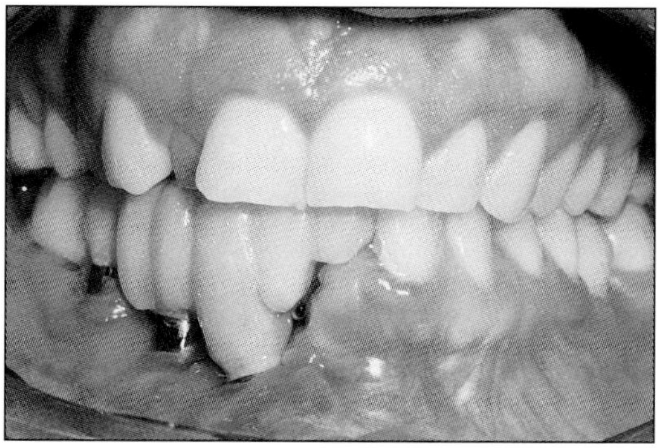

Fig 2-13f

performed, and her rehabilitation was completed by means of osseointegrated implants and a segmental prosthesis.

Another sarcoma resection patient is illustrated in Figure 2-14. Following immediate bone graft reconstruction, the patient's external contours are restored in a pleasing way, but he is still greatly handicapped by the loss of alveolar bone and teeth from almost a whole quadrant of the mouth. Implants carrying a special prosthesis provide a worthwhile solution. Mobility of peri-implant soft tissues combined with difficulty in oral hygiene caused some soft-tissue hyperplasia. This was remedied by interposing a split-skin graft along the buccal aspect of the mandibular abutments, thus providing a band of attached tissue.

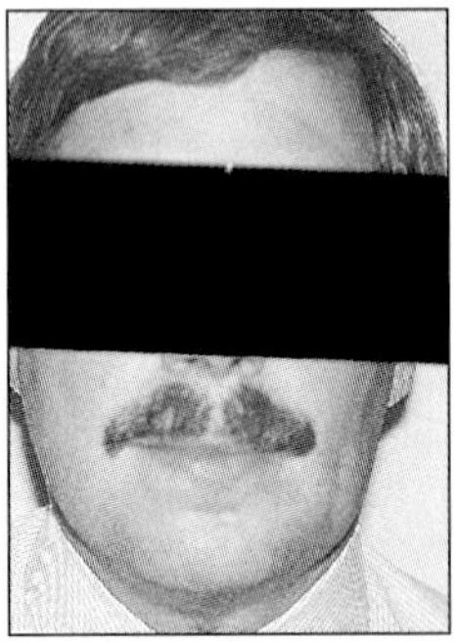

Fig 2-14a

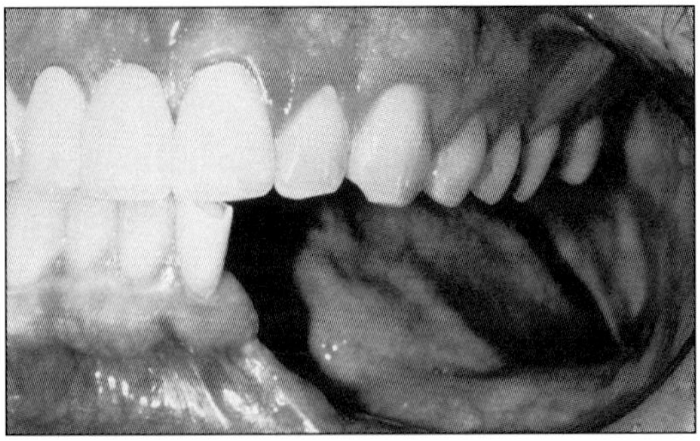

Fig 2-14b

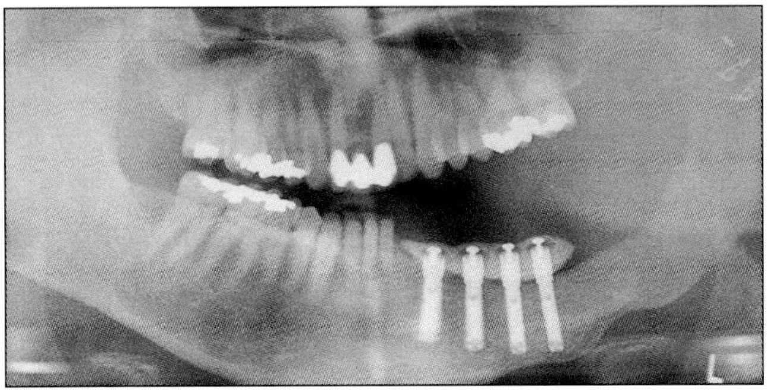

Fig 2-14c

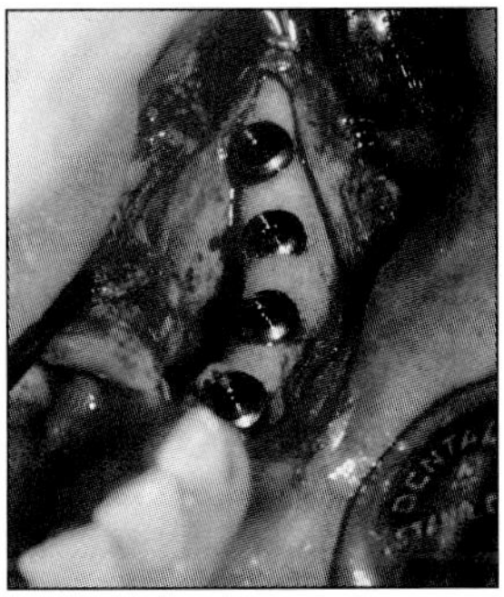

Fig 2-14d

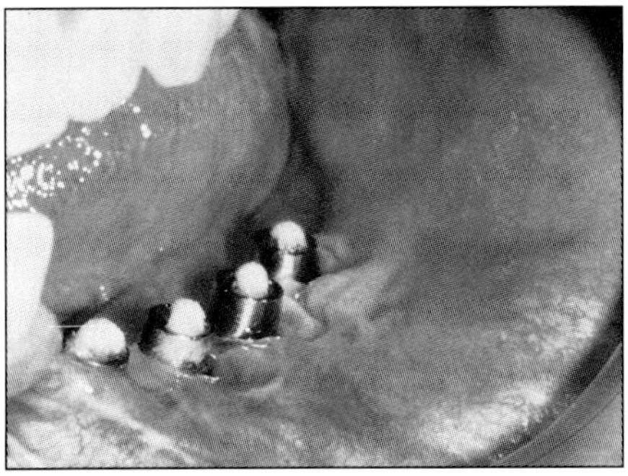

Fig 2-14e

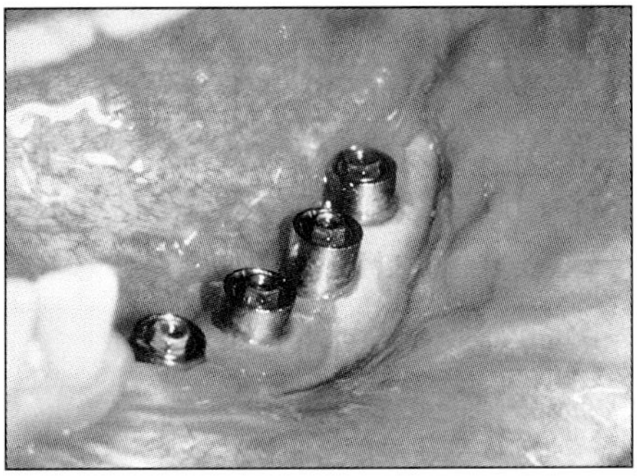

Fig 2-14f

Fig 2-14 a–g Young man with a jaw sarcoma after resection and bone graft reconstruction. External contours are well restored (a), but the patient is nevertheless handicapped by loss of alveolar bone and teeth (b). Reconstruction involved iliac bone grafting and implant replacement (c and d): the problem of soft-tissue irritation alongside the abutments was managed by means of a split-skin graft applied buccally, thus providing a band of nonmobile attached tissue (f). Mastication, speech, and appearance are vastly improved with an implant-supported prosthesis (g). (Prosthodontic treatment by Dr CL Bolender.) (From Worthington P, Brånemark P-I. Advanced Osseointegration Surgery: Applications in the Maxillofacial Region. Chicago: Quintessence; 1992. Used with permission.)

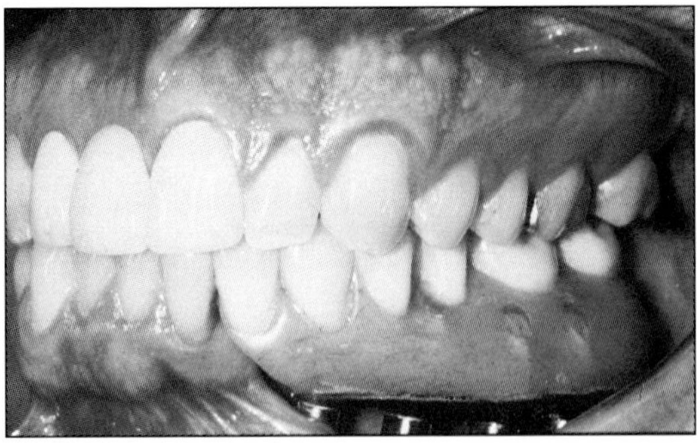

Fig 2-14g

Fig 2-15 a–d Patient after extensive ablation of mucoepidermoid carcinoma affecting the right parotid, mandible, and parapharyngeal region. Reconstruction with an iliac bone graft and implants. Note that in this case, as in many cancer-reconstruction patients, it is prudent to leave a large space beneath the prosthesis for easier access for hygiene purposes. The irradiated tissues are very injury-prone and must be handled gently. (Prosthodontic treatment by Dr J Brudvik.)

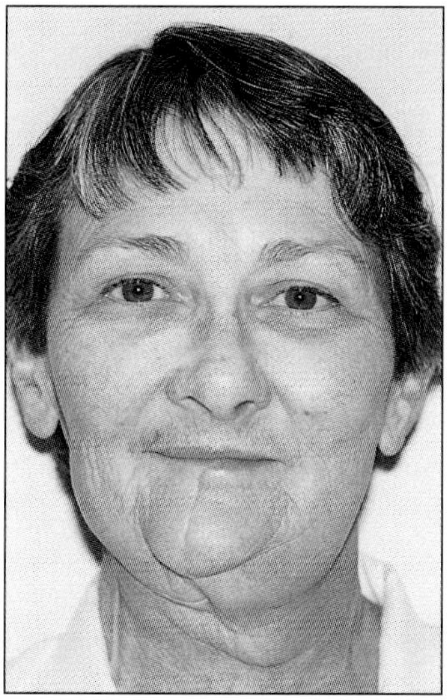

Fig 2-15a

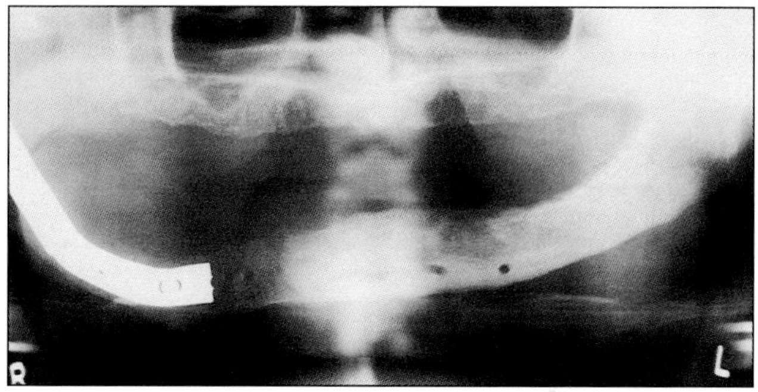

Fig 2-15b

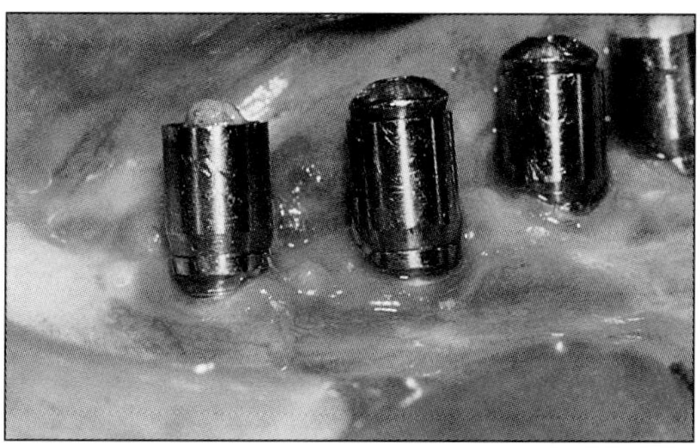

Fig 2-15c

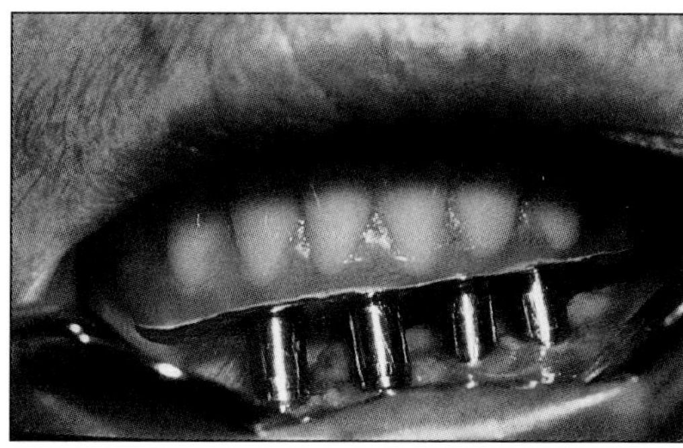

Fig 2-15d

Many previously irradiated patients have been treated successfully with implant reconstruction. However, this calls for extreme care and usually demands prior hyperbaric oxygen therapy as a prophylactic measure. Figure 2-15 shows a female patient following extensive surgery to remove a very large mucoepidermoid tumor of the right parotid and parapharyngeal region. She had had a hemimandibulectomy performed as part of her ablation and the surgery was followed by radiation. After reconstruction of the mandible with a bone graft and prophylactic hyperbaric oxygen, implants were used to stabilize a full fixed lower partial denture. Only minor complications were encountered.

Extraoral applications of osseointegration include anchorage for craniofacial prostheses. When the external ear has been removed due to tumor or trauma, a satisfactory replacement can be made by the maxillofacial prosthodontist and the artificial pinna is anchored to the temporal bone by means of special implants and a framework. Figure 2-16 shows an example. In a similar manner implants placed around the orbital rim can

Fig 2-16 a–d An artificial pinna is attached to a framework, which is in turn attached to specially designed implants placed into the temporal bone. This is far preferable to an adhesive-retained prosthesis. (Maxillofacial prosthodontic treatment by Dr J Rubenstein.)

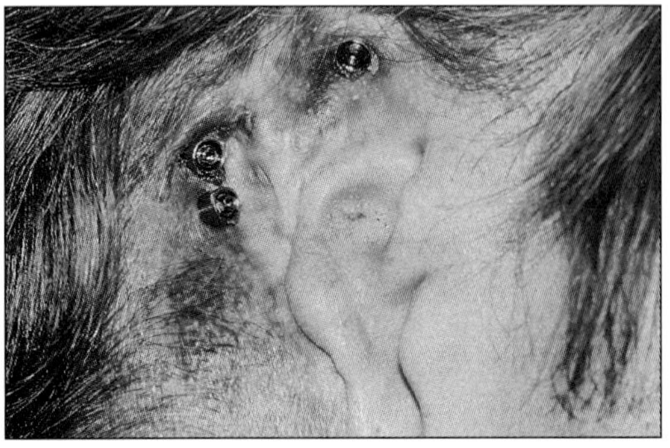

Fig 2-16a

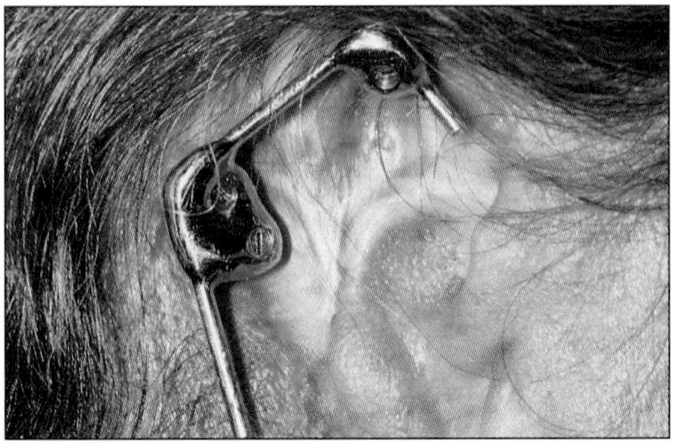

Fig 2-16b

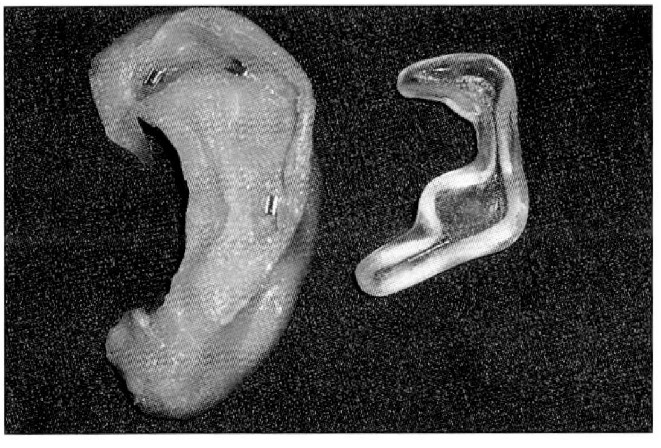

Fig 2-16c

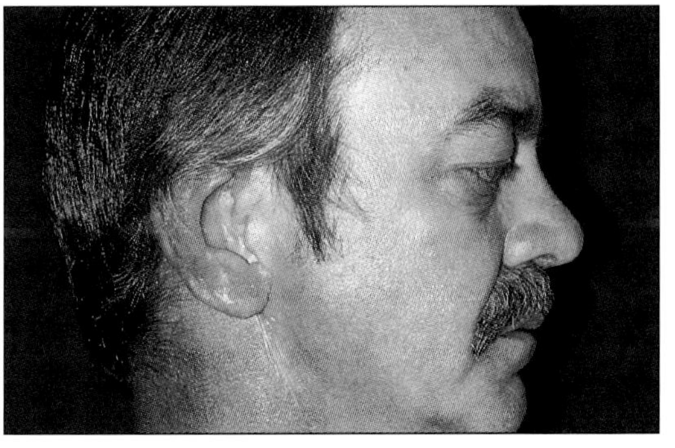

Fig 2-16d

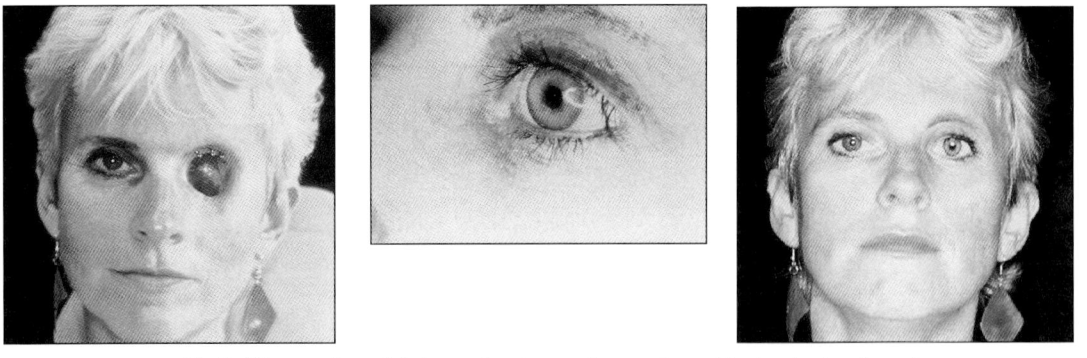

Fig 2-17 a–c An orbital prosthesis can be anchored to implants placed into the bone of the orbital rim. (Courtesy of P-I Brånemark.)

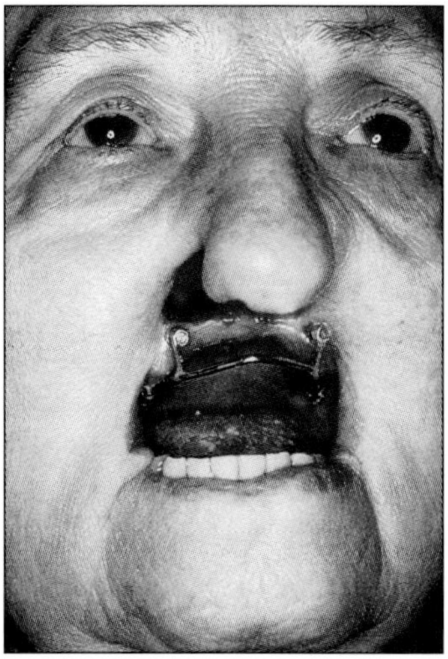

Fig 2-18a

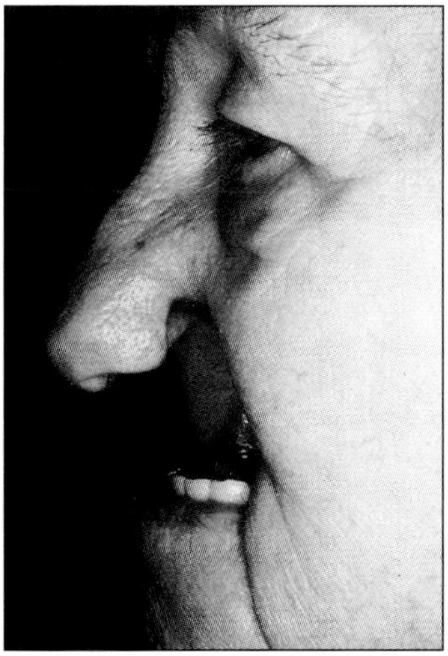

Fig 2-18b

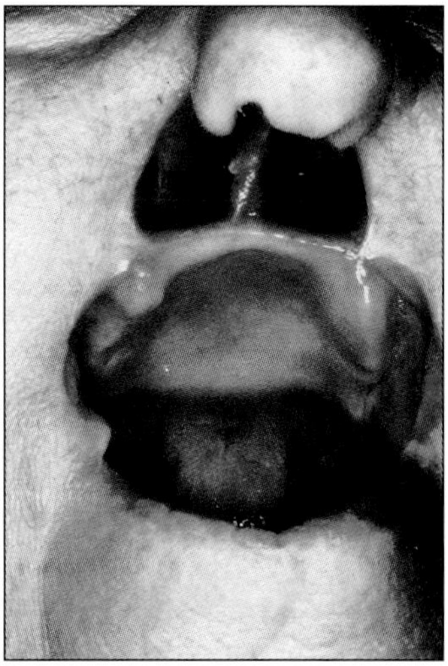

Fig 2-18c

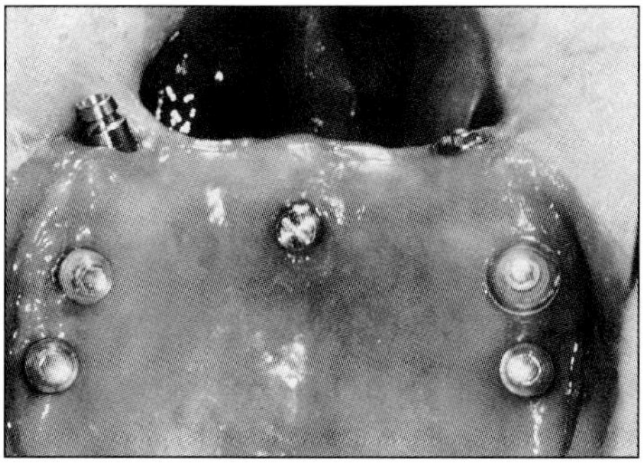

Fig 2-18d

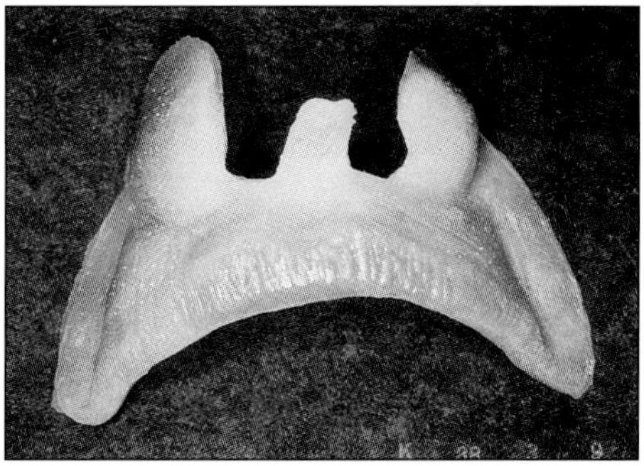

Fig 2-18e

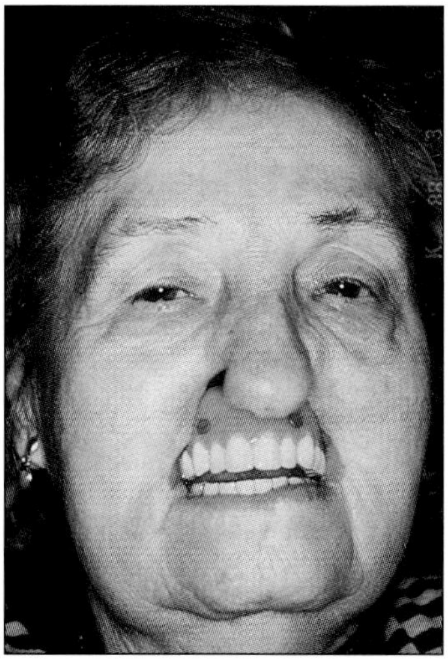

Fig 2-18f

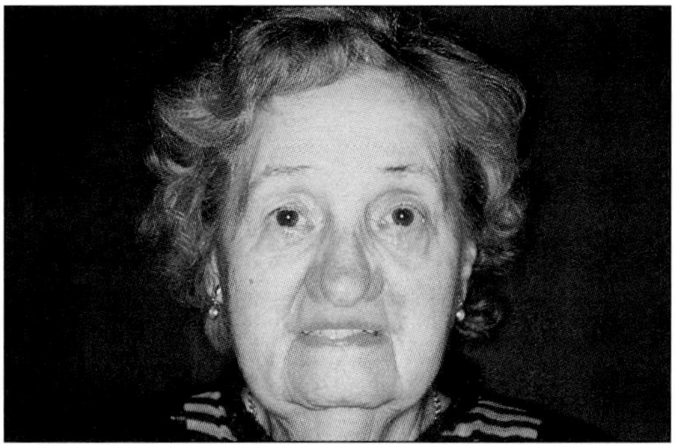

Fig 2-18g

Fig 2-18 a–g Complex problems of facial tissue loss are often amenable to management by means of implant-supported maxillofacial prostheses. This patient who suffered loss of part of her nose, her upper lip, and premaxilla (a–c) was able to return to society after she had been rehabilitated with a maxillary denture, combined with a facial prosthesis (d–g), all supported by endosseous implants. (Courtesy of P-I Brånemark.)

be used to anchor an orbital prosthesis, as illustrated in Figure 2-17. These facial prostheses are more hygienic, comfortable, and satisfactory than earlier models retained with adhesives.

Complex orofacial resections may leave a patient severely disfigured and unfit for normal society. Surgical reconstruction may be lengthy, difficult, and expensive—and may be beyond the patient's ability to undergo. By means of implant anchorage, oral and facial prosthetic components may be retained and stabilized and yet are removable for cleaning and for cavity inspections. Such a patient is shown in Figure 2-18 and is now rendered fit to rejoin society, with all the psychological benefits that that implies.

What, then, are the conclusions that we can draw from this diverse clinical experience?

1. We can say that osseointegration works predictably if the recommended protocol is followed.
2. Success depends on several factors such as the biocompatibility of titanium; the suitability of the implant bed; a gentle surgical technique; and an undisturbed healing period including controlled loading, careful case selection, and systematic oral hygiene.

It has been amply demonstrated by replication studies outside Sweden that osseointegration is a clinically useful, predictable biologic phenomenon. International, multicenter trials have produced results comparable with those from Sweden—occasionally better.[5] The success of osseointegration in the head and neck region forms a reasonable basis for believing that the same technology can be used elsewhere in the body—for joint replacement and for amputee prosthesis attachment—in fact, wherever durable anchorage is needed.

DISCUSSION

Philip Worthington I would like to respond to some extent to two of the questions that were asked previously. There was a question about bone resorption adjacent to implants and whether that had been seen in the oral field. Certainly it has. There are some occurrences for which we don't know the cause, but there have been many cases that we have been able to attribute to things such as a lack of passive fit between the superstructure and the underlying implants. This is a very critical issue.

The fit between the superstructure and a series of implants must be absolutely perfect. It is a concept of perfection that is probably difficult to achieve in reality, but where there is any degree or any detectable degree of misfit, this strains the whole system, which results in bone loss. Overload is another cause. For instance, if a lower fixed partial denture is cantilevered backward excessively beyond the implants, it may overload some of the implants. We also have the fact that Brånemark mentioned where in the preparation of the implant bed, slight thermal damage of the bone tissue had been sustained. So there are multiple factors that could be responsible for bone resorption.

We have identified a cause in most instances, but there are some that elude us.

Then there was a question about psychological acceptance of the prostheses. I suppose this is probably one of the most dramatic things about the oral application. Patients may have been struggling for years with lower dentures that float around and make them sore, are ineffective, and have poor function. Then they are a given a third dentition by means of implant-supported prostheses, and they rapidly come to regard it as part of their body. They surely don't want to take it out at night—I mean they couldn't, unless they have a special type of prostheses—but the vast majority of them regard it as part of them and never even think about it as being something artificial. So the level of psychological acceptance is extremely high.

Göran Lundborg I was intrigued by the case where you had an implant in a child who has a growing skeleton. I wonder if you had any expanded experience on this. It is a fascinating thing.

We know that an implant can be completely integrated as a part of the bone itself, but can it take part in the development and the events of the growing skeleton? Have you seen any problems with this skeletal development?

PW I am glad you brought this up because I mentioned this topic but with inadequate description. I should say first of all that this is not a recommendation that this technology should be applied to children in a sort of casual way. It needs to be done with great care and great consideration. We do not know exactly what the long-term outcome will be in this sort of situation. We take precautions, in case there was going to be any continued transverse growth of the mandible, to leave the two halves of the framework not connected to each other. Subsequent experiences are showing that that was not necessary; that was a little excessive caution.

For the same reason we did not make a single fixed partial denture linking both sides. We had a detachable overdenture that we anticipated would need to be remade periodically. But it did not take very long before it became apparent that that also was a little bit of excessive caution and the prosthodontists proceeded to transfer the child to a full fixed partial denture. The number of cases is very few. This is a child 5 years old. Prior to that my youngest patient was 13 years old, and there was no problem with transverse growth in the mandible, since growth is virtually complete anyway at that stage. So that is not a problem.

Where we do get into a problem is in the partially edentulous mandible. Imagine the mandible with a few teeth and then a space and then another few teeth. You put implants into the edentulous segment. If the jaw is growing, the adjacent parts with the teeth in will continue to grow. The alveolar bone will continue to develop, which means that the original prostheses will be left behind as the teeth and alveolar bone grow. Here the prostheses will be left behind, appearing to be somewhat submerged. So then it would need to be remade. In the case of the ectodermal dysplasia patient with the generalized hypoplasia of the jaws, there is virtually no alveolar bone. It doesn't exist because the teeth are missing and because of that there isn't any continued measurable growth in height of the bone. So in the edentulous mandible in the child or in a mandible that just has one tooth at the back, that's not a problem. The disparity between implants and teeth and growing alveolar bone is apparent in the young patient who is partially edentulous.

Dick Heinegård You mentioned that a prerequisite for a successful implantation was that you did not put load on the implant for a certain length of time. How did you define that length of time? When can you start applying the load?

PW I think the honest answer to that is only by actual experience. It's my understanding that in the early days when this method was started in Sweden, these implants were loaded and connected at a much earlier stage, and as a consequence, many were lost. This was in the early learning period, as it were. We come across the same sort of problem where a bone graft has been done to replace a resorbed alveolar ridge and implants have to be put through the bone graft as anchorage for subsequent implant prostheses' support. The question arises: How long are you going to wait before you apply any load?

Experience appears to have taught us all that the appropriate time is at about 6 months, at which time the bone graft is revascularized. The bone is in a condition where it can then undergo remodeling, providing the load stimulus is not excessive. If you wait longer than that you may reach the stage where the nonfunctional onlay bone graft begins to resorb simply because it's got nothing to do. So it's really a sort of hit-and-miss matter of experience as to know when you can load things.

There are, as you will be aware, within the whole field of oral implantology, many different opinions and there are people who have tried loading implants immediately. One cannot say it can't work. The question is, can it work as predictably as when we allow a proper healing period. There are many of us who believe—on the basis of experimental work, at least—that your predictability will be greatest if you allow a natural healing period first, before you load an implant. There are some limits to that, for instance, the temporal bone implants that really carry very little load. The hearing-aid implants have been immediately connected. But the amount of load that is applied is very little and the surrounding bone is very dense. So the immediate stability is not a problem. Whereas if you put an implant into the maxilla where the bone is spongy and not user-friendly, initial stability is hard to come by and to load those immediately would be a gross mistake.

Per-Ingvar Brånemark Dentistry has been defined as drilling-filling-billing as quickly as possible. The fact is some people have suggested a one-step procedure. But if you look behind the scene, what they do is put in the fixture and put on the abutment, which then protrudes through the mucosa. But they are not applying the functional load via a fixed partial denture until after about 2 months. And in Tjellström's studies on the mastoid bone, he put in the fixture and the abutment at the same operation but didn't apply the hearing aid until after 6 to 8 weeks. I believe that the initial period of revascularization and adaptation of the bone is critical and I believe so far that this so-called one-stage procedure has actually been a 1.5-stage procedure.

Björn Rydevik Professor Lundborg, I think you have some experience with the graded loading of implants.

Göran Lundborg Yes, I would like to comment on the loading issue and maybe question your concepts a little bit. In my field of hand surgery, we use the osseointegration principle for anchoring artificial joints. There is no skin penetration in these cases and it is impossible to unload a hand for more than a couple of weeks, so we start moving the joints after 5 days with a carefully controlled program. I was always struck by the fact that even when we had very bad conditions, our implants almost always became integrated. I have placed more than 300 fixtures, many in rheumatoid patients receiving heavy steroid medication and heavy cytostatic medication. We have obtained osseointegration in nearly 100% of the fixtures, even when we start loading them early. To be provocative, I even thought that this occurred perhaps *because* we started loading the implants so early.

PW Well, there are several things to say about that. First of all, it's not true to say that the only way you can load a dental implant is by having it penetrate the mucosa and to have something attached to it. That's not true. If you have a denture over that area, which is not adequately relieved on its fitting surface, there will be pressures transmitted through the mucosa to the head of the implant. Yes, there have been studies—Brunski did some studies in the 1970s where he had animals with implants on both sides of the mandible. On one side they were protected from load, and on the other side they were exposed and allowed to be loaded. The result was that loading was associated with a less predictable degree of integration.

I think the situation with the hand is different because, if you think of the scale of things compared with the jaw implant, you've got a much larger implant with a large bone-to-implant surface and you don't have much of a problem with initial stability because you are not talking about heavy loads. These loads are relatively small compared with the jaw. You have implants that are mechanically very solid in the bone, with good cortical bone all around. I think that's why you get away with it, but I think you are "getting away with it" rather than something that can be applied in general.

BR Professor Hansson, do you have any comment from an orthopedic point of view?

Tommy Hansson Yes, I had a question related to what Dick Heinegård asked about. You said that there must be a more or less an absolutely perfect fit between the superstructures and the implants in the mouth. Otherwise you have a misfit. With all the implants installed to date, have there been any studies on what the tolerances are? How much can they deviate from a perfect fit before you have trouble? Does anyone know?

PW I am not really the right person to answer that question. I can tell you this: there have been many different schools of thought about how the superstructure should be made: whether it is possible to make a single piece casting and have it fit or whether the casting has to be sectioned and resoldered to correct any minute imperfections of fit. There are new methods of constructing what you might call the baseplate mechanism, where little blocks of titanium are placed on the implants and then welded together with a laser to try to produce something that is devoid of any misfit. But I am not a physicist, so I can't really answer your question. I can tell you it is a matter of debate on the prosthodontic side. Maybe Professor Skalak can help us with what degree a misfit can be tolerated. I don't know.

BR Do you want to comment on that, Professor Skalak?

Richard Skalak Yes, I think really there is an absence of good quantitative data. I think the principles are quite clear and as Dr Worthington said, you can generate very substantial stresses with a small degree of misfit. For example, a misfit in the micron range, maybe only 100 μ, can produce a large stress. The reason is that because the fixtures are integrated they are very stiff, so even a small amount of required motion produces a relatively large stress. I think the main principles are obvious, but we don't have what we would call controlled studies so that we could give you back a number as to how much misfit is tolerable or how much stress you can apply.

DH I think this question of mechanical stress has several aspects. The magnitude of any mechanical load will certainly influence the cells as to what to make. The nature of the load, ie, impact loading or other types of loading, may also influence the cells. I wonder if there is any information on that.

PIB Drs Higuchi and Brunski have tried cyclic loading in the dog for the orthodontic application. It is very interesting that there seems to be a possibility, as you say, to control the modeling and remodeling of bone, but it has not been tried in clinical reality: only in experimental studies in dogs in some of the orthodontic applications.

PW It is interesting that in the early stages the orthodox recommendation was that the biting surfaces of the dental prostheses should be made in acrylic because of its resilience. That would have a damping effect on the impact loading of a sudden biting. As years have gone on, the general opinion has moved away from that toward the view that it makes very little if any difference clinically whether you have a metallic biting surface or a porcelain one or acrylic. In the early days, the acrylics were used out of caution and it certainly has a damping effect on impact injury, but it seems that we can get away with a harder surface.

William Maloney I am interested in your comments on initial loads. In the hip we obviously cannot do a two-stage procedure, and despite our best efforts we can't unload the implant. Simply raising a leg may put one body weight across the hip joint. We see that in many of the implants, despite the inability to relieve stress on the interfaces, the implants become osseointegrated. It's my bias that the initial implant stability is far more important than any healing response. If you have a stable interface under load initially, I think that osseointegration will be obtained despite what you do to the bone in surgery. I know it is a different situation in the mandible because you may not have that type of implant stability initially.

PW Well, I am tempted to say that one should practice on the basis of what is right rather than what you can get away with. If you heat bones to the point where you have detectable smoking, surely you can't believe that that's a good thing to do.

WM The question is not what is right or wrong. The question is what are the critical factors in determining whether an interface become osseointegrated or not. I can tell you from our experiences allowing for an early healing response at least in the hip implants, it has no clear influence on whether the implants become osseointegrated or not.

RS I think it is interesting that you emphasize initial stability. There has been a certain amount of discussion as to whether the criterion should be stress or strain—stress being the amount of the load and strain being the amount of deformation. I think what you are hinting, which may be really a better way to look at the situation, is that absence of relative motion provides initial stability. I think the cells don't care what the stress is as long as you don't disturb the remodeling process. What we should be studying is the conditions under which fibrous layers form, because if we knew what it was that produces fibrous tissue, we could avoid it.

Dudley Childress Dr Worthington commented about the difference between the implant in the mandible and in the hand and the importance of the relative sizes of the implant and the bone. The loading of the implants in the femur is quite complex in nature. I am wondering what the results have been, especially in amputees.

BR I think that some of these mechanical aspects that you are bringing up are going to be addressed in part by Professor Skalak and maybe we could discuss them at that time.

References

1. Collins DH. J Bone Joint Surg 1954; 36B:458.
2. Southam JC, Selwyn P. Br J Oral Surg 1971;8(3): 211–221.
3. Brånemark P-I. Introduction to osseointegration. In Brånemark P-I, Zarb G, Albrektsson T (eds): Tissue-Integrated Prostheses: Osseointegration in Clinical Dentistry. Chicago: Quintessence; 1985:11–76.
4. Adell R, Lekholm U, Brånemark P-I. Surgical procedures. In Brånemark P-I, Zarb G, Albrektsson T (eds): Tissue-Integrated Prostheses: Osseointegration in Clinical Dentistry. Chicago: Quintessence; 1985:211–233.
5. Worthington P, Brånemark P-I (eds): Advanced Osseointegration Surgery: Applications in the Maxillofacial Region. Chicago: Quintessence; 1992.

Chapter 3

Biomechanics of Osseointegration

Richard Skalak

The study of biomechanics of osseointegration is currently entering a new phase. In the early phase of clinical experience with osseointegrated prostheses, the emphasis was on questions of initial stability and on achieving a functional osseointegration in the first few weeks after placement of screws or other fixation devices. At present, emphasis has shifted to the long-term performance and remodeling of bone after osseointegration has been achieved. In the dental field, the distribution of bite forces by a dental prosthesis to each of the several supporting fixtures and/or natural teeth was a problem of interest that has been largely solved both theoretically and in clinical practice. The transfer of load from a dental fixture to the surrounding bone has been described in more or less detail by theoretical studies and proven by worldwide clinical practice to provide a reliable, functional load transfer without progressive damage or loosening.

The detailed surface properties of titanium and other implant materials have been studied, and the initial and later cellular responses have been elucidated to some extent. The effects of surface roughness and coatings, such as hydroxyapatite, and porous surfaces have been subjected to primarily experimental investigation. Many such modifications have been shown to not have any dramatic advantages over machined titanium surfaces. The main fact is that direct bone attachment can be achieved in a variety of settings (see Chapters 1, 2, and 13). However, only long-term follow-up studies will enable us to tell whether these different applications will lead to functional osseointegrations like the Brånemark dental fixtures for similar long periods.

In animal studies and in limited trials in patients, the feasibility of establishing a satisfactory initial stability and a subsequent firm osseointegration has been demonstrated in hand surgery, in ankle joint replacement, and in amputees in both upper and lower limbs. It appears likely that with careful installation and a period of several weeks at zero or moderate loads, osseointegration can be achieved in most bones of the human skeletal system. Thus, the achievement of initial stability and subsequent osseointegration is, in principle, a solved problem.

The new phase of the study and practice of osseointegration has to do with long-term prognosis and design against long-time failure phenomena associated, most severely, with atrophy or resorption of bone due to stress shielding, and other long-term effects such as fatigue and corrosion failure, and inflammation and loosening due to wear products. At this time, the biophysics and biology of these long-term developments seem as difficult to define as it was to be sure how osseointegration could be achieved in the early phases of osseointegration of dental prostheses.

Here the problems of stress shielding and fatigue failures will be discussed in view of current research and clinical experience. These are biomechanical aspects of osseointegration. The effects of wear products are primarily those of inflammation and immunologic responses. While the production of wear products is a mechanical phenomenon, it does not directly involve the mechanics of the bone or implant/bone interface. The effects of particulate debris are discussed in Chapter 17.

The Dental Experience

In dental applications, the clinical experience is now sufficient in length of time and in total patient numbers to say that neither stress shielding nor fatigue appear to be limiting factors in the long-term successful functioning of dental prostheses.[1] The only regular finding suggesting that stress shielding may play a role in dental implants is the reduction of bone height around individual dental fixtures. However, this is not progressive and appears to stabilize at 1 or 2 mm after about 1 year. Moreover, there is an unresolved issue as to whether this loss of bone height is due to stress shielding or microdamage due to overstressing. Recent studies in animal models by Hoshaw et al[2] suggest that a maximum stress is found near the surface of the bone at the point of entry of the fixture. The applicability of these short-term tests to long-term results in patients needs further study for us to be sure of the clinical relevance.

The dental experience has provided insights into basic requirements for the achievement of successful osseointegration. One requirement is careful surgery with low drilling speeds and temperature control to minimize damage of the remaining bone. Another requirement is provision of a mechanism of initial stability. In this regard, screw shapes provide a secure and clinically proven means to provide initial stability and a high pull-out strength.[3]

The dental experience has demonstrated that after an implant is placed, a period of no loading or very light loading is necessary to achieve osseointegration. Excessive early loading leads to the formation of a fibrous capsule. While these facts are well known and currently well accepted, quantitative data are scarce, and no quantitative mathematical theory is available that allows clinical prediction of the loading and timing that is permissible and will still result in osseointegration.

The obverse side of these phenomena is also unknown quantitatively. It has been suggested that the production of bone or fibrous tissue in fracture healing is similar to the processes of tissue differentiation during embryonic development. Bone is produced if the shear and hydrostatic stress are within a certain range, whereas larger cyclical stresses produce fibrous tissue or cartilage, dependent also on the degree of vascularity.[4] Another view of fibrous tissue formation at an implant interface is that it is sensitive to the relative micromotion of the implant relative to the surrounding bone. At present, there are no comprehensive data or theories that allow prediction of the rate of growth of a fibrous capsule. However, the dental experience in clinical practice is sufficiently extensive to demonstrate the long-range possibilities of osseointegration. Experience in maxillofacial and orthopedic applications shows that osseointegration can be reliably achieved in regions outside the oral cavity. From a clinical viewpoint, the more pressing problems are possible long-term resorption due to stress shielding and related problems.

Stress Shielding

The basic ideas supporting the concept of stress shielding are well known and generally accepted. The first idea is that of Wolff's law (1892), which postulates that bone remodels in response to the stress applied to it.[5] The

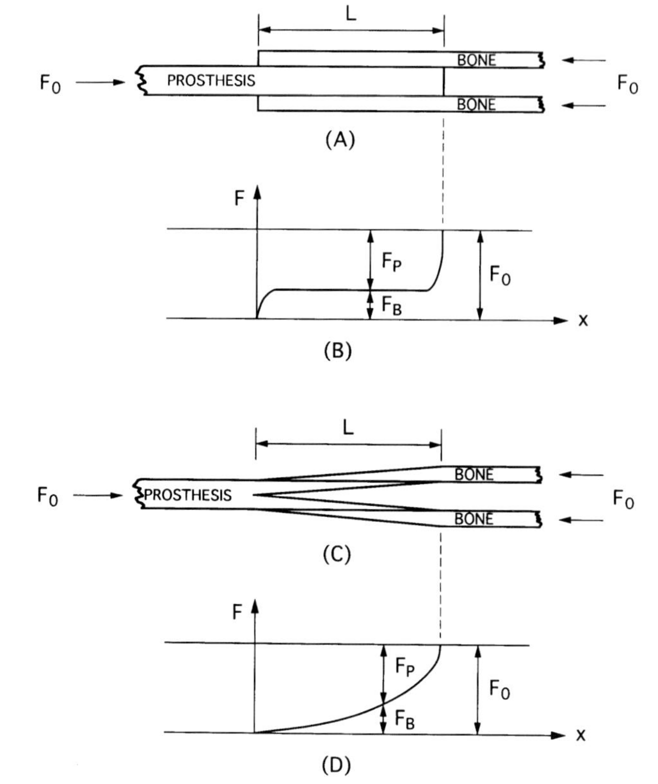

Fig 3-1 Idealized metal implants in the medullary cavity of a long bone. (A), The implant is represented by a solid cylindrical rod, and the bone is represented by a uniform circular cylinder of cortical bone. (B), The portions of the total load, F_O, carried by the bone, F_B, and by the prosthesis, F_P, in (A) as a function of the axial distance, x. (C), A similar prosthesis as in (A), but both the bone and prosthesis are assumed to be tapered, as shown. (D), The portions of the total load, F_O, carried by the bone, F_B, and by the prosthesis, F_P, in (C) as a function of the axial distance, x.

second idea is that in many prosthetic designs of connections of prostheses to bone, some parts or regions of the bone may not be loaded to their full normal level of stress. The end result is that in regions of low stress, the bone may atrophy, which is clinically observed as bone loss, usually in the proximal region of an implant. While these basic premises may be generally accepted, their detailed, quantitative formulation is still under development, and their application to the detailed design of orthopedic implants is not yet being utilized in a systematic, deliberate way so that long-term bone resorption can be accurately predicted and controlled.

The general phenomenon of stress shielding may be illustrated by a simplified model of an implant fixed centrally in a bone, as in a hip replacement prosthesis. Figure 3-1(A) shows that model of the bone as a uniform circular tube and the prosthesis stem as a uniform circular rod. Figure 3-1(B) shows the variation of the axial force along the length of the contact in the two components. The total axial force, F_o, is divided between the prosthesis and the bone in a proportion that depends on the stiffness of each component. If the prosthesis stem stiffness is greater than that of the bone, most of the load will be carried by the prosthesis (F_P) for most of the length, as shown in Figure 3-1b. Most of the load is then transferred to the bone (F_B) at the distal end. Hence the proximal portion of the bone is shielded from the full stress, and this may lead to bone resorption.

A somewhat more favorable distribution of stress in the bone may be achieved by tapering the bone and the prosthesis, as shown in Figure 3-1(C). In this case, the load transfer is more gradual, as shown in Figure 3-1(D). There is still a tendency for more of the stress to be transferred to the bone toward the distal end of the prosthesis, but due to the taper of the bone, the stress in the bone will be somewhat more uniform than in the case shown in Figure 3-1(A). The taper of the bone suggested in Figure 3-1(C) resembles the results found clinically in cases in which the atrophy of bone due to stress shielding produces a similar region of tapered bone at the distal end of a bone.

It is theoretically possible to achieve a uniform distribution of stress in the bone and the prosthesis if the axial stiffness of the bone and implant are equal and each is tapered as shown in Figure 3-1(C). The axial stiffness is the product cross-sectional area and the elastic modulus of each material. Thus, if the implant has a Young's modulus that is five times that of the bone, the cross-sectional area of the implant should be one fifth of the cross-sectional area of the bone to match stiffness. These simplified model considerations serve to illustrate the basic phenomena of stress shielding and suggest the direction of possible improvements in implant designs. However, in clinical practice, the geometry and loading patterns will be more complex, and the resulting stress distributions are not simple. Hence more detailed analyses are required before implant designs can be based directly on theoretical analysis. Progress has been made in this direction by use of finite element analysis of stress distribution and subsequent remodeling of bone. Some of these results are outlined in the following paragraphs.

To generate a realistic model of bone resorption due to stress shielding, it is necessary to postulate some criteria of the loading and the response of bone to stress. The forces applied to any bone will generally vary in magnitude, direction, and duration depending on many factors of body form and activity. At this time, it is not precisely known what loading characteristics are important to bone remodeling. An average over time and directions is probably significant, but peak values and rates or frequency of application likely play some role.

After some selection of the loading characteristics is made, the finite element method or other methods may be used to compute stress distribution in any particular geometry of an implant and bone. With current computer capacities, three-dimensional models and stress distributions can be efficiently and accurately computed.

At this point, some law of bone remodeling must be adopted to represent the response of bone to the stress applied to it. Judging from the homeostasis observed in many situations, it is generally assumed that bones will not atrophy or augment if the stress applied to them is within some normal range. Under lower stresses bones may atrophy; for higher stresses they may augment. Of course, if the stress is sufficiently high, failure may ensue. If there is bone atrophy or growth, the stress distribution will change and may lead to further changes of form or density. Such progressive alterations may be modeled in computer simulations by iterative computations in which bone properties and form are updated by small increments and the stress field is recomputed based on the new form. This cycle is repeated until either a steady state or failure is reached.

The particular characteristics of the stress at a point leading to specific atrophy rates in bone are not precisely known. Candidate features might be peak stress, or some average over the various directions at a point. One such average is the strain energy, W, which represents the energy stored by the elastic deformation of the bone. Other criteria may be based on strain rather than stress. The various possibilities have been reviewed by Cowin.[6] Another factor that is unknown is the influence of different patterns of stress as a function of time. The same average stress may be delivered by forces acting steadily, or at different frequencies and peak amplitudes. Data on such frequency dependence have also been reviewed by Cowin.[6]

The response of bone to applied stresses may take two forms, which are described as internal remodeling and surface remodeling. Internal remodeling implies a change of density or porosity or rearrangement of internal geometry such as trabeculae, without change of external form of the bone. Surface remodeling involves the removal or addition of bone to the periosteal surface so that the cross section of the bone is altered. Various simulations have been carried out assuming either only internal remodeling, only surface remodeling, or both. The merits and disadvantages of several formulations have been reviewed by Huiskes and Hollister[7] and Cowin.[6]

An example of three-dimensional finite element simulation of stress shielding involving both internal and surface remodeling has been reported by Weinans et al[8] with comparison between animal experiments and the computer simulation. The experiments were noncemented hip replacements in dogs in which bone density and cortical bone area were measured over 2 years. The stems of the implants were titanium alloy with plasma-sprayed coatings of titanium.

The finite element method used was applied to both the prosthesis and the contralateral femur. The finite element meshes and forces loading the models are shown in Figure 3-2. The remodeling stimulus was the strain energy per unit mass, S = W/r, where *W* is the strain energy per unit volume and *r* is the mass density. It is assumed that there is a normal range of S in which there is no remodeling. The normal range is centered on an assumed reference value designated S_{ref}, and no response of the bone is assumed for the range of S equal to ±35% of S_{ref}. Larger values of S result in increased bone density, and smaller values result in atrophy. At the periosteal surface, the rate of change of bone mass is used to compute the movement of the external surface in the perpendicular direction to the periosteal surface. In the cortical zone, the bone density is assumed to be constant (r = 1.73 g/cm^3), so apposition or resorption of bone can occur only on the surface (external remodeling). In the medullary zone, the bone apposition or resorption is also assumed to be at the surfaces of the trabeculae, so the mean density may vary (internal remodeling).

The flow chart of the computational procedure is shown in Figure 3-3. The basic loop is a computation of stresses that give rise to density and geometry alterations Δp and Δx (see Fig 3-3) over a small time step Δt. The stresses are then recomputed taking these changes into account, and the cycle is repeated to simulate the progressive remodeling.

The results of the computed simulation and animal experiments were compared on the basis of two measures of remodeling as shown in Figure 3-4. The first is the medullary bone area fraction, which is the percentage of the medullary area of each cross section (B through J in Fig 3-2) occupied by bone trabeculae. The second is the percent change in cortical bone area in the same cross sections. In both the experimental and computational results as shown in Figure 3-4, the cortical bone area decreased progressively with increasing time. This is the stress-shielding effect. At the same time, the medullary bone fraction increased, particularly near the distal end of the prosthesis. This represents bone augmentation due to the high stresses found in the medullary regions.

While the comparisons of experimental results and theoretical computations as shown in Figure 3-4 are encouraging, there are a number of parameters that may be adjusted

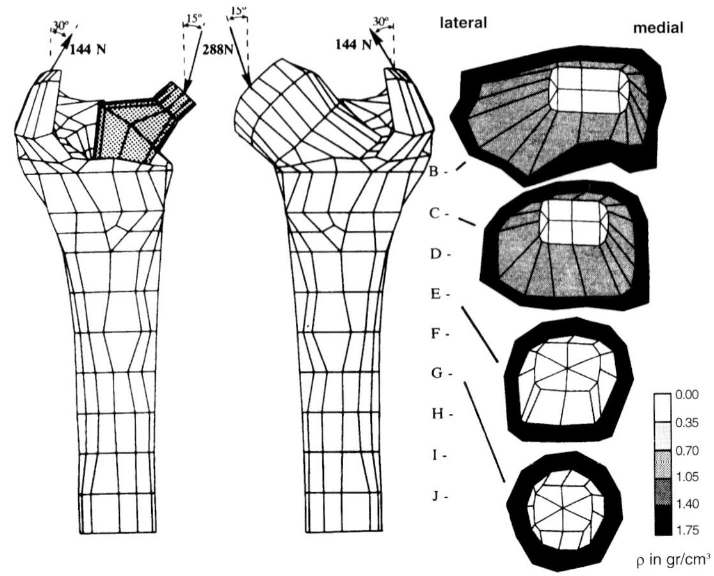

Fig 3-2 Finite-element meshes of the femur of a dog with a prosthesis and the contralateral femur; the initial density distribution at different cross sections of the finite-element model with a prosthesis is shown at the right. The density of cortical bone was taken to be 1.73 g/cm³ in all sections, the maximal density value for bone in this model. The densities of cancellous bone in the cross sections were taken to be 0.40 g/cm³ (B), 0.37 g/cm³ (C), 0.24 g/cm³ (E), and 0.09 g/cm³ (G); all are in agreement with the measurements in medullary bone area fraction of the experimental contralateral configuration. (Figs 3-2 to 3-4 from Weinans H, Huiskes R, van Rietbergen B, Sumner DR, Turner TM, Galante JO. Adaptive bone remodelling around bonded non-cemented total hip arthroplasty: A comparison between animal experiments and computer simulation. J Orthop Res 1993;11:500–513. Used with permission.)

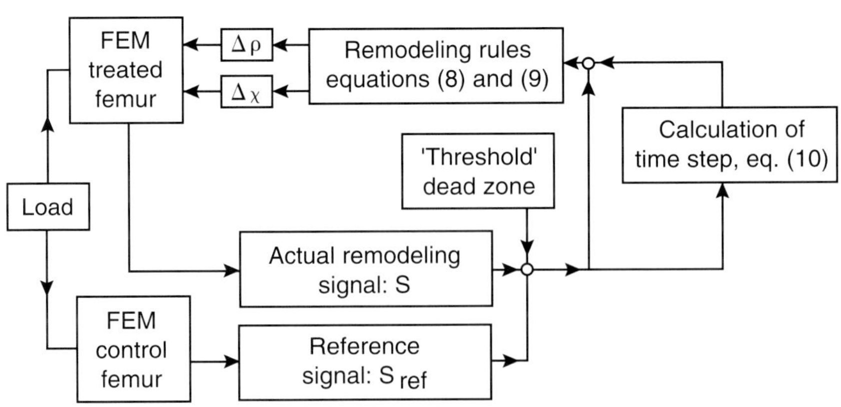

Fig 3-3 Schematic representation of the incremental computer simulation model.

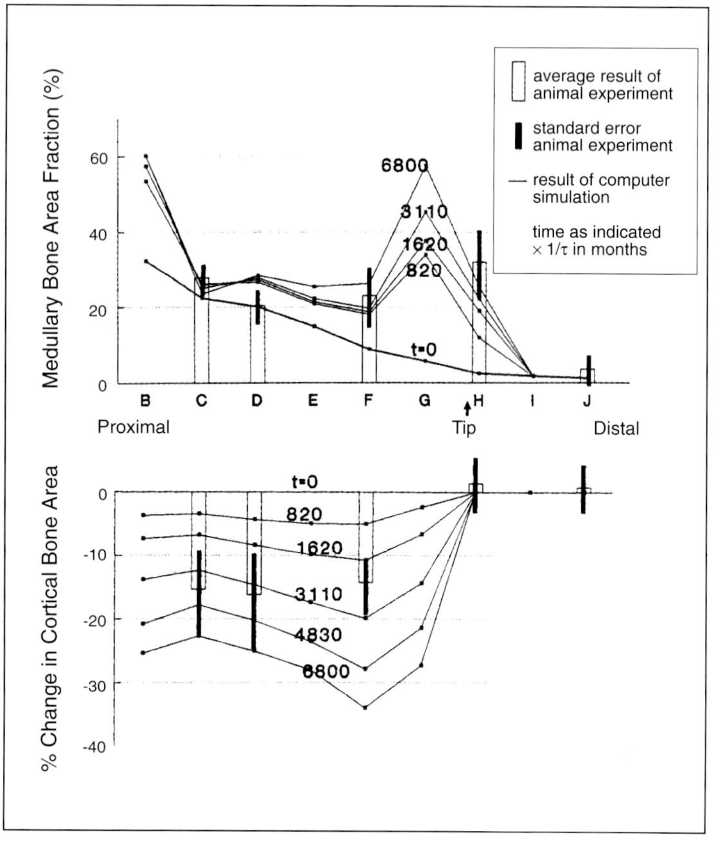

Fig 3-4 The medullary bone area fraction and the change in total bone area per cross section of the animal experiment immediately postoperatively (t = 0) and after 2 years, and computer simulations after different intervals from t = 0 to (6800/t) months. The best fit of the experimental data was found to correspond to the constant t equal to 129.6. (From Weinans H, Huiskes R, van Rietbergen B, Sumner DR, Turner TM, Galante JO. Adaptive bone remodelling around bonded noncemented total hip arthroplasty: A comparison between animal experiments and computer simulation.

in theory. The values of these parameters may vary for different bones, age, and conditions. These uncertainties prevent the use of such computations as exact predictions in specific clinical cases. The critical parameters that are not entirely known concern mainly the sensing and response functions of bone: for example, the value of the reference state of stress, S_{ref}, and the range of strain energy, S, in which there is no response of the bone. For the results shown in Figure 3-4, this range was ±35% of S_{ref}, but other studies suggest that this range may be ±50% or even ±75% of S_{ref}.[9] The rate constants of bone resorption and apposition are also not entirely known. These likely have large variations also with age and anatomic site. While the rate of resorption may not affect the final outcome, it of course controls how soon the end state is reached.

Fatigue and Remodeling of Bone

Most nonbiologic materials will rupture under repeated stresses at lower stress levels than if the same peak stress is applied only once. This phenomenon is known as fatigue. The peak stress that produces a fatigue failure decreases with the number of cycles of loading and can be a fraction of the static single cycle strength after thousands or millions of cycles. In inert materials, fatigue failure proceeds by the formation of microcracks within the material or on its surface. The growth of such cracks in time leads to the eventual failure. Fatigue of implanted prostheses can be a clinically significant problem in metal or ceramic components. Bone cement may also be subject to fatigue.

The usual concept of fatigue failure may also be applied to bone if it is removed from the body and tested as a nonliving material. However, in vivo, bone has the capacity to repair itself. Taking this into account, a homeostatic state is possible in which the rate of production of damage by stress is balanced by the rate of repair by the cells of the bone. These concepts may be expressed in a mathematical form as a theory of bone adaptation by damage accumulation and repair.[10] The result is an alternative theory of remodeling of bone due to stress, rather than a simple fatigue failure. However, if the rate of microdamage exceeds the capacity of the living bone to repair itself, then a fatigue failure in living bone is possible. Below the levels and repetitions of stress that lead to failure, the microcracking and repair provide a detailed mechanism of strain-adaptive remodeling of bone that can lead to optimized, stable bone structures. At present, the details of how stress or strain are sensed by bone and the metabolic pathways leading to remodeling are not sufficiently well known to allow reliable predictions of bone response in particular prosthetic designs.

One aspect of bone remodeling that has been demonstrated to influence the results of simulations, but is not yet well defined physiologically, is the distribution and range of action of sensors and responding elements in bone. It has been suggested that osteocytes being regularly distributed and interconnected may be the mechanoreceptors in bone.[11] A detailed theory of excitation of osteocytes by mechanical loading has been developed by Weinbaum et al[12] and is also discussed in Chapter 4 by Cowin. Due to their interconnection, the osteocytes may respond to stresses applied over some finite region, rather than to the stress at the cell body only.

Similarly, bone remodeling by osteoblasts and osteoclasts will have some finite region of response rather than only at one point. This notion of range of sensing and response has been shown to be important to the results of simulations of bone remodeling.[13] In modeling the development of trabecular morphology, it is shown that the size and branching of trabeculae depends on the spacing of the mechanotransducers and the range of influence of the repairing elements. In view of these complexities of the bone responses and the paucity of detailed knowledge of these processes, prediction of long-term outcomes of particular cases and new prosthetic designs is not yet possible with reliable accuracy.

The present level of technology in computation of stresses and knowledge of material properties are adequate to design prosthetic elements of sufficient strength and fatigue life under prescribed loading conditions and degree of osseointegration. However, if the long-term response of bone leads to major resorption, the loading of the prosthesis may change sufficiently to produce fatigue failure.

Similarly, the surgical techniques and knowledge of biomaterials behavior are currently adequate to ensure osseointegration in any skeletal component. Stress distributions in bone for complete osseointegration can be reliably computed. However, the limits of stress and strain that may be applied without interfering with the development of osseointegration are not precisely known.

The study of the mechanisms and quantitative aspects of bone remodeling is still in a research and development stage. It is clear that bone loss due to stress shielding observed clinically may be reproduced theoretically by assuming that understressed regions of a bone will be resorbed by the body over time. The quantitative predictions of resorption due to stress shielding depend on several parameters of the sensing and response processes, which are not yet known with sufficient accuracy and generality to permit reliable predictions in particular cases, although the general trends to be expected are clear and should be taken into account in prosthetic designs.

A future stage of development of individual patient design based on prediction of long-term bone remodeling is likely to be achieved as knowledge of the processes of transduction and growth in bone is established. With successful initial osseointegration and controlled long-term resorption, the prospect of permanent (life-long) osseointegrated implants is a most likely possibility. Such long-term stability has already been achieved in osseointegrated dental fixtures.

DISCUSSION

William Maloney You suggested that the threshold for allowable relative motion of an implant to obtain osseointegration was somewhere around 100 μm. I would suggest to you that this figure may be an order of magnitude too high. There is some work, some practical work, that has been done by Dennis Burke of Harvard. He has designed a system that actually can stress a pin that is otherwise unloaded. Then one looks at tissue differentiation at the interface. At 100 μm, 100% of the cases developed a fibrous tissue interface. To reliably obtain an osseointegration interface, you had to have 40 μm or less and probably maybe more on the order of 25 μm.

Richard Skalak I think we agree on the order of magnitude. It could be that low. I do not know the particular work to which you refer.

WM I stated the number, so let me clarify it. If I recall, it was someone in Stephan Perren's group who gave the 100-μm figure. This was for the development of the fibrous layer rather than the criterion for osseointegration, which would suggest that there is a range below this for which the results are different.

RS I think that we certainly have a correct order of magnitude. It is somewhere below 100 μm.

Göran Lundborg You showed a nice picture of a screw and bone indicating that bone integration did not occur all around the screw, and you said that it was not necessary to have such integration all the way around. What are the limits here? Is it enough to have osseointegration in the small area once it occurs because it gives sufficient stability? How much of the area of the screw do you think has to be integrated to give a good result?

RS My opinion is that it can be quite a small percentage. I think for instance in the dental case, probably the capacity of the system is the factor of something in the order of two to three over the load that actually is placed on it by the patient, particularly if he or she is careful. So that means that in fact if you had only 30% surface integration you would not see failures. Structurally it is not necessary to have that support and contact at every point. You know we build all kinds of open trusses that are only supported at distant points. Mathematically speaking you get the stability you need if you have just three points that are not in a straight line fixed. So that if there were theoretically just a few points that were well fixed, you have stability, but capacity for load is more or less proportional to the area. I think many of these fixtures could operate with as low as 30% to 50% contact and be perfectly satisfactory clinically. I think Dr Worthington may say you are pushing what you can get away with rather than what's desirable or reliable, but I think the practical answer to your question is that a fairly low percentage is actually satisfactory.

Per-Ingvar Brånemark I agree with you completely. I do not believe that it is a particular percentage that is required, but it is the capacity of the bone to carry load and remodel that is important. I hope that Dr Hansson will comment later on this. I believe very strongly that we should really define what tissue that we have there and what would come later. So I believe surgery is not a piece of carpentry; bone is much more a question of highly differentiable tissue. So sometimes we are surprised that so little bone can carry so much.

Another comment: You mentioned this very interesting item of the long-term situation, and you had as you said a simple prosthesis in the tubular bone where there was thinning out distally. If you thinned out as you were, the proximal prosthesis would thin out as well. That is exactly what we have observed in some of the bones, of course. Would you please comment on the possibility that you establish harmonic balance so that it is then a dynamic equilibrium, or is there a risk that resorption will propagate?

RS There is a risk that it will propagate. I am sorry to tell you that this is in the theory and you can see it in Dr Huiskes' computation. If you have this mismatch of stiffness, the initial distal absorption will move proximally and there is nothing to stop it. It may shorten its length, but the stress distribution will still be similar. What the resorption does is produce a very small area with a very steep front. Ideally you should transfer the load directly to the cut surface because then the bone beyond that cut would receive its usual load. So in fact the natural progression in these tubular implants is to work back toward that situation.

Richard Coutts I want to mention several things. First, with regard to the issue of how much motion it takes or how much you can get away with and still have bone attached to your implant, I think we can draw on the experience of the research that was done on porous ingrowth, which has what has really been used for osseointegration in the hip joint. There were three things that were noted to be necessary to achieve integration. One was that you had the proper size voids in your porous surface, somewhere between 100 to 500 μm in average pore size. You had to have no gaps or to minimize the gaps. There is the so-called jumping distance that osteoblasts are capable of traversing in open space and it requires no motion. I think that fits with what you were saying as to the necessity for minimizing the motion, but I agree with Dr Maloney that your figures are too high for the long bone situation.

There has been an interesting experiment going on in hip replacement. There is fully porous coated stem that was actually the first one used in this country, and it is still being widely used. It does cause stress shielding, particularly in the larger sizes. The interesting thing is that the stress shielding progresses to a certain point and then stops. It seems to progress somewhere between 1 and 2 years and then it stabilizes. It shows you how smart the bone really is. I guess it has figured out that it can reduce itself to a certain size but that it needs to be present at a certain minimum mass. There have been efforts made to try to transfer load at the proximal end of the stem by just *putting the porosity up at the upper end* and then having the rest of the stem smooth. The smooth portion of the stem gives initial stability by virtue of a tight fit in the bone, but the ultimate fixation is by bony ingrowth into the porous surfaces proximally. It is a fact that with those designs there is less stress shielding.

From our perspective, a final point I wish to make is that stress shielding is a concern, but it is not what we consider to be the absolute harbinger of defeat. It is the wear products that are a major concern to us. In the joints we are talking about, we have not only the issue of the stem being implanted, but we have a wear surface that the joint works against, and that is what we are really struggling with and what we will be talking about in a later session.

RS Thank you for these comments. I did not remark at all on the pore size or roughness effects, but that has to be related to the cell size and what the cells will do. That is a very real effect. The motion question is a little different for deep pores and for a smooth surface. Clearly, if the pores are of the order of 100 µm, and you move 100 µm, you shear off whatever was there. So in that case, any movement needs to be well below 100 µm in relative motion. That the stress shielding does stop at a certain point could be explained in terms of the fact that there are bending moments as well as axial forces. This theoretical argument of not stopping applies in the case where you have only axial loading. But if you have bending moments, they stress the bone differently and that leads it to a functional minimum.

RC There are also torsional forces.

RS Yes, there are torsional stresses also.

RC The other interesting fact is that the percentage of ingrowth that you get in one of these prostheses is relatively small. It could be as little as less than 10%, and yet those are well-fixed implants that function quite satisfactorily.

RS Yes, I think that corresponds also to my comments on the screws. You don't really need 100% contact to have something that is quite stable and effective. I did not know the ingrowth was that low a percentage, of only 10%.

RC Well, it varies all over the map. If you have 20% ingrowth, that would be considered very high.

Björn Rydevik Dr Skalak, could you comment on the issue that was brought up before, namely the anchoring capacity of bone, particularly in relation to the size of the various implants? This was brought up by Dr Childress. Are there any possible relations between maximal loads and the dimensions of the fixtures in the skeleton, for example, the lower extremity and the loads applied in various situations?

RS Again, without having exact figures, I think that if you reduce the different situations to local stress intensities, the capacities of the various bones are not so different. In the hip joint, you have quite large loads but normally you also distribute them over a larger area. The long bone is proportionately larger. So I don't see any reason to expect any strong size effects.

In observing the work over the years, it seemed to me one should be able to scale up to orthopedic values, including long bones of legs and arms, from the dental experience. We call this *scale up* and we are accustomed to it, especially in structures. You build the model structure of an airplane and then you build a big one. We are able to correlate the different sizes. So I don't expect any striking structural effects from the differences in sizes. I think the permissible stress levels will turn out to be the same and the responses will be turn out to be similar.

I would like to throw out a question to the medical and biological people. That is—perhaps there are differences in bone? For example, I do not think the skull bones have as much stress in them as the leg bones, but the skull bones do not atrophy. This raises another issue that was not in my talk, but I talked to Dr Eriksson and he is planning to bring it up. It has to do with the genetics of the situation. I think the molecular genetics people should be in on these discussions. Molecular genetics controls development and may control differences in different sites. Maybe we could take advantage of them.

Tommy Hansson There is some work in a turkey-wing model in which the investigators stimulate with vibration and measure bone remodeling in different places. They are putting forward a hypothesis that we are actually helping to test, namely that small vibrations with very low amplitude can have an effect. Vibration of the skeleton can mean that strain levels maybe less than 15 microstrains could remodel the skeleton, which should mean that the skeleton could have its own *remodeling environment*. So that is why we do not lose it, so to speak.

RS Yes, that is a factor that I did not mention in my talk, but it is in the manuscript. We really don't know in detail what the differences in responses are for a steady load versus an oscillating load, and that may be quite significant. That is another aspect that needs exploration.

TH But I have not seen anyone so far who has completely lost their skeleton. Not even space astronauts lose their skeleton; I have not seen anyone without a spine, literally at least.

PIB I am pleased you brought this up, because when we applied the bone-conduction hearing aid by a fixture in the bone behind the ear, we vibrated the entire brain and scalp. So that might be suggested as prevention when it comes to disappearance of the scalp bone.

BR Thank you very much. You have all received a transcript of the various aspects of definition of osseointegration. May I suggest that this could be the subject of informal discussions or comments during the workshop and also that this could be brought up at the final general discussion.

References

1. Worthington P, Brånemark P-I. Advanced Osseointegration Surgery. Chicago: Quintessence; 1992.
2. Hoshaw SJ, Brunski JB, Cochran GVB. Mechanical loading of Brånemark fixtures affects interfacial bone modelling and remodelling. Int J Oral Maxillofac Implants 1994;9:345–360.
3. Brunski JB, Skalak R. Biomechanics of osseointegration and dental prostheses. In Naert I, van Steenberghe D, Worthington P (eds): Osseointegration in Oral Rehabilitation. Chicago: Quintessence; 1993:133–156.
4. Carter DR. The regulation of skeletal biology by mechanical stress histories. In Skalak R, Fox CF (eds): Tissue Engineering. New York: Alan R. Liss; 1988:173–180.
5. Roesler H. The history of some fundamental concepts in bone mechanics. J Biomech 1987;20: 1025–1034.
6. Cowin SC. Bone stress adaptation models. J Biomech Eng 1993;115:528–533.
7. Huiskes R, Hollister SJ. From structure to process, from cell to organ: Developments of F–E analysis in orthopaedic biomechanics. J Biomech Eng 1993; 115:520–527.
8. Weinans H, Huiskes R, van Rietbergen B, Sumner DR, Turner TM, Galante JO. Adaptive bone remodelling around bonded noncemented total hip arthroplasty: A comparison between animal experiments and computer simulation. J Orthop Res 1993;11:500–513.
9. Weinans J, Huiskes R, Verdonschot N, van Rietbergen B. The effect of adaptive bone remodelling threshold levels on resorption around noncemented hip stems. Adv Bioeng 1991;20:303–306.
10. Prendergast PJ, Taylor D. Prediction of bone adaptation using damage accumulation. J Biomech 1994;27:1067–1076.
11. Cowin SC, Moss-Salentijn L, Moss ML. Candidates for the mechanosensory system in bone. J Biomech Eng 1991;113:191–197.
12. Weinbaum S, Cowin SC, Zeng Y. A model for the excitation of osteocytes by mechanical loading-induced bone fluid shear stress. J Biomech 1994; 27:339–360.
13. Mullender MG, Huiskes R, Weinans H. A physiological approach to the simulation of bone remodelling as a self-organizational control process. J Biomech 1994;27:1389–1394.

Modeling of Ingrowth into Implant Cavities Along the Bone-Implant Interface

Stephen C. Cowin and Gangming Luo

The topic addressed in this study is the influence of local bone remodeling at the bone-implant interface on the long-term global stability of the implant. To emphasize local events on the implant interface and their relation to the global stability of the implant, the study outlined here consisted of three parts: local models, global models, and the local-global feedback mechanism. The words "global" and "local" are used to distinguish between events associated with the entire bone-implant system and those events associated with small regions of the implant interface.

A set of interchangeable local computational models of contact surface structures that permit the modeling of different interfaces has been constructed to represent the local level. The influence of remodeling is evaluated at all the local levels, that is to say, at the bone-implant interface and on the bone surfaces not contacting the implant, over time and under specified load conditions. The local models are used to determine the ingrowth and effective friction coefficient at the interface employed in the global model.

The results of the local analysis are used as input in the global model, which is used to determine the overall strain distribution. Since the overall bone remodeling, in turn, influences the local interface conditions, it is necessary to make the models interactive, the output of one being the input for the other. Therefore a computational model has been developed for the interaction between the time dependent global bone-implant structure and the time-dependent local interface conditions. The models have been used to determine the long-term stability and load-carrying capacity of the implant.

One feature of this computational model is a variety of bone-implant interface structures that permit the modeling of different bone-implant interfaces, including screw and sintered bead. The global model contains adjustable or selectable parameters, which include the mechanical loading, the initial interference or fit, and the significant characteristics of the bone-implant interface. The local bone-implant surface friction can be varied to reflect different implant surface coatings, such as a form of the mineral hydroxyapatite, one of the two basic constituents of bone tissue that has been found to enhance bone tissue attachment and growth.

Many noncemented prostheses have slots or ridges designed for bone to grow into. The purpose of these slots is to create a strong mechanical interlock for effective load transfer. In this work, the effects of implant surface coatings have been mechanically modeled, on the local level, as frictional boundary conditions at the bone-implant interface. When the implant is placed in the bone, a new strain field is created in the surrounding bone tissue, and the bone tissue subsequently changes its structure to accommodate the new strain field. This study uses a surface bone remodeling theory and the boundary element method for local remodeling, and the finite element method for global remodeling, to calculate the structural changes induced in the bone due to implant placement.

We consider the model as predicting the final stable shape of the remodeled mature bone. Thus, when we compare with experimental results, it is necessary to compare with the final remodeled ingrown shape. Comparison of our predictions with experimental results can only be done qualitatively because the type of experimental data needed is not available. The type of experimental bone ingrowth information needed is the kind that documents the history of ingrowth as well as the mechanical loading and strain history at the ingrowth site. This type of study has not been accomplished due to the great technical difficulties in recording these variables in a long-term animal study. However, although the type of data needed for the verification of the present model is nonexistent, we have compared the calculated predictions of the model with the bony ingrowth observations and experiments we found reported in the literature, and found that our calculated predictions are qualitatively consistent with these data.

Interference and Initial Ingrowth

An implant is generally placed in bone with an initially tight fit. In this work, this tight fit is called the initial interference where interference is a quantitative measure of tightness. The *interference* is a length equal to the excess of the cross-sectional dimension of the implant (or implant plus cement in a cemented hip prosthesis) over the cross-sectional dimension of the bone cavity in which the implant is currently residing, both cross-sectional dimensional lengths being measured in the same direction.

Interference can vary with anatomic site due to the shape mismatch of the prosthesis with the bone cavity into which it is placed. There are two major reasons that the loosening might occur. First, there is the possibility of too small an initial interference.[1] Second, if significant initial interference exists, the amount of the interference may diminish in time due to bone loss at the interface. In cementless implants, nonuniform bone-implant surface contact at the initial stage of placement occurs frequently because it is almost impossible to accomplish uniform contact in the placement surgery. This initial nonuniform bone-implant surface contact leads to localized areas of low and high stresses.

A second concept that is used in this study and which might be confused with interference is *initial ingrowth*. To describe the initial state of the implant on certain bone interfaces, it is assumed there is some initial bone ingrowth into the implant cavities, such as screw threads or pores on a porous surface, before any external load is applied. The initial ingrowth can be viewed as the effect of wound healing. When an implant is placed in a femur, the initial reaction of bone is to fill the cavities of the coated implant. The numerical remodeling process does not start until there is some calcified bone filling the cavities due to wound healing. This initial bone ingrowth is called *initial ingrowth* and is not the same as interference. The lack of interference and ingrowth, initial or evolved, leads to excessive *micromotion* (relative displacement between bone and implant surfaces), which can stop the formation of bone tissue and create a structurally insubstantial fibrous tissue in its place.

Surface Remodeling Theory

Surface bone remodeling is the deposition and resorption of the surface layer of bone tissue. The *stimulus* for bone remodeling is defined as that particular aspect of the bone's stress or strain history that is used by the bone to sense its mechanical load environment and to signal for the deposition or resorption of bone tissue. The true biologic stimulus, although much discussed, is unknown.[2] The case for strain rate as a remodeling stimulus has been building over the last quarter century. The animal studies of Lanyon,[3] Rubin and Lanyon,[4,5] McLeod and Rubin,[6] and others have quantified the importance of strain rate over strain as a remodeling stimulus. The studies directed at the understanding of the cellular mechanism for bone remodeling have suggested that the prime mover is the bone strain rate–driven motion of the bone fluid whose signal is transduced by osteocytes. These include the model of Weinbaum et al,[7] where the shear stress from the bone fluid flow over the osteocytic processes in the canaliculi is a cellular mechanism–based model suggesting strain rate as a stimulus.

In the surface remodeling theory, the bone is assumed to be a linear elastic solid with an additional constitutive equation that specifies the velocity of movement of the exterior bone surfaces. The remodeling rate equation is the equation that governs the velocity U of the free surface movement as a function of the remodeling stimulus S at the surface at the time t on the remodeling time scale. In the present work, a strain rate stimulus is used.[2,8] Our implementation of the strain rate stimulus is accomplished here for the simple case of constant load frequency. The constant load frequency is assumed to be that of normal activity such as walking, say 1 to 2 Hz, and the shape of the applied load cycle is assumed to be the normal or reference shape; thus the surface bone remodeling strain rate stimulus is twice the trace of strain amplitude. The surface bone remodeling equation is written for a range of remodeling equilibrium stimuli as:

$$U = K_1(S - S_{o-}), \text{ if } S < S_{o-},$$
$$U = 0, \text{ if } S_{o-} < S < S_{o+},$$
$$U = K_2(S - S_{o+}), \text{ if } S_{o+} < S,$$

where S_{o+} is the value of the stimulus above which deposition occurs, and S_{o-} is the value of the stimulus below which resorption occurs. The stimulus domain, $S_{o-} < S < S_{o+}$, is the remodeling equilibrium domain. A graphical representation of the equations given above is shown in Figure 4-1. In this analysis, the remodeling rate constant $K_1 = K_2 = K$ is equal to 1 mm per unit of time where the unit of time is unspecified. The value of K only affects the rate at which the final remodeled equilibrium shape is attained, not its form.[9] In the calculations reported, bone was assumed to be an isotropic linear elastic solid with Young's modulus E (18 GPa), shear modulus G (7031 MPa), and Poisson's ratio υ (0.28). In the computational procedure, the stimulus is not evaluated at surface points but at points near the surface for accurate strain values, as explained by Sadegh et al.[9]

Based on the work of Lanyon[3] and Rubin and Lanyon,[4,5] who could only measure the peak value of one tangential strain (the tangential axial strain in a long bone, say ε_t), bone resorption will occur at strains ε_t greater than about −0.001, and bone deposition will occur for strains ε_t less than −0.003. Because the surface bone remodeling strain rate stimulus is twice the trace of strain amplitude, while the Rubin and Lanyon data are in terms of ε_t only, there is an adjustment that must be made in the determination of the numerical value of the limits S_{o-} and S_{o+}. Assuming the uniaxial loading condition in the Rubin and Lanyon experiments, the other two normal strain components ε_s and ε_n are equal to each other and also to $-\upsilon\varepsilon_t$. Thus these data should be adjusted by multiplying the Rubin and Lanyon measured ε_t by $(1 - 2\upsilon)$, where Poisson's ratio υ is 0.28; thus

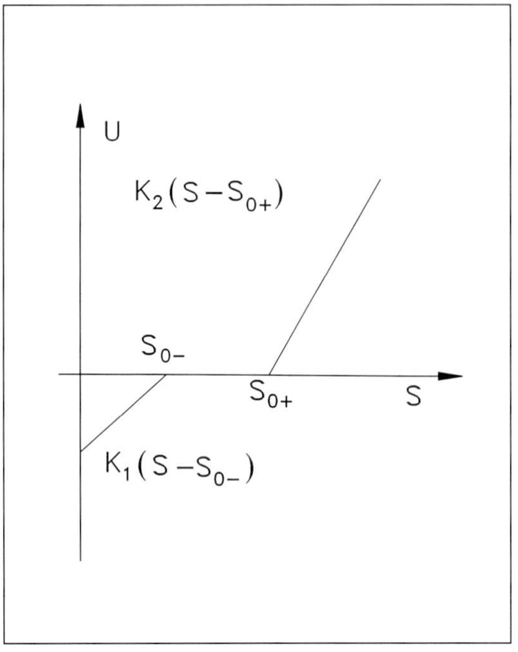

Fig 4-1 A plot of relationship between bone deposition and/or resorption and the stimulus. The bone deposition and resorption are represented by positive and negative values of the remodeling velocity U. This relationship is used to determine the velocity of the free bone surface movement as a function of the stimulus.

$S_{0-} = 0.00044$ and $S_{0+} = 0.00132$. Bone is in remodeling equilibrium between those two strain rate stimulus limits.

Local-Global–Feedback Model

The surface bone remodeling theory just described is used in this study of the long-term stability of the total hip implant. The primary information required for implementation of the surface bone remodeling theory is the strain distribution on all remodeling bone surfaces. It is not feasible to construct a mesh for the whole total hip implant while keeping all the local structural details. The strategy to circumvent this restriction is to divide the problem into three parts.

The first part consists of local models that are used to study the surface bone remodeling under given loading conditions for certain local structures, ie, screw thread or bead models. The second part consists of global models that are used to find strain and displacement distribution near the bone surface of the global models without keeping local details. The local remodeling is reflected on the global model by changing the bone-implant interface condition and bone geometry. The last part is the interaction between the local model and the global model, ie, the passage from the global level strain and displacement distribution to the local level and the reciprocal influence of the remodeling at the local level on the global level response.

Local Model of a Screw Thread

To achieve sturdy bone-implant contact, the implant surface is roughened so that a strong mechanical interlocking of the bone at the implant will develop. From the viewpoint of the modeling done here, this roughing is on the global scale; the actual surfaces of screw threads may be either smooth or rough on the local scale. The rough surface (on the global scale) may be manufactured by means of sintered beads, wire mesh, notches, or by coatings of various types. When the implants are placed in the bone, a tight fit is essential for a successful mechanical interlocking. The bone at the implant interface is thus subjected to multi-axial load. A number of different local interface models are considered in Sadegh et al,[9] including slots, beads, and screw thread. The two-dimensional screw thread model is described here. Plane strain is assumed in the calculation.

To understand the influence of the applied load on the final ingrowth, a two-dimensional model, as shown in Figure 4-2, of a typical region of the bone around a screw thread was used. In this model, the thread angle of the screw is denoted by 2θ, the total height of the screw tooth is H, the pitch of the screw is $2H\tan\theta$, and the stress applied by the surrounding bone tissue is denoted by σ_x and σ_y. The initial penetration depth of the screw into the bone is denoted by h_i, the depth in the middle by $h(t)$, and the final depth in the middle by h_f. The final penetration depth h_f is called the *final ingrowth*. The surface of the bone tissue represented by the linear domain BC is initially in contact with the implant. The bone boundary represented by the domain AB is a free-moving boundary, and it may move into the triangular region below AB as determined by the remodeling process governed by surface bone remodeling theory. As the ingrowth continues, the moving face AB of the bone becomes smaller and closer to the vertex between the teeth, and the domain of bone-implant contact, BC, increases.

The boundary element method (BEM) is used for the local stress analysis.[10] To obtain accurate numerical results, different boundary element meshes are used for three different ranges of $h_b(t)$. The BEM model is one-half of Figure 4-2. On boundary BC, a smooth contact boundary condition is applied for a smooth implant surface and a fixed interface condition is applied for an HA-coated implant surface. On boundary CD, a uniform normal stress σ_x is applied, and on boundary DE a uniform normal stress σ_y is applied. The roller boundary condition is applied on surface EA.

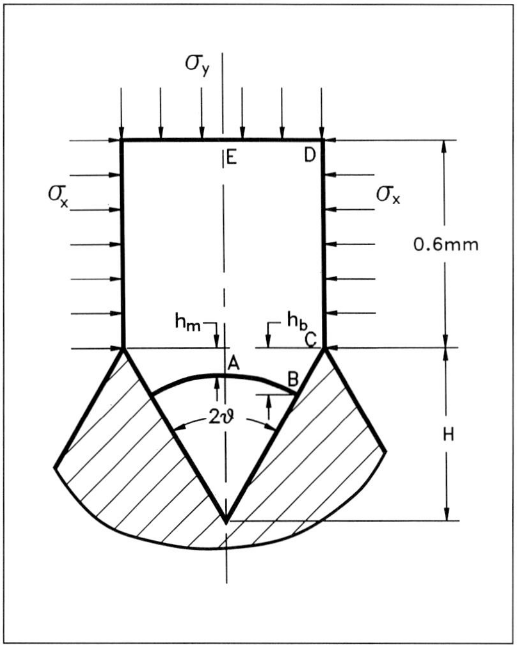

Fig 4-2 The screw thread model. The cross-hatched area represents a portion of a screw thread; block ABCDE represents a piece of bone tissue. The bone–screw thread interface is indicated by segment BC.

Results for the Local Screw Thread Model

For the case with smooth contact interface BC, the variation of the normalized final ingrowth depth, h_f/H, as a function of the load $\sigma_x = 0$ and $\sigma_y = \sigma$ for thread angles of $2\theta = 60°$ is shown in Figure 4-3. This curve shows that the final ingrowth is almost complete if the compressive stress σ is greater than 23 MPa, while stresses less than 10 MPa produce incomplete final ingrowth. The depth of the bone final ingrowth is a function of the applied compressive stress $\sigma_x = 10$ MPa and $\sigma_y = \sigma$, as suggested by the three panels of Figure 4-4. The curves illustrating the progressive ingrowth of bone into the thread tooth in Figure 4-4 have significance only with regard to the method of calculation.

A similar calculation was also performed for fixed interface BC. The depth of the bone final ingrowth is a function of the applied compressive stress $\sigma_x = 10$ MPa and $\sigma_y = \sigma$ for thread angles of $2\theta = 60°$, as suggested by the three panels of Figure 4-5.

Comparison of Local Model Results with Experimental Results

The results show that for the given boundary condition, increasing loading results in more

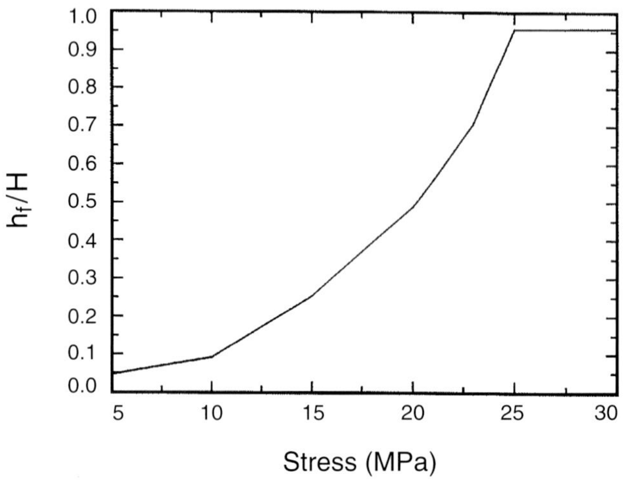

Fig 4-3 Plots of the curves representing the non-dimensionalized final ingrowth between the screw threads, h_f/H, as a function of the applied stress σ for the 60° thread angle considered.

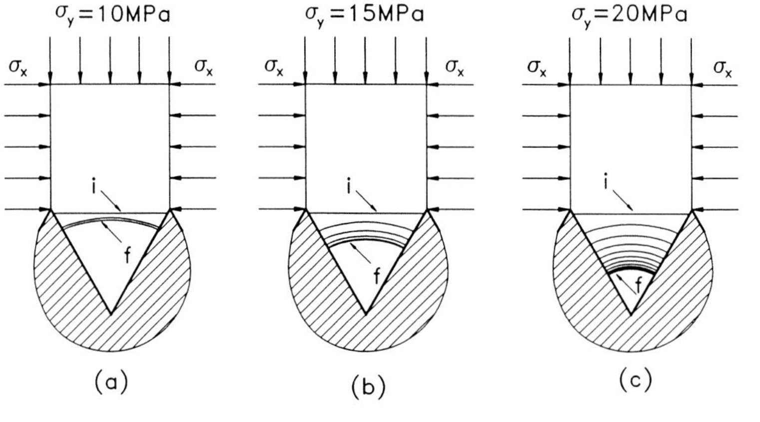

Fig 4-4 Illustration of the depth of bony ingrowth into screw thread with smooth contact bone-implant interface (as indicated by segment BC in Fig 4-2) for three levels of applied stress (a–c). The "i" and "f" stand for initial and final, respectively.

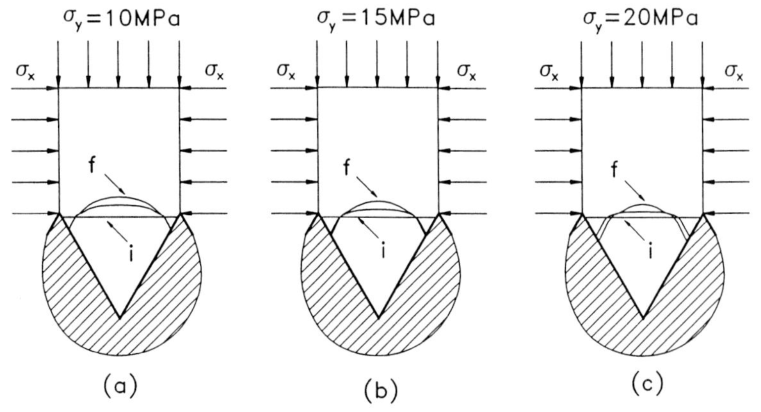

Fig 4-5 Illustration of the depth of bony ingrowth into screw thread with fixed bone-implant interface (segment BC in Fig 4-2) for three levels of applied stress (a–c). The "i" and "f" stand for initial and final, respectively.

bone ingrowth into screw threads and reducing loading results in less bone ingrowth into screw threads. An interesting observation is that, for a given loading condition, the numerical model predicts less bone ingrowth into screw threads for HA-coated surfaces than for smooth surfaces. It is important to note that the results predicted are based on a surface bone remodeling theory for mature bone instead of the results of the healing process and/or trauma reaction.

Carlsson et al[11] inserted 60 implants of three designs into the proximal tibial or distal femoral metaphysis of six dogs. A screw (12-mm long, 3.75 mm in diameter) was inserted in a pretapped hole (n = 48). A double cylinder with an irregular surface (length 24 mm, 8 mm in diameter) was inserted into a preformed defect (n = 6). A T plate was placed with the transverse part of the T (length 18 mm) on the bone surface, and the 8-mm-long shaft was inserted into a preformed slit in the cortical bone (n = 6). Their results show that the titanium screws were well integrated into the bone. At the cortical passage, bone with haversian systems filled the threads of the implants entirely, and there were no differences between the level of bone contact after 3 months and that after 14 months.

The comparison of our numerical prediction of bone ingrowth into screw threads and Carlsson's experimental results shows that under certain conditions bone may fill the screw threads. Due to the lack of loading history data on the screws and adjacent bone, only qualitative comparison is possible.

Local-Global–Feedback Mechanism

The connection between the local model (interface) and the global model (whole bone-implant system) is achieved by passing the stresses and the displacements determined by the global model (Fig 4-6) to the local model, and the local model in turn feeds back the surface movement and the

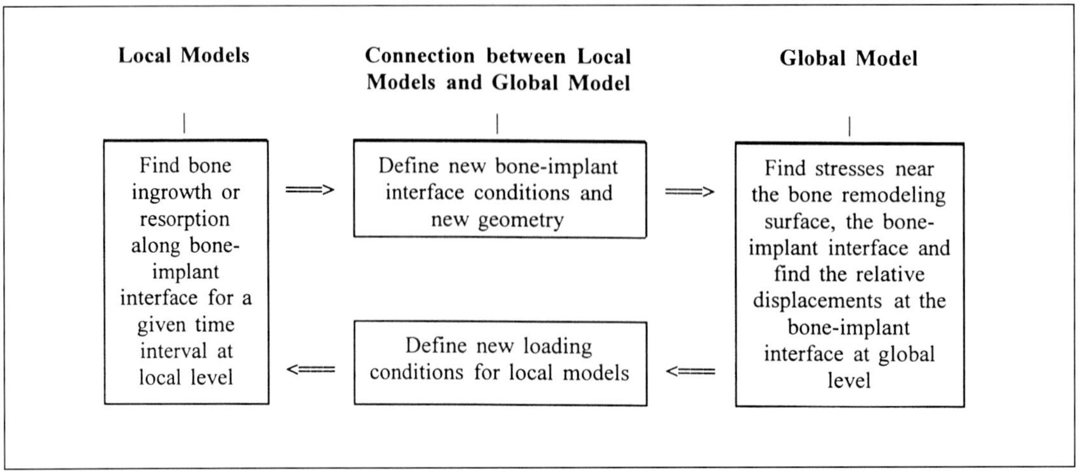

Fig 4-6 A chart illustrating the local-global feed-back process.

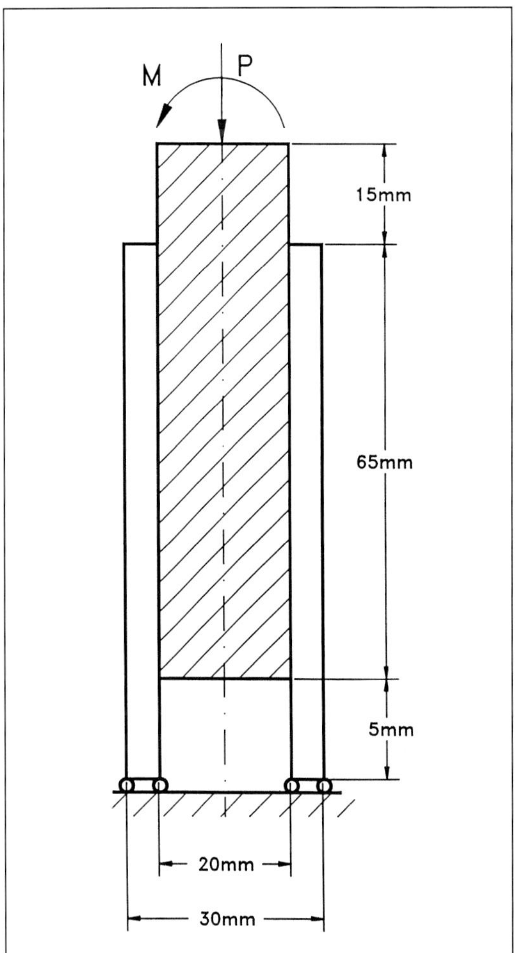

Fig 4-7 The global model with its loading and boundary conditions.

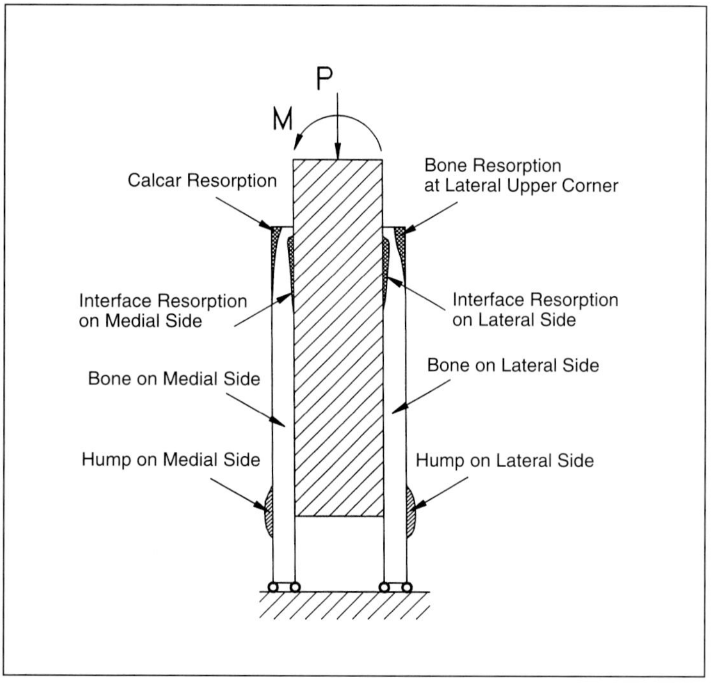

Fig 4-8 A general representation of local-global feedback model results.

Coulomb friction to the global model (Fig 4-7). The process repeats itself until a criterion for stopping is achieved. The criterion for stopping is the remodeling equilibrium on all bone remodeling surfaces or the failure of mechanical equilibrium. The steps of the local-global computational process are as follows:

Step 1. Set initial values: (a) initial interference associated with the force fitting, (b) initial ingrowth into the implant's porous surface before any external load is applied, (c) applied loading, and (d) initial distribution of the effective Coulomb friction coefficient along the interface.

Step 2. Global analysis: find the stress states and relative displacements along all the bone remodeling surfaces given the applied loading, the present interference state, and geometry of the global model using ABAQUS (an FEM package).

Step 3. Mechanical equilibrium check: mechanical equilibrium checks are made at each iteration loop. If mechanical equilibrium is not satisfied (ie, the implant cannot support the external load), the implant-bone structure has failed, and the calculation stops.

Step 4. Remodeling equilibrium check: each local site on all remodeling surfaces is tested to determine if the stimulus on the free surface of the current ingrowth has reached remodeling equilibrium. If all the local site remodeling processes stop, the program stops.

Step 5. Local analysis: the ingrowth during the time interval Δt at each node of the bone-implant interface under given stress states is determined from the appropriate data base.

Step 6. Based on the local remodeling results, the new bone-implant interface is defined. The calculation procedure then returns to step 2 and the iteration continues until the criterion for stopping is achieved.

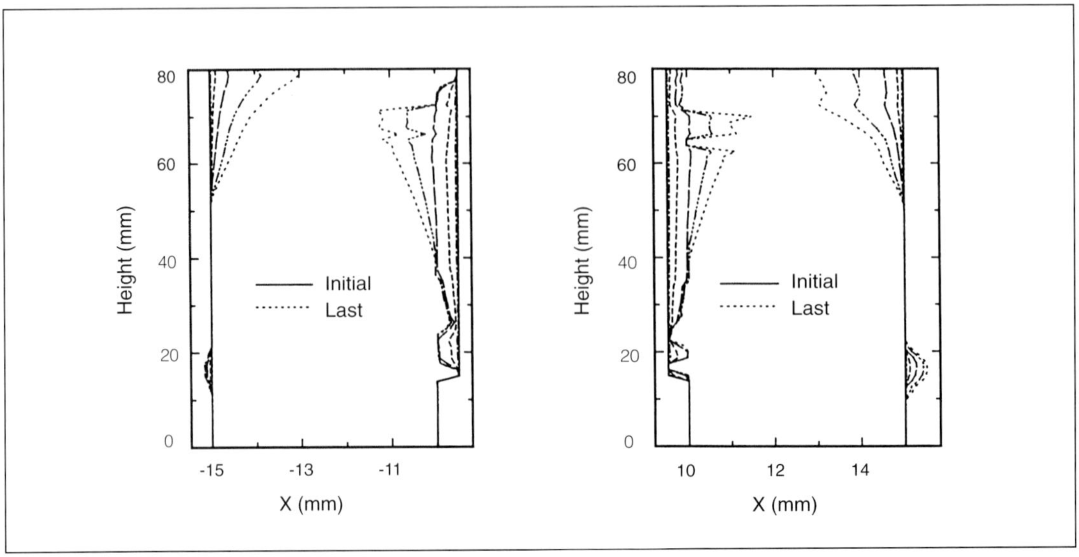

Fig 4-9 Remodeling predictions for a titanium implant fully covered with screw threads. The initial ingrowth was 0.48 mm and the initial interference was 0.01 mm. The bone remodels along the bone-implant interface as well as on the endosteal and periosteal surfaces on the medial side *(left)* and the lateral side *(right)*.

Local-Global–Feedback Results

A general chart of the predictions of the local-global feedback model is given in Figure 4-8. A specific example is given in Figure 4-9. The local-global–feedback model predicted the following:

1. Calcar resorption and that the resorption area increases as the stiffness of the implant increases.
2. Interface resorption, as shown in Figure 4-8, on both the medial and lateral sides is predicted in all cases. The area of resorption increases when the coated area of an implant increases.
3. Humps on the periosteal surface corresponding to the location of the distal end of the implant (see Fig 4-8). Increasing the initial interference will increase the hump size on the medial side and reduce the hump size on the lateral side, if other conditions are unchanged.
4. Since interface resorption is predicted in all cases and resorption starts at the very beginning of remodeling, uniformly distributed initial ingrowth and initial interference are not necessary to get the results as predicted in this study, as long as there is enough initial ingrowth and initial interference in both the upper and lower parts of the bone-implant interface on both medial and lateral sides.
5. The resorption in the greater trochanter area is predicted in the model if the abductor force is not simulated. In the model with the simulated abductor force, no resorption is predicted in the region of the greater trochanter.

Comparison of Local-Global–Feedback Model Results with Experiment

The comparison of our numerical results with clinical studies shows qualitative consistency. Calcar resorption is predicted by the numerical models and is observed clinically. Both numerical models and clinical studies find that the greatest bone resorption occurred in the proximal third and that implants having a more extensive porous coating and a larger-diameter (stiffer) stem cause more-extensive resorptive bone remodeling. Although bone resorption near the bone-implant interface has not been measured quantitatively, it is visible from radiographs and is mostly proximal. Humps on the periosteal surface or cortical thickening are predicted by the numerical models on both medial and lateral sides distally. The size of these humps depends on the initial condition. If the distal end of an implant initially contacts tightly with the endosteal surface on the medial side, a hump on the periosteal surface on the medial side will be predicted. If the distal end of an implant initially contacts the endosteal surface on the lateral side, not necessarily tightly, a hump on the periosteal surface on the lateral side will be predicted.

Cook et al[12] found that cortical thickening occurred only on the medial side. An explanation based on our numerical model is that there is a tight contact between the distal part of the implant and the endosteal surface on the medial side but not on the lateral side. Cortical thickening was observed on the lateral side in the clinical study. Due to the lack of knowledge of how the distal part of an implant initially contacts the endosteal surface on the medial and lateral sides, comparison is made based on the final numerical results and clinical observations.

Two local-global–feedback models, one without abductor force and one with abductor force, have been studied. By comparing the results obtained with these global models, the following conclusions may be drawn:

1. The one without abductor force predicts remodeling patterns that are consistent with clinical observation except in the proximal end of the model (near the greater trochanter). This is because the model retains the basic features of the load applied to the hip implant.
2. The one with abductor force predicts remodeling patterns that are consistent with clinical observations even in the greater trochanter area. This is because the model maintains the features of load applied on both the hip implant and the region of the greater trochanter.
3. The results obtained with these global models can be compared only qualitatively with clinical observations, due to the simplification of geometry, material properties, and loadings.

Closure

The long-term stability of the hip implant was studied using the local-global–feedback model. The surface bone remodeling theory with strain rate stimulus was used to predict surface bone remodeling for the interface as well as the bone surfaces that are traction-free. The advantages of the local-global model are the following:

1. Remodeling at both the local level, which is in a scale less than 1 mm, and at the global level, which is in a scale greater than 30 mm, can be considered at the same time.
2. Due to the creation of data bases for the local models, accomplishing the feedback model consumes minimal computer time.
3. The local models are interchangeable and a combination of different local models for an implant is easily accomplished. Thus the local-global feedback may be used to design an implant with different interfaces in different locations.

4. Although the models used in this study are two-dimensional, theoretically there is no barrier to extending the local-global–feedback models to three-dimensional models. However, a powerful computer is needed for three-dimensional calculation.

At the local level, the microstructural bone ingrowth remodeling into screw threads was investigated with computational models. Our four principal conclusions are the following:

1. The boundary conditions representing the type of contact at the local level (ie, bonded vs smooth interface) at the bone-implant interface have a significant effect on the final ingrowth distance of the remodeled bone tissue.
2. The bonded interface associated with HA-coated surfaces may result in a very different remodeling pattern from a smooth interface. We found that, for a slot, a bonded interface increases bone ingrowth into the slot, while for a screw thread, a bonded interface resists bone ingrowth into the screw thread.
3. The load applied on bone adjacent to a slot or screw thread also has a significant effect on the final ingrowth distance of the remodeled bone tissue. Increasing the load results in more bone ingrowth and vice versa.
4. The initial ingrowth depth has a significant effect on the final ingrowth distance of the remodeled bone tissue.

The following are the limitations on the comparison of the numerical results and clinical observations:

1. The simplified global models only can simulate the stress distribution in the middle plane in the medial and lateral directions and no out-plane effects can be considered.
2. Homogeneous and isotropic material properties are used in the numerical calculation, while in the femur the material properties are neither homogeneous nor isotropic, especially in the proximal femur.
3. The load applied on a bone depends on an individual's activity. Many muscles attached to the bone make the loading more complicated. The loads considered in this numerical study were only the 1- or 2-Hz components of the walking load. The higher frequency loading due to muscle firing was not accounted for. Lack of knowledge of the loading history and exactly what aspect of the loading history influences bone remodeling is a significant problem.
4. Comparison of the numerical results with clinical observations can be made only qualitatively due to the lack of data on the loading history in these experiments. Comparison of the numerical results with the animal experimental results at the local level shows that the numerical results are qualitatively consistent with experimental results; however, quantitative comparison is not possible at this time.

The models used in this study are useful for gaining understanding of bone remodeling at both the local and the global levels and for studying the long-term stability of hip implants. The models can be used for numerically testing the stability of an implant and as aids in choosing good implant designs. Using numerical models to design animal experiments will be very cost-effective, because computer analysis is much less expensive than animal experiments.

DISCUSSION

Dudley Childress Dr Cowin, qualitatively your model seems to work. These are mathematical models. How do these compare with physical reality? How do the results compare, or do you have results with respect to biological experiments?

Stephen Cowin You can go through the literature and you can go through the radiologic collections of surgeons and you find excellent qualitative comparison. You can see that the models succeed. You find these humps, you find the resorption, of course. The calcar resorption has long been known. But you can't quantitate this precisely yet because we do not have a good idea of what aspect of the mechanical load history influences remodeling. That is to say, of all the mechanical loading events that a bone experiences, what aspect of the event is recognized and responded to by the cellular mechanosensory system? The mechanical loading events bone experiences have range of force magnitudes up to fracture strength, variable duration, and a considerable range of frequency. We really need to do very careful biologic experiments to try and determine exactly what the signal, the mechanical signal, is that excites the osteocyte.

Shu Chien I am very interested in your comments on the similarity between the effects of shear stress on the osteoblasts and the endothelial cells in the blood vessel. The magnitude may be very similar, but there may be a difference in the temporal pattern; the frequency spectrum must be different. The circulation is always subjected to stress. The blood is normally flowing all the time, at least in the arterial system. Many genes are turned on by the shear stress with a short time constant. But in the bone, the loading or the shear stress is not constant, and the time constant of the response may be important.

SC That's right. It is intermittent.

SCh Since it is intermittent, it has a different implication in terms of gene expression because many of these genes would turn on for about an hour and shut off very quickly. So either in circulation, the straight part of the vessels with a relatively constant flow would not be turned on. But in the bone, there may be a different situation.

Another comment is the spatial influence. In the circulation the shear patterns seen by the branch point are different from that in the straight part. The branch point orifice is where the atherosclerosis is formed. In the bone, do you have spatial variation in shear stress? Is it similar to the straight part of the blood vessels?

SC Yes, we have spatial variation because when the femur is bending (the primary load is bending), it means that at the outer edges you are going to get greater deformation. The fluid flow will be greater in interior regions. You will get higher stresses there because the fluid is driven across the neutral axis. There is higher stimulation of the cells at the center because that is where the fluid speed is greatest. At the external surfaces, you get lesser stimulation.

I would like to hear your thoughts because we realize that we have to have some spatial integration. We think that these bone cells communicate over a certain distance and make group decisions as to whether to deposit or restore. The question is, how many of their neighbors do they sample? Do they sample out to 100 or 200 μm? This is the sort of thing we are trying to build the model for—to figure out over what distance they sample. I think that is a spatial aspect similar to what you are citing.

SCh In the vascular tree, you have stagnation points, flow reversals, and other conditions of disturbed flow. Do you see that in bone tissue?

SC Not in the bone. You have such an elegant control in the cardiovascular system when the shear stress stays the same, through ten generations. So we have not seen anything like that in bone at all. In regard to the question about the top of the skull, though, I like to answer with the frequency discussion. Our friend, Melvin Moss, will tell you that the masseter muscle really excites the top of the skull. He really likes the results of Ruben and McLeod because they are at a frequency that is reasonable and consistent with his idea about functional matrices.

Richard Skalak You are starting to look now at the sensory mechanisms involved. Do you have any information or intuition on the mechanism of generating signals that are interpreted by the brain so that there will be a sense of touch or pressure to the patient?

SC We just had a discussion on proprioception in teeth. The innervation of teeth is with pain receptors, not with probe receptors, and there is some debate along that line. But if you look at the structure of teeth, the nerve wraps right around and goes deep into the tooth. You don't see such nerves in bone. Not to the same extent certainly, so I don't think it is a similar situation.

Göran Lundborg We will address this question later on, I think, but there is no question that there are nerves in bones. This has been demonstrated with immunohistochemical techniques. There seem to be nerves in the outer layer in the periosteum and in the endosteum. We know from patients having bone-anchored prostheses that these patients not only achieve proprioception but to some extent also learn discrimination and detection of the shape of items.

Björn Rydevik This is for amputation prostheses, not for joint prostheses, is that correct?

GL Yes, I am talking about amputees. This is a very fascinating issue, which Dr Brånemark has named osseoperception. There are nerves, but the question is, what are the receptors for touch and discrimination? Your thoughts about redistribution of fluid in the canaliculi are very interesting because if that is so, perhaps receptors peripheral in the bone may be reacting to the distribution of fluid inside the bone tissue.

Richard Coutts I am intrigued by this concept that it is the deformation of the osteocyte lacuna and then transmission of fluid that generates a signal. Have you actually been able to measure the deformation of a lacuna, and have you measured fluid flow in a canaliculus?

SC The first question is answered by an easy calculation using the properties of bone. We know that the overall tissue deforms, and we can calculate the deformation. I don't know that anyone has measured it. The fluid flow in the canaliculus has not been measured, but there is a report that is about to come out from Elizabeth Berger's laboratory about the sensitivity of osteocytes, and the report says that the osteocytes respond to shear stresses of a half a Pascal but to hydrostatic stresses of 13,000 Pascals. So they are very sensitive to shear stresses and relatively insensitive to hydrostatic stresses, which for us is a very happy indication.

RC Where would the fluid in the canaliculi flow if the system is closed?

SC The fluid flows back and forth due to cyclic bending loading. At one point in the cycle it is likely to drain into the vascular porosity, then mix with bone fluid rich in nutrients from the blood, and then be driven by fluid pressure gradient back into the canaliculi to repeat the process. The bone fluid velocity will approach zero as the bone fluid approaches or leaves the vascular porosity and it will achieve its maximum value at points in the canaliculi furthest from the vascular porosity.

RC So it is an oscillation, as opposed to a unidirectional flow.

SC That is correct, but I believe that there are also enhancements; that the convection enhances the diffusion. That is part of the loading and part of the mechanisms by which the osteocytes get their nutrients. The result is not just back-and-forth motion of the same fluid; at each back-and-forth motion you mix up a little bit and then diffuse out again.

RS I would like to comment on the comparison to the cardiovascular system again. This system is also very sensitive to shear stress, such as a few dynes per square centimeter. However, pressures of thousands of dynes per square centimeter don't seem to do anything.

SCh I agree with you partially. There is this difference in order of magnitude regarding sensitivity to the two kinds of stresses. But in terms of molecular programming, the time scale is different. Shear stress seems to turn on the gene, as I previously mentioned, for a short time, and then shut off. On the other hand, the pressure seems to have its effect for a long time. If you elevate the pressure in the cardiovascular system, for example, the monocyte chemotactic protein gene is turned on and does not turn off if the pressure is kept elevated. So there is a difference in the time scale. As Dr Skalak said, cells are more sensitive to shear stress, but the existing shear stress is also much lower than the pressure loads.

Geert Schmid-Schönbein I have a fundamental question. What are the molecular mechanisms by which new bone is generated? Is it a new cell that is incorporated into bone? Are the osteoclasts capable of resorbing bone? How do these cells do that?

SC Osteoclasts resorb the bone. I think there are others here who can speak on this issue better than I. Our model consists of a sequence of steps in which the mechanical loading is the source of a signal for the bone to model or remodel. First, the mechanical loading sets up a fluid motion in the bone. Second, this fluid motion is sensed by the osteocytes. Third, the information concerning the mechanical loading is sent to the surface lining cells from the osteocytes. When the surface lining cells or osteoblasts receive this information, they have at least three choices. First, they can deposit bone. Second, they can also change the parathyroid level in the circulating blood and can call osteoclasts to their neighborhood to resorb bone. Third, they can decide the signal was not strong enough and do nothing.

GSS Does it require living cells to resorb bone? If I die, my bones can lie in the ground and they do not resorb anymore. Is this because I have no more living osteoclasts? Do I need a new cell in the bone for resorption?

Tommy Hansson You can't always answer such questions in a simple way, but these processes always start with resorption. The osteoclasts are the cells that start resorption. After a while, an osteoclast turns into osteoblast. After becoming osteoblasts, they lay down ground substance, which is then calcified. So it is always a cyclic behavior—first the resorption and then the regeneration or both. There can't be another way, apparently.

SC New additional bone can be formed on a bone surface. You can have formation without resorption. You can have additional bone layered on top of old bone.

TH Yes, you can have surface apposition with or without resorption. The cycle is resorption, then an apposition which as a rule can be positive. I mean you can lay down more than what is absorbed, but I do not think you usually have direct apposition of bone.

SC I certainly have seen sections of bone where there are osteoclasts lying superficially and where there is no resorption. You don't necessarily see a debit in the bone. I have seen osteoblasts just lying on what appears to be a normal bone surface, and I think that apposition of new bone could begin in response to a change in the load environment. If they sense the need for bone, osteoblasts would build on top of the existing bone. I don't believe that you have to have resorption in the trabecular bone, but you do in cortical bone because you don't have any space to build new bone, unless you remove the old. You can have new bone formation on the periosteal side without resorption.

Dick Heinegård I am more and more impressed with what is being learned about the molecular mechanisms. Bone resorption always triggers the cells to make new bone also. You also need to know when to remove old fatigued bones, and that is part of the resorption cycle. The initiating event, I think, in part has to be a stimulant to parathyroid hormone and vitamin D3 and so forth. The initial event must be very focal because if you look at where this starts, it is very focal. One of the events is that cells are triggered to remove the existing cells at the same time that the cells are stimulated to deposit new mineralized contents onto the surface. Osteopontin is an excellent recruiting factor. It is a surprising protein and stimulates all kinds of osteoprecursor cells. They are integrated on the cell surface and are triggered to polarize and to form the resorbing osteoclasts. It actually seems like the osteoclast resorbs and secretes some of this acid phosphatase, and that the osteoclast has lost its content. Thus, there is no more support of the binding of the osteoclast, and it can go somewhere else to do something else. The riddle is then, why would the osteoblast come back in? What is it recognizing?

Because cells need a surface to migrate upon and something to bind to, it's not clear why the osteopontin-covered surface does not support the osteoblast, but the surface that the osteoclast has left supports binding. Then you have a repopulation of cells on this cold or naked surface by the osteoblasts that are triggered to deposit new matrix. It is a balance to remove all bone and put new bone down, but you can always trigger these cells by giving them the right growth factors.

BR Thank you, Professor Heinegård.

Per-Ingvar Brånemark That also goes for situations where you have synthetic material and bone tissue at that actual interface because it has, as suggested, osteopontin as a component at least in some regions.

DH Yes, but I mean osteopontin is also present as the fully phosphorylated osteopontin, which permits binding in the osteoclasts. The fully dephosphorylated osteopontins are not very well delineated yet. Then there are fragmentations of osteopontin and of course it is a rather complex situation.

Acknowledgment

This investigation was supported by NSF Grant No. BSC-9103236. This research was also supported (in part) by grant nos. 662319 and 662306 from the PSC-CUNY Research Award Program of the City University of New York.

References

1. Ling R. Observation of the fixation of implants to the bone skeleton. Clin Orthop 1986;210:80–96.
2. Luo GM, Cowin SC, Sadegh AM, Arramon Y. Implementation of strain rate as a bone remodelling stimulus. J Biomech Eng 1995;117:329–338.
3. Lanyon LE. Functional strain as a determinant for bone remodeling. Calcif Tissue Int 1984;36:S56–S61.
4. Rubin CT, Lanyon LE. Regulation of bone formation by applied dynamic loads. J Bone Joint Surg 1984;66A:397.
5. Rubin CT, Lanyon LE. Osteoregulatory nature of mechanical stimuli: Function as a determinant for adaptive bone remodelling. J Orthop Res 1987; 5:300.
6. Rubin CT, McLeod KJ. Endogenous control of bone morphology via frequency specific, low magnitude functional strain. In Odgaard A, Weinans H (eds): Bone Structure and Remodeling. Singapore: World Scientific; 1995:79–90.
7. Weinbaum S, Cowin SC, Zeng Y. A model for the excitation of osteocytes by mechanical loading-induced bone fluid shear stresses. J Biomech 1994;27:339–360.
8. Luo GM, Cowin SC, Sadegh AM, Arramon Y. Strain rate and strain as bone remodelling stimuli. In Odgaard A, Weinans H (eds): Bone Structure and Remodeling. Singapore: World Scientific; 1995: 189–199.
9. Sadegh AM, Luo GM, Cowin SC. Bone ingrowth: An application of the boundary element method to bone remodelling at the implant interface. J Biomech 1993;26:167–182.
10. Luo GM. A new boundary element method coupled with FEM packages. Commun Appl Numerical Methods 1989;5:365–371.
11. Carlsson L, Rostlund T, Albrektsson B, Albrektsson T, Brånemark P-I. Osseointegration of titanium implants. Acta Orthop Scand 1986;57:285–289.
12. Cook SD, Barrack RL, Thomas KA, Haddad RJ. Quantitative analysis of tissue growth into human porous total hip components. J Arthroplasty 1988;3:3–15.

Chapter 5

Matrix Molecules in Connective Tissues with a Potential for Use in Guiding Osseointegration

Dick Heinegård, Björn Klinge, and Mikael Wendel

In recent years, the use of implants in orthopedics has increased dramatically. The design of most of these implants has involved prioritizing the *mechanical* properties of the implant. The interface to the surrounding tissue has often been created by including a cementum layer to fill any sterical defects. In the process of polymerization, the temperature increases and will most likely permanently cause damage to some of the surrounding cells, therefore hampering remodeling.

A problem that may arise is that the implant is not interacting with the surrounding tissue to become an integrated part of the organ. Attempts are continuously being made to modify the surface of the implant to achieve this integration. In the past, this has been attempted by attaching a hydroxyapatite layer onto metal implants. Still, complications are frequent, with loosening of implants and stress fractures due to the relative mechanical incompatibility of the implant with the surrounding bone.

For major progress to be made, prerequisites are that the mechanical properties of the implant can be adjusted to mimic the properties of the surrounding bone. Implants should also be designed so that they become integrated with the bone and the two structures will form a unit. The potential use of different types of polymer materials and alloys may provide implants with adequate mechanical properties.

To create an interface to the bone, one should consider modifying the surface of the implant so that it can directly integrate into the bone. We suggest that there are two avenues that should be attempted. One is to select cells that bind to the surface by strong interactions and can then lay down a bone matrix with adequate composition, structure, and function. These cells should also have the capacity to remodel the surrounding matrix to best meet the needs of the system. Another approach would be to create a surface that interacts with major structural components in the surrounding bone matrix. Examples would be molecules that connect to the fibrillar collagen network that apparently guides the deposition of the hydroxyapatite mineral crystals in bone. Important prerequisites for this approach are that we have an adequate knowledge of factors that selectively bind bone cells as well as of components that interact with bone matrix structural molecules.

What follows is an account of our current knowledge on bone matrix macromolecules and their interactions among themselves and with the cells in the tissue.

Over recent years, an increasing number of bone matrix proteins have been isolated and characterized (Tables 5-1, 5-2). In many cases, their primary structure has been determined with complementary DNA cloning and sequencing.[1-3] Among these molecules are those that are primarily found in bone.

Table 5-1 Matrix Proteins That Interact with Different Integrins

Protein	*Cell type bound*	*Integrin (site in tissue)*
BSP	Osteoblasts	$\alpha_{IIb}\beta_3$ and $\alpha v\beta_3$ (depends on surface) early in bone cartilage interfaces
OPN	Osteoclasts (does not bind osteoblasts)	$\alpha v\beta_3$ on osteoclasts
Chondroadherin	Chondrocytes	$\alpha_2\beta_1$ and $\alpha_5\beta_1$
Collagen I/II	Osteoblasts (many cell types)	$\alpha_1\beta_1$ and $\alpha_2\beta_1$
Fibronectin and other integrins	Many cell types	$\alpha_5\beta_1$

Abbreviations: BSP = bone sialoprotein; OPN = osteopontin.

Table 5-2 Matrix Proteins with Strong Interactions with Other Matrix Constituents

Protein	*Properties*
Collagen I	Early calcium-phosphate deposition Tensile strength
Fibromodulin	Binds collagen I/II fibers, TGF-β
Decorin	Binds collagen I/II fibers, TGF-β
Lumican	Binds collagen I/II fibers
Thrombospondin	Regulation of cell growth/migration
Fibronectin	Collagen, heparan sulfate

Abbreviation: TGF = transforming growth factor-beta

Examples are osteocalcin, bone sialoprotein (BSP), the 85-kDa protein osteoadherin, and bone-specific alkaline phosphatase. A number of proteins produced by cells in the bone are found with a wider distribution among tissues, but are most prominent in bone. These include osteonectin, osteopontin, and the 63-kDa protein nucleobindin. Interestingly, there are other proteins that appear to be synthesized elsewhere, eg, in the liver, and then primarily deposited in bone tissue. An example is α_2HS glycoprotein, which may be identical to a prominent component in bone. One should not forget the major component of the organic matrix in bone, which is collagen I. This molecule is abundant in many tissues, but a large proportion of the total body content of collagen I is actually found in bone.

Information on the structure and the properties of some of these molecules will help us shed light on their functions. Also, this is important background information in any future attempts to provide a special function to an implant surface by providing properties mimicking or depending on the presence of distinct matrix macromolecules.

The collagen molecules are assembled by a very precise mechanism into large fibers forming the primary scaffolding of the tissue. Each collagen molecule is made up of three chains, where almost the entire molecule is tied together in a triple helical structure. The molecules are secreted in a proform with extension peptides both in the N- and the C-terminal parts. These peptides have to be cleaved off from the major triple helical region after the proform has been secreted into the matrix to allow precise assembly into the fibers. The collagen fibers are arranged in parallel in some tissues where they are essential contributors of tensile properties, a major function of these molecules. Examples are tendons and ligaments.

In bone, the fibers appear to have an important role in regulating and providing the primary site for the nucleation and growth of the hydroxyapatite crystals. Interestingly, the collagen fiber surface properties are modified by a coat with several other matrix constituents. These include a family of small proteoglycans that have a prominent large and heavily negatively charged side chain of chondroitin sulfate. It is possible that one role of this charged molecule is to regulate and influence mineral deposition and growth.

These small proteoglycans are part of the family of proteins usually containing ten leucine-rich repeats (LRR), each consisting of 25 amino acid residues. As stated, some of these molecules have the capacity to specifically bind to different collagens. Other properties include binding of members of the TGF-superfamily, notably TGF-β1, -β2, and -β3. It is therefore possible that they bind growth factors that are released when the proteoglycan is degraded in the remodeling processes, thereby inducing a cellular response.

An example of a bone molecule that is more restricted in its distribution and that has important functional properties is BSP. This molecule contains an RGD-sequence that can mediate binding of cells via integrin receptors. It appears that there is some preference for BSP binding to osteoblasts among bone cells. In addition, BSP contains polyglutamic acid sequences, which may have a role in binding to hydroxyapatite. Interestingly, BSP has a rather distinct localization in bone at the interface region between growth cartilage and bone. Its synthesis is most prominent in this particular region. Thus, the protein appears to be used in nature to provide an interface between these very different tissues. This could be of major relevance in promoting and guiding reactions and tissue structure at bone-implant interfaces.

Another molecule with a specific capacity to bind certain cells is osteopontin. It primarily binds osteoclasts via their $\alpha_V\beta_3$ integrin. The protein is also enriched at the mineralization front in bone and may therefore have a role in calcium deposition. It has been shown that the protein can modulate and inhibit hydroxyapatite nucleation and growth in vitro. Thus, this protein or fragments of this protein can be used to modulate the mineralization and perhaps also to regulate cell attachment to implant surfaces.

In this context it should be mentioned that collagen will also have a capacity to bind cells via a different set of integrins, primarily $\alpha_2\beta_1$ and $\alpha_1\beta_1$. Other molecules with different specificity for integrins are laminin, fibronectin, vitronectin, and some of the members of the leucine-rich repeat family.

There are a number of molecules that primarily will bind a distinct set of integrins. Since cells of different lineages and different phenotypes express distinct patterns of integrins, it should be possible to design protocols where a given surface exposes a particular ligand and thereby recruits a particular set of cells. Using this approach, it should be possible to guide matrix deposition at the interface. A schematic illustration of this approach is provided in Figure 5-1, where the ligand is immobilized, eg, by covalent linkage to the surface of the implant. There it will serve to attract particular cell types, which then may elaborate the matrix.

Another approach to modulating the character of the interface would be to provide a coat with matrix molecules with distinct properties. Thus, it should be possible to promote calcification by combinations of certain collagen arrangements as well as by components of BSP that have been shown to promote hydroxyapatite nucleation and growth. Alternatively, one could make use of the interactions in the matrix by providing molecules that show distinct interactions with other major structural constituents in the bone matrix. Thus, the surfaces of molecules that bind hydroxyapatite could be of interest in linking the mineral to the surface. Another attractive approach would be to make use of molecules that can bind the collagen fibers,

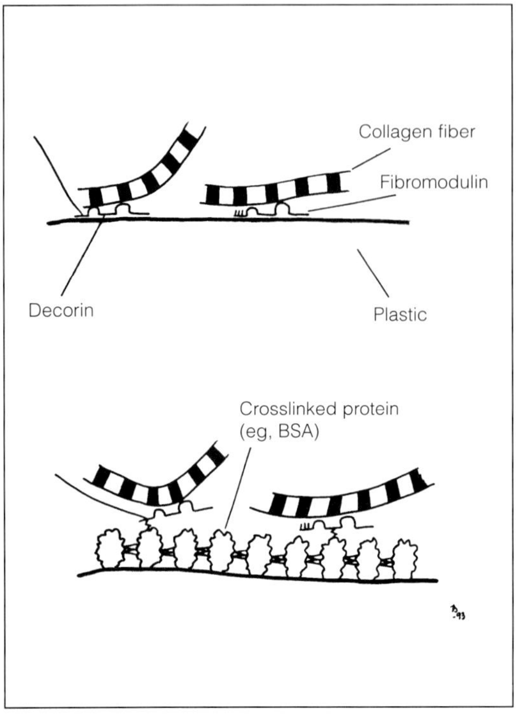

Fig 5-1 Matrix collagen attaches to the surface. *Top*, A surface is passively coated with a collagen-binding protein. *Bottom*, Alternatively, such proteins are covalently attached to a continuous coat of the surface. The new surface created can then form strong interactions with specific molecules or structures, in this case collagen fibers.

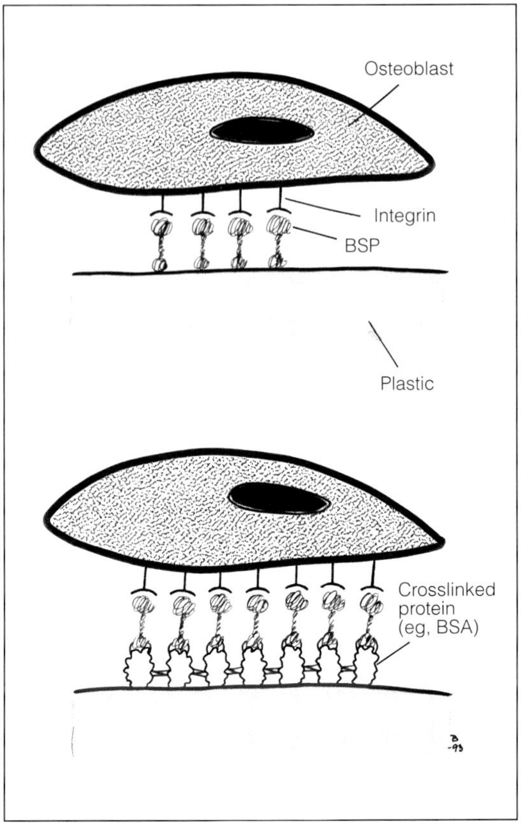

Fig 5-2 Cells attach to the surface. *Top*, A surface is passively coated with a specific cell binding (to integrins at the cell surface) protein. *Bottom*, Alternatively, such proteins are covalently bound to a continuous coat of the surface. The new surface created can then attract and bind select cells that are induced to form a new matrix.

eg, leucine-rich repeat proteins, other collagen-binding matrix proteins, collagens themselves, and fibronectin fragments. This approach is schematically outlined in Figure 5-2, where a molecule with the capacity to bind collagen has been immobilized on the surface. This could then interface to the newly formed bone matrix.

In summary, currently available bone matrix macromolecular constituents have a potential use in tailoring the reaction and tissue composition at bone implants. In this process, a given molecule or combination of molecules can be immobilized and bound to an implant surface, and the development of an interface matrix upon implantation can be monitored. In principle, one advantage of such a system is that the interface is guided via interactions to the specific matrix proteins or analogs of these and should thereby be independent of the properties of the implant itself. This would open a potential to design implants of materials that have not been feasible to this point. The different compositions of bones at different locations may require use of different modifications depending on the site for implantation.

DISCUSSION

Shu Chien Thank you very much for this very interesting presentation. You mentioned that various kinds of cells, such as osteoblasts and osteoclasts, have surface receptors that interact with the matrix and that there were initial signaling molecules. Eventually, I suppose that is what directs the cell to proliferate or to resorb. What are these signaling pathways for the different kinds of integrins? Are they different? What makes the cell proliferate or go the other way?

Dick Heinegård That has really not been very well mapped out. There are definitely phosphorylation cascades involved, but the pathways may be different and have not been identified.

SC I have another question on the production of a strong mechanical bonding between the collagen and an artificial surface. Do you see any way that could be achieved?

DH I believe that nature provides a clue for this. If you look in tendon, the collagen fibers are coated with a number of proteins/proteoglycans apparently cross-linking the fibers. If you remove the side chains that appear involved, which we can do enzymatically, we change the properties of the tissue. I believe that one approach is to provide a coat to the implant with such collagen-binding molecules. We will of course have to identify the right ones because not all of the side chains show the same strong binding. I suggest that the collagen fibers of the continuous collagen fiber network in the tissue can be tied down to the surface by using such collagen-binding molecules.

Steven Garfin How easy or difficult is it to get these binding proteins to attach to a metal, to an implant? I understand how they bind to the biologic tissues, but how do you get them to bind to the metal, or is there a natural affinity to that? A second question related to that is, how stable are these proteins? Would they be something that you could apply commercially, so to speak, and have them on the shelf so when you take the implant out of the box, they would be biologically active?

DH Let me answer the second question first. When we purify these molecules, we have to be very harsh to the tissues. We use extreme, denaturing conditions because we have to break up the network that ties the molecules down. Thus, it is not very likely that they can be damaged by storage. There are many ways of preparing the proteins and stability should not be a problem. Can they bind to surfaces? I believe that you can achieve this in many ways. This is not a major problem in our model systems. For instance, we have used a surface with a thin gold layer. You can then use thiol chemistry to attach your ligands to this surface. In many other cases, we know that the proteins can be passively or actively bound to the surface. Thus, it should be possible to attach proteins to any kind of material. It should also be possible to design procedures so that they retain their stability.

Elof Eriksson This is a very interesting topic. Where are you experimentally in relation to this? Most of this I assume is done in vitro. Can you use these molecules in vivo?

DH The last set of slides showed materials that were actually implanted into a dog mandible to try to modulate what is happening at the interface. However, most of our work is still aiming at attempts to learn the functions of these molecules. That provides the background for selecting various molecules in a combination of in vivo and in vitro work.

Åke Carlsson Concerning in vivo events: a recent study has shown that osteopontin is laid down very prominently at the interface between the untreated titanium implants and the bone in a rat model. This is the work of McKee and others. I wonder if you would speculate on the possible role of osteopontin, in view of its binding to osteoblasts and how it may be related to implant surgery.

DH I am not convinced that this is actually osteopontin. They may well be looking at BSP. When you study where these proteins are made—by, eg, in situ hybridization—you tend to find primarily BSP at these locations. One of the problems with antibodies to osteopontin and BSP is that many of them do show cross-activity.

Another complicating factor in evaluating what cell ligands to use is the often different binding characteristics of fragments. So even though osteopontin of the type that is isolated from bone will primarily bind to osteoclasts, a differently phosphorylated osteopontin or a fragment of osteopontin will show a different binding specificity. This protein is also a product of activated T cells. It has been suggested to have a role in macrophage-tumor interactions. It is relevant in many systems, but the fine specificity may be determined by the post-translation modifications.

SG Dr Rydevik.

Björn Rydevik If you are thinking of the in vivo situation, then you would like to coat a certain implant surface with various components to attract certain cells. The rate of release from the coating may be of some concern. In other words, consideration of the healing process is necessary if you want to influence the first year or the later phases of the integration process.

DH It should become possible to design it into the system. It would be relevant to create a permanent coating. I think it is something that should be explored. In this it should be possible to prevent degradation by cross-linking the molecules such that they retain function. The amount of the coating may be very important, ie, the density of binding sites. You may actually prefer to use a combination of various coats. It may turn out that you need a two-phase coat, with an early phase that guides early events and is then removed by the integration process. Function is later taken over by a second-phase, permanent coat. There still are many variables that need to be tested. I am convinced that the possibilities and the potentials are there for us to become more active in guiding what is happening at the implant.

SG We talked yesterday about the fibrous layer around some of the implants. Can these mediators work through the fibrous layers or immature fibrous tissue? Can they convert some of these cells to bone cells?

DH It could well be. Throughout life you have stem cells in most tissues. These cells may be triggered to go on to form osteoblasts or chondrocytes or muscle cells depending on how we trigger them. This could become a very intriguing way of learning how to modulate what is happening at various surfaces and, of course, to induce repair.

Richard Coutts I sense from what you are saying is that in addition to being a binding site, which to me is a passive type of event, the cells somehow also find that there is a signaling component, that the presence of these binding sites is somehow attracting these cells to the site. Is that your hypothesis?

DH Yes, clearly when cells bind to many of the cell-binding proteins, intracellular signals are elicited. For instance, cells can be induced to migrate.

RC Is that random or is that directed?

DH It could be directed along a concentration gradient of a protein. Furthermore, when cells bind they can start making growth factors of various kinds that will attract other types of cells. It is quite clear that when the cells bind to different molecules they will respond and do things or they may just sit still. We clearly need to find out more about these systems.

References

1. Heinegård D, Oldberg Å. Glycosylated matrix proteins. In Royce PM, Steinmann B (eds): Connective Tissue and Its Heritable Disorders. New York: Wiley-Liss; 1993:189–209.
2. Williams DC, Frolik CA. Physiological and pharmacological regulation of biological calcification. Int Rev Cytol 1991;126:195–292.
3. Young MF, Kerr JM, Ibaraki K, Heegaard AM, Robey PG. Structure, expression, and regulation of the major noncollagenous matrix proteins of bone. Clin Orthop 1992;281:275–294.

The Impact of Bone Marrow Biology on Skeletal Reconstruction and Osseointegration

Dennis G. Osmond and M. Dwayne Rahal

Within virtually every living bone lies a complex and highly active tissue, the bone marrow, from which throughout life the cells of the blood and immune system are continuously replenished. Prostheses implanted into bones in most cases must also transgress the associated bone marrow cavity, often to an extensive degree. The implantation of nonbiologic prostheses in orthopedic and dental practice aims to achieve a "true and lasting tissue integration with minimal risk of local or general tissue reactions."[1] In this context, while *osseointegration* of the implant is clearly of prime importance to give long-lasting mechanical stability, a satisfactory state of *myelointegration* is equally of great biologic significance for both the health of the recipient and the successful retention of the implant.

The term *myelointegration** may be defined from various viewpoints and with respect to several scientific scales of interest. From the viewpoint of the patient, a fixture is myelointegrated if it provides a prosthesis without discomfort, inflammation, or other injurious effects on health over the lifetime of the patient. From a viewpoint of macroscopic and microscopic biology and medicine, myelointegration of a fixture in bone marrow is defined as the close apposition of new and reformed bone marrow with no perturbation, either locally or systematically, in immune or hemopoietic mechanisms. Myelointegration requires that at the light microscopic and electron microscopic levels, the identifiable components of tissue adjacent to a fixture surface are identified as *normal* bone marrow constituents: this implies that lymphohemopoietic tissue is in contact with either the fixture surface itself or with intervening newly reconstituted bone and bone-lining cells, re-creating a normal lymphohemopoietic microenvironment.

Prolonged perturbations of lymphohemopoietic cells by intrusion of foreign materials into the bone marrow cavity could potentially have deleterious local and systemic consequences, of which the possibility of predisposing to neoplastic dysregulation is perhaps of most concern. Yet, the nature of the implant–bone marrow interface and the tissue reactions to it have been little studied.

The purpose of the present article is to describe the response of bone marrow to pure titanium implants, using an experimental model in mice. This is preceded by a brief background of relevant bone marrow biology to note the intimate relationships between bone and bone marrow that should be disrupted by the presence of an implant, and the main cellular elements in bone marrow

*Adapted from the "Definition of Osseointegration" at the 2nd International Workshop on Osseointegration in Skeletal Reconstruction and Joint Replacement. Rancho Valencia Resort, Rancho Santa Fe, CA; October 27–29, 1994.

that could be perturbed by implant reactions, taking B-lymphocyte production as a particular example.

Bone–Bone Marrow Interrelationships

The bone marrow is normally inextricably associated with the surrounding bony tissue, upon which it depends for the establishment of a microenvironment conducive to hemopoiesis.[2] A large organ with a total mass of about 3 kg in adult humans, the bone marrow is scattered throughout the skeleton, about half being hemopoietically active red marrow located in the proximal parts of the long bones and the axial skeleton, while the remaining bones contain fatty, yellow marrow. Bone marrow develops wherever bone is formed, including ectopic sites and grafted bone.

In ontogeny, the onset of hemopoiesis in individual bones is related to the particular time at which ossification begins. A prehemopoietic marrow stroma is formed by the migration of mesenchymal cells from the periosteum into the forming marrow cavity, coinciding with the appearance of osteoclasts. Following vascularization of the marrow cavity and dilatation of the vessels to form thin-walled sinuses, hemopoiesis begins in the third month of intrauterine life in the long bones, and is widespread by the fifth month. In addition to erythroid, granulocytic, monocytic, and thrombocytic precursor cells, lymphoid cells are particularly prominent.

Bone and bone marrow continue to provide each other with cells throughout life. Local ablation of bone marrow is repaired by the reconstitution of a supportive stroma from mesenchymal cells in the surrounding bony haversian system, subsequently seeded with hematopoietic stem cells from other bone marrow sites via the bloodstream. On the other hand, osteoclasts are formed by the fusion of hematopoietic cells of the monocyte-macrophage lineage within the bone marrow.

Intercommunications between the blood circulations of bone and bone marrow appear to be important in defining the functional microenvironment near the bone–bone marrow interface and in establishing a gradient of hemopoietic activity maximal in the subosteal region of the marrow[3,4] (Fig 6-1).

In the shaft of long bones, the small nutrient artery, after traversing the bone cortex, divides into longitudinal branches running centrally along the long axis of the marrow cavity, giving off radial branches that pass outward to a subosteal capillary plexus. Here, some capillaries form loops in the surrounding haversian canaliculi, while the subosteal capillaries are joined by perforating vessels through osteal channels from the rich periosteal capillary network. Large sinusoids form a rich anastomosing network converging from the subosteal plexus to a central venous sinusoid that finally drains via the nutrient vein into the systemic blood circulation. The remarkable richness of the bone marrow circulation results in a substantial blood flow, estimated to be 3% to 4% of the cardiac output, and a rapid "washout" of protein-bound or diffusible substances introduced into the medullary cavity with a half-clearance time of only 20 to 30 seconds. The walls of the sinusoids, consisting solely of a layer of thin endothelial cells, an incomplete basement membrane, and a partial covering of reticular cell processes, freely transport even large molecules to and from the blood plasma.

The blood supply to the subosteal area of bone marrow represents, in effect, a portal system. Blood from vessels draining the periosteum and bone is joined by arterial blood from the nutrient arterioles to supply the bone-lining zone. Physiologic parameters including blood gaseous tensions and pH may thus be distinctive in this zone. More especially, however, the vascular arrangements permit a local network of cytokines and other

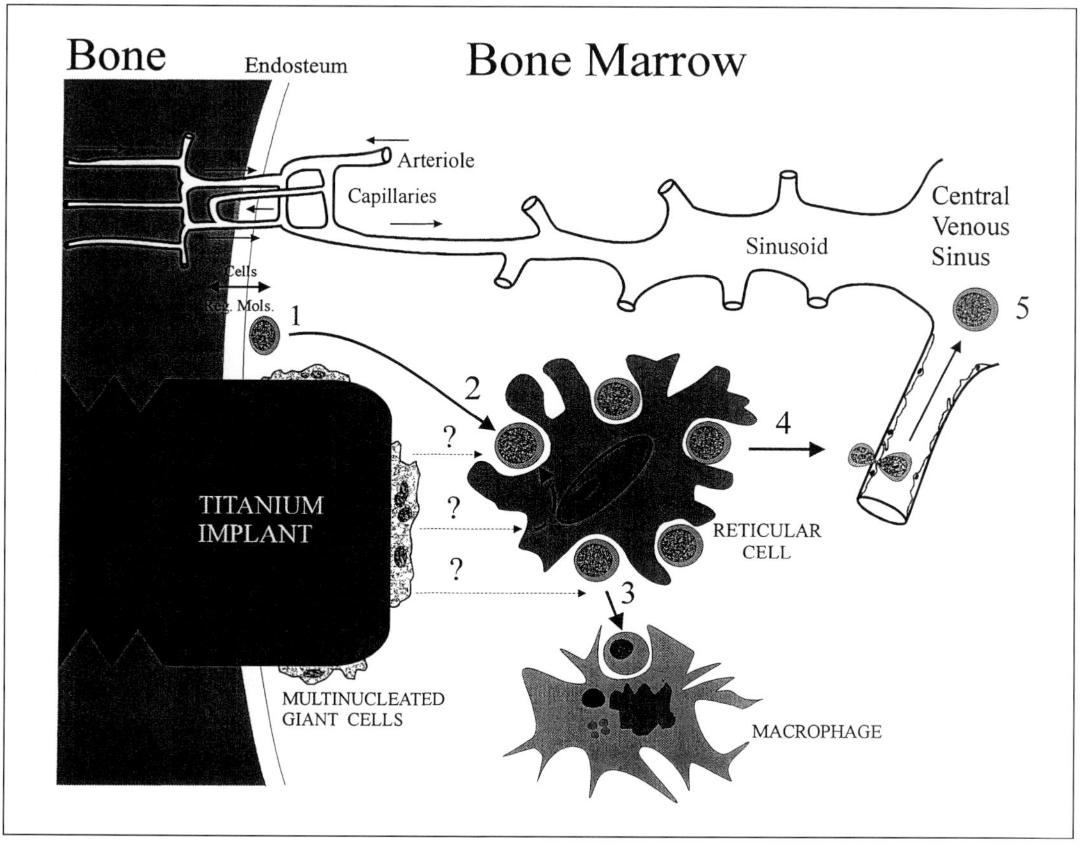

Fig 6-1 Model of bone–bone marrow interrelationships, lymphostromal associations involved in successive phases of B lymphocytopoiesis (1–4), and the titanium implant–bone marrow interface, as seen in part of a transverse section of a mouse femur extending from the bone cortex to the central venous sinus. Broken line arrows represent possible action sites of factors released from multinucleated giant cells applied to the implant surface.

regulatory molecules to operate between the bone and bone marrow. Bone vascular endothelium constitutes an active organ capable of producing many growth factors (CSF-G, -GM, -M; fibroblast growth factor), cytokines (IL-1, IL-6), arachidonic acid metabolites, and a variety of small peptides (endothelin 1) that may pass into the adjacent bone marrow.[4] On the other hand, the bone vascular endothelial cells themselves constitute targets for various factors (IL-1, TGF-ß, TNF, PGE-2, IGF-1 and -2, PDGF) and hormones (parathormone, estrogen E2, progesterone), so that the cells seem likely to be responsive to both local and systemic regulatory signals.[5]

Replacement of some of the surrounding bone by a nonbiologic implant, while of little consequence as a space-occupying lesion, will nevertheless deprive the immediately adjacent bone marrow microenvironment of the blood flow, cells, and regulatory products normally derived from the bone and its vasculature. More particularly, the normal bone interface will be replaced by the physicochemical properties of the implant. Titanium implants in vivo develop a surface layer of titanium oxide to which various biomolecules may become bonded and possibly modified.[6]

In the case of interfaces with bone tissue, the implant surface develops an afibrillar covering that seems to consist of proteoglycans[7] and be rich in glycoconjugates, including osteopontin and 2 HS-glycoprotein.[8] This has suggested that the host response to biomaterials may be similar to osteogenic processes occurring at natural bone interfaces. Implant surfaces in the marrow cavity, however, are exposed to a quite different and highly aqueous molecular microenvironment. The molecular nature of the bone marrow–implant surface and the extent to which it may either resemble or differ from the natural local environment remain to be elucidated.

Bone Marrow Lymphohemopoiesis, Stromal Cells, and the Molecular Environment

Of the various cell lineages generated in the bone marrow, this article will concentrate on B-lymphocyte production, a process that is liable to perturbations and that has been the subject of study in our laboratory for many years.[9-13]

The B-cell lineage that mediates humoral immunity stems from the bone marrow throughout postnatal life. A continuous stream of newly formed virgin B lymphocytes representing a wide diversity of antigen-binding specificities passes from the bone marrow to the spleen, lymph nodes, and other lymphoid tissues. Here, the cells may respond to antigens resulting in the formation of memory B cells, which can recirculate between blood and lymphoid tissues, including the bone marrow, and of antibody-secreting plasma cells. After booster antigen exposure, plasma cells migrate back into the bone marrow. This becomes the major site of sustained secondary antibody production. The bone marrow thus contains subsets of blood-borne plasma cells and recirculating B lymphocytes, but by far its largest B-lymphocyte lineage component consists of differentiating precursor B cells. Thus, factors causing defective differentiation of these cells within the bone marrow could have systemic consequences, the defects becoming evident at later stages in the complex life history of the lineage.

The peripheral region of bone marrow adjacent to and within the bone-lining cell layer appears to provide a particularly important microenvironment for the initiation of lymphohemopoiesis (see Fig 6-1). Early precursor B cells, hemopoietic progenitors (colony-forming units), and highly proliferative cells are most concentrated in this subosteal region.[14-19] Postirradiation regeneration and the activation of yellow marrow begin near the bone interface. This is also the location of active subsets of stromal reticular cells characterized by cytokine (IL-1) receptors and surface determinants shared by the follicular dendritic cells of peripheral lymphoid tissues.[20] Between the cells, electron microscopy reveals a particularly thick electron-dense extracellular matrix of as yet unknown composition.

Differentiating precursor B cells proliferate actively and move centrally in the bone marrow, closely adherent to stromal reticular cells (see Fig 6-1).[17,21] B-cell development at this stage depends on close stromal cell contact and the action of the cytokine, interleukin-7 (IL-7), a stromal cell–derived protein that binds to receptors and stimulates mitosis of precursor B cells.[22] Synergistic interactions, notably involving c-kit ligand (stem cell factor) and insulin-like growth factor (IGF)-1, enhance the effect of IL-7. The result is a considerable proliferative expansion of each B cell precursor into a clone of B lymphocytes, each expressing surface immunoglobulin (Ig)M molecules of identical antigen-binding specificity.

The Ig specificities are determined during the differentiation of individual precursor B cells by the formation of genes for the variable (v) regions of IgM heavy (H) and light (L) chains.[23] Recombinant enzymes cut DNA between multiple gene segments and rearrange them in a variety of sequences to form functional Ig genes encoding for a wide diversity of antibody specificities. The rearrangement of V_H and the expression of μ heavy chains precede V_L rearrangements and the expression of the complete IgM molecule. Early precursor B cells, named pro-B cells, are in the process of rearranging V_H genes before the expression of the μ heavy chain, while transiently expressing intranuclear terminal deoxynucleotidyl transferase (TdT).[24] This enzyme inserts additional nucleotides at gene segment joints, contributing to antibody diversification. Later precursor B cells, named pre-B cells, having rearranged V_H, express free μ chains within their cytoplasm, while proceeding to rearrange V_L. Finally, on completion of Ig gene rearrangements and the synthesis of both heavy and light chains, complete IgM molecules are expressed as cell surface receptors, giving the cell the hallmark of a definitive B lymphocyte. Mitoses cease and the cell enters a resting noncycling state during the pre-B stage of development.

The processes of DNA recombination represent a phase of genetic instability, making precursor B cells prone to genetic errors. These may lead to nonproductive Ig gene rearrangements[23] and to more gross mistakes, including chromosomal translocations, which may place Ig genes in proximity to oncogenes. As a result, the great majority of precursor B cells never complete their development, but instead undergo a programmed cell death (apoptosis).[10,13] The apoptotic cells are rapidly recognized, ingested, and degraded by the resident macrophages that pervade the bone marrow, especially in its more central regions (see Fig 6-1).[13,25] This process may be viewed as an important quality-control mechanism, culling ineffective, dysregulated, and some potentially self-reactive B cells, and preventing their dissemination throughout the immune system.

Viable precursor B cells that successfully complete their development migrate across the endothelium of certain sinusoid segments and are flushed from the lumen into the systemic blood circulation (see Fig 6-1).[18]

The B-cell lineage is prone to a number of disorders that may stem from events initiated in the bone marrow. Differentiation blocks may be associated with humoral immunodeficiency, while the development of self-reactive Ig specificities may cause autoimmune disease, and oncogene dysregulation may lead to a variety of B-cell neoplasms (leukemias, lymphomas, and myeloma).

Oncogenesis results from the accumulation of several genetic errors within an individual cell, causing overexpression of either cell cycle genes or cell survival genes. Although a B-cell neoplasm may show the phenotype of a later stage of the lineage, the first genetic error that initiated the whole oncogenic process may have occurred early in the development of the cell. In some cases, it appears clear that the oncogenic process can be initiated by mistakes in Ig gene segment joining during precursor B-cell development in the bone marrow.[26]

Animal models have considerably illuminated the process of B-cell oncogenesis. Mice bearing a c-myc transgene linked to the Ig heavy chain enhancer (Eμ-myc mice) all eventually develop fatal B-cell lineage neoplasms.[27] During a prolonged pretumorous phase, however, the outstanding effect is a marked expansion of proliferating precursor B-cell populations and excessive degrees of apoptosis in the bone marrow.[25,27,28] The insertion of cooperating oncogenes as transgenes greatly accelerates the onset of tumors. This concurs with the view that the long time lag between the Eμ-myc dysregulation and the appearance of a monoclonal tumor represents the development of further mutations in individual cells. Thus, the initial transgenic event is not itself oncogenic but, by increasing the rate of proliferation of precursor B cells, increases the statistical probability of further genetic errors, leading eventually to neoplastic dysregulation.

Other processes that either stimulate precursor B-cell proliferation or interfere with the selective death of aberrant cells also predispose to B-cell neoplasms. Mice bearing an IL-7 transgene show intense precursor B-cell activity,[29] while bcl-2 transgenic mice show enhanced B-cell survival, each case leading eventually to B-cell neoplasia (Janani R, Osmond DG, unpublished data, 1996).

Even nonspecific inflammatory stimuli appear to be capable of stimulating precursor B-cell activity. The stromal reticular cells of the bone marrow can be considered to form part of the cytokine network of the body. In addition to the constitutive production of a range of hemopoietic growth factors (colony stimulating factors, IL-4, IL-6, IL-7, TGF-ß), they possess receptors for circulating cytokines, which can modify their behavior.[20] Thus, IL-1 induces the production by stromal cells of IL-4, IL-6, IL-7, and colony-stimulating factors, while IL-6 production can be induced by IL-1, IL-7, TNF, and TGF-ß.

We have found that products of activated macrophages, even at distant sites, notably the spleen, can increase precursor B-cell proliferation in the bone marrow,[30] probably acting indirectly by stimulating the activity of the stromal reticular cells.[20] This has led us to propose that conditions associated with prolonged excessive macrophage activation, leading to high levels of inflammatory cytokines, may predispose eventually to neoplasia in the B-cell lineage.[31] It thus becomes a matter of practical concern to question the possible effects of disrupting the normal microenvironment of the bone marrow by the insertion of nonbiologic prosthetic implants that may remain exposed to myeloid tissue for several decades of life.

Bone Marrow–Titanium Reactions

To examine the impact of endosseous titanium implants on the bone marrow, an experimental system has been developed using miniaturized implants in the femurs of inbred mice (see Fig 6-1).[32] In this species, immunologic reagents are highly developed and the processes of hemopoiesis and B-cell lymphopoiesis have been well characterized. Commercially pure titanium implants (Avesta Jernverks AB, Avesta, Sweden) in the form of smooth-ended annular-ridged cylinders (1.0 mm x 0.58 mm), have been produced (Institute for Applied Biotechnology, Gothenburg, Sweden). Recipient C3H/HeJ male mice have been used at 4 weeks of age, a time when the femoral diaphysis has a clearly defined medullary cavity. With the mice under anesthesia, implants have been inserted into the mid-diaphysis and medullary cavity of the surgically exposed femur, closely fitting a hole in the bony cortex produced by gentle drilling and cooled by local irrigation. After surgical closure, the implants and bone marrow have been examined at six intervals ranging from 1 week to 78 weeks later.[32]

Ground sections (10 to 15 µm thick) of the femurs and implants, 4 weeks after implant placement, demonstrate a state of apparent osseointegration by morphologic criteria, the cortical bone having grown into the annular serrations on the implant without evidence of intervening connective tissue.[32] The mouse appears, therefore, to provide a satisfactory experimental model for work on the biologic impact of osseointegrated titanium implants.

Implants have been removed from Epon-embedded perfusion-fixed femurs to examine the bone marrow–titanium interface with light microscopy and electron microscopy.[32] The implants project through the normal bone–bone marrow interface and endosteum. Neither reconstitution of the bone-lining cell layer nor a deposition of bone or connective tissue have shown any signs of taking place over the medullary surface of the implant. Instead, the extensive titanium surface remains apparently in direct contact with the bone marrow elements. Following the initial disruption produced by drilling and inserting the implant, the regenerating bone marrow intimately surrounds the implant. Many hemopoietic cells, including granulocytes, appear to be in direct contact with the titanium surface.

The most striking feature has been, however, that a considerable part of the bone marrow interface becomes formed of prominent multinucleated giant cells with voluminous cytoplasmic processes containing phagocytic inclusions (see Fig 6-1). Such cells are not seen in normal bone marrow. Their numerous cytoplasmic vesicular organelles and complex microvillous processes facing the bone marrow parenchyma suggest considerable molecular synthesis and traffic. The giant cell layer has proved to be a persistent feature from 18 to 78 weeks after implantation.[33]

The actual interface between titanium implants and the giant cells has not yet been directly observed. The fracture technique used to separate the implant from the marrow appears to shear off the surface membrane of the giant cells, suggesting a tight apposition and spreading of the cells on the titanium surface. The molecules responsible for this surface recognition and adhesion remain a matter of speculation. In addition to the metallic surface and its possible ionic discharge, the local marrow microenvironment is rich in growth factors, cell adhesion molecules, and extracellular matrix (including fibronectin, laminin, collagen type IV, heparan sulfate, chondroitin sulfate, and hyaluronic acid). Some of these molecules might have become adsorbed to the oxidized titanium surface and possibly modified to become recognized by binding sites on the surface membrane of the giant-cell precursors. Further studies need to be directed at this molecular interface.

The multinucleated interface cells have the morphologic features of inflammatory giant cells.[32] These are characteristically formed by the fusion of monocyte-macrophage lineage cells in response to insoluble and poorly removable foreign material. Strong attachment to the foreign material induces cell spreading and fusion with adjacent cells to form the typical macrophage polykaryons.

The functional role of the multinucleated interface cells remains conjectural. The cells might exert a beneficial barrier function, separating the hemopoietic cells from the ions and molecules associated with the implant surface. Alternatively, the morphologic evidence of secretory activity raises the question whether the giant cells may be producing factors that might perturb local lymphohemopoiesis (see Fig 6-1). Such perturbations, although affecting only a small volume of bone marrow, could be long-lasting and might theoretically predispose to the emergence of a dysregulated clone of lymphohemopoietic cells. Such a risk might be compounded if the bone marrow were also exposed to ionizing irradiation or myelotoxic drugs. So far, however, in recent and current work, the results appear to be reassuring in this respect.[33]

Suspensions of cells from the entire femoral diaphyseal marrow cavity show no significant differences in cell numbers or proliferative activity in the populations of pro-B cells, B cells, and B lymphocytes in titanium-implanted femurs compared with sham-operated controls. Detailed radioautographic histologic mapping of the cellular organization throughout the bone marrow cavity shows no evident perturbation near the implant surface in the distribution of precursor B cells labeled for the expression of the B lineage–associated surface glycoprotein, B220 (CD45RA), and no increase in the incidence of proliferating cells of all lineages labeled by the uptake of the DNA precursor, ^{3}H-thymidine. Further long-term studies are required to determine the molecular nature of the titanium-implant surface exposed to bone marrow, the expression of regulatory molecules by the interface giant cells, single-cell genetic effects in the vicinity, and the possibility of modifying the implant surface to achieve the most harmonious state of myelointegration.

Summary

Bone and bone marrow are normally closely associated with each other by an exchange of blood circulation, cells, and regulatory molecules, which establish a microenvironment conducive to lymphohemopoiesis. This microenvironment may be disrupted by the intrusion through the bone of a nonbiologic implant. The extravascular compartment of the bone marrow between large venous sinusoids is the site of production of B lymphocytes and the various cells of the blood, supported by stromal reticular cells that supply essential growth factors. Early hemopoietic progenitor cells together with a subset of active reticular cells are concentrated near the bone-lining cell layer. B-lymphocyte precursors may be particularly susceptible to local perturbations. They differentiate in apposition to reticular cells while dividing and undergoing Ig gene rearrangements, influenced by reticular cell growth factors and cell adhesion molecules.

Many precursor B cells undergo ineffective or aberrant Ig gene rearrangements and, together with certain cells that may have generated self-reactive Ig specificities, undergo an apoptotic programmed cell death and are rapidly culled by local resident macrophages. Viable surviving B lymphocytes traverse the sinusoidal endothelium and pass via the bloodstream to the peripheral immune system, where they may mediate primary and subsequent humoral immune responses to antigenic challenges.

Perturbations of the B-cell lineage can cause immunodeficiencies, autoimmunity, and oncogenesis. Dysregulation of oncogenes produced by mistakes in Ig gene recombination, or prolonged proliferative stimulation of the precursor B cells by overexpressed growth factors and inflammatory cytokines, may all predispose to the development of monoclonal neoplasms of the B-cell lineage, after a long time lag. Insertion of a titanium implant through the bone cortex to impinge upon the bone marrow cavity, as seen using miniaturized implants in mouse femurs, is followed by a vascularized regeneration of the disrupted hemopoietic tissue up to the surface of the implant without intervening formation of an endosteal cell layer or bone tissue.

The bone marrow–titanium interface is characterized by the presence of certain large multinucleated giant cells, apparently tightly apposed to the implant surface and persisting for prolonged periods. The nature of the local biomolecules that may be bonded to the titanium oxide surface remains unknown. Results to date reveal no significant changes in the local incidences of B-cell precursors and proliferating cells. Further studies are required to elucidate the molecular basis of myelointegration of prosthetic implants, and to rule out the possibility that products released from the interface giant cells might eventually predispose to dysregulation of local lymphohemopoietic cells.

Acknowledgments

The authors' work referred to in this article was supported by grants from the Medical Research Council of Canada and National Cancer Institute of Canada (DG Osmond) and the Institute for Applied Biotechnology, Sweden (MD Rahal).

DISCUSSION

Åke Carlsson I would like you to tell a little bit more about this chronic macrophage activation, because such a reaction is regularly observed around loose orthopedic implants and when there is a production of wear debris. In spite of this common phenomenon, oncogenesis has been reported in extremely few cases.

Dennis Osmond This is admittedly a hypothetical issue. The reason we have some concern relates to experimental data. If we inject some fairly nonspecific foreign agent, for example sheep blood cells, into mice, we find a wave of increased proliferation among early precursor B cells in the bone marrow. A series of experiments has shown us that this is dependent upon the activation of macrophages located in the spleen of the animal. If the animal has been splenectomized, this reaction does not occur. If the macrophages are functionally depressed by giving silica particles, the reaction does not occur. And in transfer experiments, we find a macrophage product seemingly to have this effect.

Now in itself, of course, there is no problem of having a wave of B-cell genesis. On the other side of the coin, however, it is shown by a number of transgenic systems that if, in the marrow, you have a heightened proliferation of early precursor B cells that is intense enough and goes on long enough, there is a heightened incidence of B-cell tumors. For example, in mice where the c-myc oncogene is upregulated in the B-cell lineage, all the mice eventually die of B-lineage tumors. But for a number of months, the only finding in the bone marrow is a greatly increased proliferation of precursor B cells and increased apoptotic deletion. Eventually, possibly through a second mutation, a single cell is able to survive and to give rise to a monoclonal tumor. We see the same sequence of events in interleukin-7 transgenic mice, producing excessive precursor B cells.

We have also looked at a chronic stimulation from oil-induced granuloma. These mice eventually develop monoclonal plasmacytomas. Although I cannot say that there is a cause-and-effect relationship, these mice also show prolonged proliferative aberrations of precursor B cells in the bone marrow before developing tumors. So, at the moment, it is simply a cautionary tale, and I have no clinical evidence that orthopedic implants are followed by increased incidence of B-cell mutations or tumors.

The striking thing in the present work is the persistence of giant cells in direct contact with the titanium implant adjacent to what we believe is the site of early B lymphocytopoiesis in bone marrow. So far, however, the results are reassuring in that there is no evident perturbation of B lymphopoiesis in the bone marrow cavity as a whole.

Göran Lundborg It is fascinating to listen to your discussion of the bone marrow as such an active and dynamic organ. I wonder how it may differ in the different parts of the body. We often have to graft bone marrow from one part of the body to another; for instance, in rheumatoid patients, we take bone marrow from the iliac crest and bring it into the hand before we introduce our implants. So, I wonder if the interaction between bone marrow and the implant that you have discovered would also hold when you implant titanium into a grafted bone marrow. Would that bone marrow survive when you graft? If you had filled the rat femur with grafted bone marrow before you introduced the screw, would you expect the same observation?

DO Bone marrow varies regionally and with age. In human infants, the marrow cavity is full of hemopoietically active red marrow. By adulthood, about half the cavity contains red marrow, the other half being yellow. In the example you quoted, the iliac crest would contain red, active marrow, while the marrow that you are grafting into would be yellow, fatty, inactive marrow. The transplanted marrow will assume the marrow characteristics of the recipient region. This is thought to be related to local physiologic parameters, notably, the local bone marrow circulation. Classic experiments showed that after rat tails were implanted into the abdomen, the rat vertebrae become hemopoietically active. This was thought to be a temperature effect, probably inducing changes in the bone marrow circulation. I may add that the adult mouse is more like a human infant than a human adult because the skeleton is full of red marrow.

GL That was a great talk. May I comment on the oncogenicity or potential oncogenicity of implants? As you probably know, in soft tissue, there is an interesting experiment that nature has undertaken. A chronic wound, say a burn wound in a young individual, typically doesn't heal because of lack of skin or repeated trauma from motion. Over time, there would be new plasia in that wound. Typically, after about 15 years, you will start seeing tumors. These are sarcoma cancers first described by a Frenchman named Marjolin. He might be very interested to hear the hypothesis that this is caused by chronic macrophage activation. I found your hypothesis to be very interesting.

DO Thank you. The hypothesis certainly could have a wider application.

Lars Magnus Bjursten I just want to comment on these findings of the giant cells close to the implant. That is a regular finding in other soft tissues, like connective tissues, where you have these giant cells present at all times. So the real question here, I think, is if you have a proliferation of such cells all the time or not. That may vary depending on different materials in other tissues. What is the experience in the bone marrow?

DO Some giant cells are there for very long periods of time. We have not yet explored whether they are the same cells, or cells continuously being recruited and turning over at the interface. That is a very good point. What would you suspect?

LMB Looking at your pictures, they look like fairly inactive cells.

DO From a proliferative point of view?

LMB Yes.

DO We see no evidence that they are turning over, but we haven't examined this point specifically.

References

1. Brånemark P-I. Introduction to Osseointegration. In Brånemark P-I, Zarb GA, Albrektsson T (eds): Tissue-Integrated Prostheses: Osseointegration in Clinical Dentistry. Chicago: Quintessence; 1985:11.
2. Trubowitz S, David S. The bone marrow matrix. In Trubowitz S, Davis S (eds): The Human Bone Marrow: Anatomy, Physiology and Pathophysiology, vol 1. Boca Raton, FL: CRC Press; 1982:43.
3. Brånemark PI. Vital microscopy of bone marrow in rabbit. Scand J Lab Invest Suppl 1959;38:2.
4. de Bruyn PPH, Breen PC, Thomas TB. The microcirculation of bone marrow. Anat Rec 1970;168:55.
5. Collin-Osdoby P. Role of vascular endothelial cells in bone biology. J Cell Biochem 1994;55:304.
6. Kasemo B. Biocompatability of titanium implants: Surface science aspects. J Prosthet Dent 1983;49:832.
7. Linder L, Albrektsson T, Brånemark P-I, et al. Electron microscopic analysis of bone-titanium interface. Acta Orthop Scand 1983;54:45.
8. Nanci A, McCarthy GV, Zalzal S, Clokie CML, Warshawsky H, McKee MD. Tissue response to titanium implants in the rat tibia: Ultrastructural, immunocytochemical and lectin-cytochemical characterization of the bone-titanium interface. Cells Mater 1994;4:1.
9. Osmond DG. Population dynamics of bone marrow B lymphocytes. Immunol Rev 1986;93:103.
10. Osmond DG. B cell development in the bone marrow. Semin Immunol 1990;2:173.
11. Osmond DG. Insights into normal myelopoiesis. Can J Infect Dis 1992;3(suppl B):69.
12. Osmond DG. The turnover of B lymphocyte populations. Immunol Today 1993;14:34.
13. Osmond DG, Rico-Vargas S, Valenzona H, et al. Apoptosis and macrophage-mediated cell deletion in the regulation of B-lymphopoiesis in mouse bone marrow. Immunol Rev 1994;142:209.
14. Osmond DG, Fahlman MTE, Fulop GM, Rahal MD. Regulation and localisation of lymphocyte production in the bone marrow. In Microenvironments in Haemopoietic and Lymphoid Differentiation. CIBA Foundation Symposium 84. London: Pitman; 1981:68.
15. Dedlar A, Lewis H, Weiss L. Studies on the organisation and regeneration of bone marrow: Origin, growth, and differentiation of endocloned hematopoietic colonies. Anat Rec 1985;214:187.
16. Lord BI, Testa NG, Hendry JH. The relative spatial distribution of CFU-s and CFU-c in the normal mouse femur. Blood 1985;46:65.
17. Jacobsen K, Tepper J, Osmond DG. Early B lymphocyte precursor cells in mouse bone marrow: Subosteal localization of B220+ cells during post-irradiation regeneration. Exp Hematol 1989;18:304.
18. Jacobsen K, Osmond DG. Microenvironmental organisation and stromal cell associations of B lymphocyte precursor cells in mouse bone marrow. Eur J Immunol 1990;20:2395.
19. Osmond DG, Kim N, Manoukian R, Philips RA, Rico-Vargas SA, Jacobsen K. Dynamics and localization of early B-lymphocyte precursor cells (pro-B cells) in the bone marrow of scid mice. Blood 1992;79:1695.
20. Fauteux L, Osmond DG. Interleukin-1 as a systemic modifier of B lymphopoiesis: IL-1 alpha binds to stromal cells and sinusoid endothelium in bone marrow and perturbs precursor B cell dynamics. J Immunol 1996;156:2376.
21. Weiss L, Sakai H. The hemopoietic stroma. Am J Anat 1984;170:447.
22. Kincade PQ, Lee G, Pietrangeli CE, Hayashi SH, Gimble JM. Cells and molecules which regulate B lymphopoiesis in bone marrow. Ann Rev Immunol 1989;7:111.
23. Rolink A, Melchers F. Generation and regeneration of cells of the B-lymphocyte lineage. Curr Opin Immunol 1993;5:207.
24. Park YH, Osmond DG. Dynamics of B lymphocyte precursor cells in mouse bone marrow: Proliferation of cells containing terminal deoxynucleotidyl transferase. Eur J Immunol 1989;19:2139.
25. Jacobsen KA, Prasad VS, Sidman CL, Osmond DG. Apoptosis and macrophage-mediated deletion of precursor B cells in the bone marrow of Eμ-myc transgenic mice. Blood 1994;84:2784.
26. Bertoli LF, Kibagawa H, Borzillo GV, Borrows PD, Schreeder MT, Carroll AJ, Cooper MD. Bone marrow origin of a B-cell lymphoma. Blood 1988;72:94.
27. Langdon WY, Harris AW, Cory S, Adams JM. The c-myc oncogene perturbs B lymphocyte development in Eμ-myc transgenic mice. Cell 1986;47:11.
28. Sidman CL, Shaffer DJ, Jacobsen K, Rico Vargas S, Osmond DG. Cell populations during tumorigenesis in Eμ-myc transgenic mice. Leukemia 1993;7:887.
29. Valenzona HO, Pointer R, Ceredig R, Osmond DG. Preneoplastic lymphoproliferation in the bone marrow of interleukin-7 transgenic mice: Precursor B cell dynamics, microenvironmental organization and osteolysis. Exp Hematol (in press).
30. Park YH, Osmond DG. Regulation of early precursor B cell proliferation in mouse bone marrow: Stimulation by exogenous agents mediated by macrophages in the spleen. Cell Immunol 1991;135:168.
31. Osmond DG, Priddle S, Rico-Vargas SR. Proliferation of B cell precursors in bone marrow of pristane-conditioned and malaria-infected mice: Implications for B cell oncogenesis. In Potter M, Melchers F (eds): Current Topics in Microbiology and Immunology, vol 166. Heidelberg: Springer-Verlag; 1990:149.

32. Rahal MD, Brånemark P-I, Osmond DG. Response of bone marrow to titanium implants: Osseointegration and the establishment of a bone marrow-titanium interface in mice. J Oral Maxillofac Implants 1993;8:573.

33. Rahal MD, Delorme D, Clokie C, et al. Distribution of B lineage and dividing cells in the mouse bone marrow around titanium implants. In Program book of the Tenth Annual Meeting of the Academy of Osseointegration, October 22–29, 1994, Chicago. Academy of Osseointegration; 1994.

Chapter 7

Skeletal Mechanics as Related to Bone Mineral Content and Bone Remodeling

Tommy Hansson

More than 100 years ago, Wolff proposed that functional loading influenced the remodeling of the skeleton.[1] In spite of a general acceptance of this hypothesis and numerous subsequent studies exploring the hypothesis, it has not been possible to identify any specific factor controlling the adaption of bone to mechanical loadings. This is, however, a very important aspect affecting the long-term success of osseointegration in orthopedic applications. Hence, any information on the processes involved in bone remodeling and means of monitoring bone regrowth and mineralization is useful background, as well as an aid in evaluating possibilities and the condition of bone surrounding osseointegrated implants.

In this chapter, the means available for assessing bone mineral content, factors that affect it, and some clinical findings are reviewed with this purpose in mind. The relatively recent advent of an accurate noninvasive technique for determinations of the amount of bone mineral in the skeleton, the absorptiometric technique, allows new possibilities for bone remodeling studies in humans and other mammals (Table 7-1). This technique has made bone mineral determination possible in the peripheral and central axial skeleton.

Biomechanical testing of human lumbar vertebrae has revealed a close correlation between the noninvasively determined amount of bone mineral and the compressive strength of the specimens.[2] Several other studies have confirmed this close relationship, and found that the variance in compressive strength to around 80% is explained by variations in the amount of bone mineral.[3–8] Other studies of the bone mineral of the spine have revealed that it also predicts fatigue properties of the vertebral bodies of

Table 7-1 Types of Bone Mineral Determinations

Type of bone determinator	*Energy(ies)*	*Suitable for part of skeleton*	*Bone close to implants*
Single photon	Gamma	Peripheral	No
Double photon	Gamma	Central	Yes
Quantitative CT	X-ray	Peripheral and central	Yes
DEXA*	X-ray	Peripheral and central	Yes
TRIXA†	X-ray	Peripheral and central	Yes

*DEXA= dual-energy x-ray absorptiometry †TRIXA = triple-energy x-ray absorptiometry

the human thoracolumbar spine.[9,10] The amount of bone mineral, at least in the spine, has also been found to accurately predict mechanical properties of the intervertebral disk as well as of spinal ligaments.[11,12] We found that the relationships between the amount of bone mineral and the biomechanical properties differed from one tissue type to another.[12] In cases of linear relationships, the slope of regression lines representing those relations differed between the different tissues tested.

The close relations between the bone mineral and major biomechanical properties of not only skeletal but soft-tissue structures strongly suggest that the concept of a skeletal adaptation to load has even wider implications. From our studies, the bone, ligament, cartilage, and muscle structures obviously belong to the same functional entity; eg, the spine is subject to a remodeling process that is generated through the same, or fractions of the same, mechanical stimuli.

Absorptiometric technique also has allowed studies of the effect on the human spine of extremely low as well as extremely high loads. Strict and literally almost complete immobilization of young women in bed for up to 8 weeks disclosed considerable bone mineral losses from the fourth lumbar vertebra.[13] When the same population was reexamined 5 years after the immobilization period, it was found that actual mineral losses were regained and well compensated for in only around 50% of the women studied.[14] The remaining young women apparently had sustained permanent bone loss in their spine.

Other spinal studies also have revealed that much less extreme physical inactivation can be detected through bone mineral measurements. This was demonstrated among subjects who had retired prematurely from normal everyday work. They had significantly lower bone-mineral content than age- and gender-matched normal working populations.[15] These findings make it clear that reduction of the spinal loading will cause lowering of the amount of bone mineral, and consequently reduce the mechanical competence, not only of the bone, but also of the surrounding soft tissue. An increase of the spinal loading will, on the other hand, have the opposite effect. In studies of world championship weight lifters, we found that frequent and extreme practicing caused subsequent incorporation of bone mineral in the spines of the lifters.[5] An exponential relationship was found between annually lifted load and the amount of bone mineral in the lumbar spine.

We found a linear positive relation between practicing hours and the amount of bone mineral in the lumbar spine in other types of athletes, such as long distance runners, chosen to represent much lower spinal loading than the previously mentioned power lifters. The results from these studies clearly demonstrated that the adaptation of the bony structures in the spine was affected by at least two spatial components of its mechanical load exposure, ie, magnitude and duration.

In an additional study, the bone mineral was determined in the lumbar spine among presently and previously active top male athletes (Olympic wrestlers and heavy weight lifters) between 16 and 35 years old, and 35 to 50 years old, respectively.[5] Their (L_3/the third lumbar vertebra) bone mineral values were compared with the corresponding values in an age-matched male population cohort. Considerably higher bone mineral values were found among the active top athletes of all included ages in comparison to the males in the cohort. The bone mineral in the former top athletes was lower than among the presently active ($P<.01$). This difference remained even after correction for age and weight differences between the two groups. When the former athletes were compared with the age-matched males in the control group, and corrections were made for body weight differences (the former athletes were heavier), there was no difference between the groups. These results made it clear that the remodeling of the human skeleton is a very adaptive process, however, with a short memory.

On Load, Stress, and Strain

A load (force) on a skeletal tissue deforms or strains the tissue. The deformation or strain stretches the loaded tissue, which resists it with an elastic force called a stress. Strain rather than stress seems to control and generate the stimuli that control the biologic reactions to mechanical loads. It is quite likely that large strains affect the tissue (bone) more than small strains, and that many have more influence than few.[16,17]

The magnitude, peak, rate, average, duration, frequency, number of peak cycles, etc, of the load facing the bone might alone or combined be the factors that determine the remodeling process. The temporal integration of all the factors that might influence the remodeling process creates the concept of *strain history*.[18] The decisive details of this concept are still missing, however. So far, no definition of strain history has been able to explain bone morphology. It is apparent, however, that the tissue aims at optimizing morphology. The consequence of optimal morphology will be a minimizing not only of the strain but, in addition, a minimizing of the amount of bone mineral.[19]

According to Frost,[20] remodeling takes place through the BMU (basic metabolic units). The sequence of the BMU is an orderly cascade of events including activation-resorption-formation (ARF). Activation involves the initiation of the resorptive ability of the BMU. Resorption creates cavities when bone matrix is removed. Formation is the repair phase when the cavities caused through the resorption are filled with new matrix. The resorptive or osteoclastic activity is coupled to the subsequent formative or osteoblastic activity.[21] The final result of this ARF sequence, again according to Frost,[20] is a BMU. It is estimated that the skeleton in a normal adult will create many millions of new BMUs annually.[20] The time frame of an ARF cycle is around 4 months. The bone replaced is estimated to be about 0.05 mm^3.

There are three possible outcomes of one ARF sequence: same amount of bone, less bone, or more bone. Remodeling occurs on periosteal, cortical-endosteal, haversian, and trabecular surfaces.[22] It is hypothesized that during normal conditions, an ARF sequence can only be positive (more bone is formed than resorbed) on the periosteal surface. On the cortical-endosteal and trabecular surfaces, resorption is believed to exceed formation, resulting in a loss of bone. Every BMU on these surfaces could lose 0.003 mm^3 during one ARF sequence.[23] The expansion of the marrow cavity with age in all long bones is one expression of this apparently lifelong negative process. This "interior excavation" is one of many processes that might threaten the longevity of manmade implants anchored within the marrow cavity of bones.

Microdamage and Remodeling

Repetitive loading of the skeleton can cause both macro- and micro-fatigue fractures.[9,10,18,24-27] Microfractures due to fatigue have been verified in the central as well as peripheral human skeleton.[27,28] It is supposed that the microdamage is usually repaired, preventing the development of an ultimately gross fracture. The repairing process is said to take place as a removal of the damaged bone and a replacement with new bone. Overload, such as around the tip of the stem of an endoprosthesis, can accumulate microdamage. The subsequent reparative or remodeling process will sometimes be visible even on routine radiographs.

Strain Levels

A human long bone fractures at a deformation-strain of around 25,000 µE (micro strains). Normal deliberate physical activities have been estimated not to exceed a longitudinal strain of around 3,000 µE. For loads causing less strain than 2,000 µE, the normal repair processes are adequate. When the strain exceeds 4,000 µE, microdamage can accumulate to an extent to cause trabecular as well as whole bone failures. Frost suggests that in the range 2,000 to 4,000 µE, a relatively small increase in strain can increase microdamage severalfold. Bony implants might cause an increase of the strain within this range and in such a way induce an accumulation of microdamage.

Cellular Response and Remodeling

Evidently, there is a close relationship between the strain history and the adjustments of the bones' architecture and its content of mineral. The mechanisms by which mechanical events are transformed into biologic responses at the bone cell level are still, to a great extent, unresolved. As previously mentioned, the present definition of strain history has not made it possible to explain bone morphology. The strain measured as peak, shear, energy, etc, in, for example, a vertebral body or a femur, is far from uniformly distributed in the bone. The relatively homogenous distribution of bone mineral in the cortices of the vertebral body or diaphysis of the femur suggests, however, a more homogenous distribution of the mechanical signals than what could be expected if the only strategy would be to minimize strain/minimize bone mineral. It has been suggested that the osteocyte, with its "strategic" placement within the bone matrix in response to their strain environment, could be a mediator of a uniform and remote influence upon the remodeling.[29]

Experimental Influences on Bone Remodeling

Studies by McLeod and Rubin have shown that the bone is sensitive to mechanical strains up to at least 60 Hz.[30] They have also found that high-frequency (15 to 50 Hz) strains occur in the skeleton during normal activities. In vivo experiments in turkeys have revealed that the bones' sensitivity to mechanical strains was highly frequency dependent. In this way, they have shown that 1 Hz sinusoidal loading regimen of around 800 µE, applied for 10 minutes daily, maintains bone mass. Rubin and McLeod also have demonstrated that exposure to 30 Hz vibration, generating peak strains of around 200 µE and imposed during 10 minutes daily, will have the same effect. The same exposure, however, for 30 minutes daily not only maintained bone mass, but increased it.

Noninvasive Bone Mineral Determination

Cameron in the United States and Nilsson in Sweden developed the single-photon absorptiometric technique into clinically useful methods for determination of the bone mineral in peripheral skeletal parts.[31,32] Roos introduced the first double-photon absorptiometer, which allowed accurate measurements of the bone mineral in the axial skeleton, eg, the spine and the proximal femur.[33] Different energy sources have been used for these noninvasive bone mineral determinations. Some commonly used types of bone mineral determinators are presented in Table 7-1.

During the past few years, these types of noninvasive techniques have been used to determine the effect on the bone mineral in more or less direct contact with implants in the hip as well as the knee.[34,35] The results from such studies are promising. Since from a technical standpoint bone determinations

Table 7-2 Biochemical Markers of Bone Turnover

	Marker of	
Marker	*Formation*	*Resorption*
Alkaline phosphatase (ALP)	X	
Osteocalcin	X	
Procollagen	X	
Hydroxyproline		X
Bone-specific ALP	X	
Calcium		X
Pyridinoline and deoxypyridinoline		X

close to a metal surface are far from trivial, careful evaluations of both precision and accuracy are needed before their usefulness in this respect can be fully established.

Markers of Bone Turnover

Biochemical markers can also be used for studies of bone remodeling. These markers consist of substances that are released during bone resorption and formation. They can be collected and determined in serum or urine. The different molecules used as markers reflect different aspects of bone turnover. Some are byproducts formed during resorption or formation, while others reflect enzymatic activities in the bone cells. Commonly used markers are presented in Table 7-2.

The lifelong remodeling of the skeleton will also affect manmade implants. We have learned a lot about the remodeling of the human skeleton, and we can measure its effect through bone-mineral determinations. The knowledge of how the mechanical stimuli interact with a cellular response is expanding. Recent findings suggest that we might even interfere with the remodeling process by exposing bones to repetitive strains of different frequencies.

DISCUSSION

Göran Lundborg I have one question and one comment. First, when you described the bone loss in the immobilized girls, there was dramatic bone loss already within 6 weeks. Do you think that is because those bones were not loaded, or is it because there were circulatory changes? If it is because the bones were not loaded, it could certainly have implications on our previous discussion on unloading of implants.

Tommy Hansson I am convinced that it was because they were not loaded. They were immobilized, as completely as a spine can be because there is always some load on the spine. I mean, there is always some pressure. I think it is a disuse osteopenia.

GL One comment regarding bone formation and vibration. A very interesting human model on this is all the patients that have been working with handheld vibrating tools for many years. I have lots of experience with such patients. They have all kinds of problems of the vascular system and of the neurologic system, but they have minor problems of the skeletal system. It has been believed that there may be changes in the carpal region. They certainly have no indication of bone malformations in the hand bones.

TH Actually, at the moment, we are measuring truck drivers and others subjected to vibration. We want to see if they have more bone mineral in their spine, or less. According to this hypothesis, they should have a higher bone mineral content than the normal population. I mean, it is something you can find in the regular machinists or people using vibratory handheld tools.

Dick Heinegård First of all, concerning these girls who had surgery: some of them actually increased their bone mineral density afterward, and some stayed low. What is the difference between these girls?

TH The difference was that the girls who had increased bone mineral density were younger at the time of surgery. They were between 13 and 15, while the others were older. So I think they were still growing and that compensated for the losses. That is my interpretation of it.

DH Now with respect to physical exercise, like your weight lifters increasing bone strength: I think you have one variable which is rather difficult to control, and that is the effects of drugs. From my knowledge, all of them use drugs in various ways. Is there a way to control for that? Could it actually be that they induce alterations?

TH In 1984, this was not discussed, but I am pretty sure that they used them. I think most of them were disqualified for this reason some years after the championship, maybe all of them. They certainly used them, I think. They believed steroids would work, and that they might improve their capacity. Of course, you can also predict some kind of direct effect of the steroids on the skeleton. But we cannot be sure of their effects in a given case.

Bjorn Gunterberg An odd question: Is there any explanation why the osteoporosis induced in astronauts might be irreversible? Couldn't they regain bone mineral content with gravity loading or physical loading?

TH There is a classic work by Bo Nilsson, who invented the technique for determining the amount of bone mineral in the skeleton. We measured it in people who have had femur fractures several years ago and found that, at least in people treated in Sweden who had a whole leg in plaster for many weeks, the amount of bone in the femur was reduced. They found that many years after the fracture, they did not regain it, even if the functional capacity was back to normal. So, it seems that once you have lost bone it cannot be recovered, at least not easily. It is hard to understand why, when I am talking about the training effect. It seems that if you demineralize your skeleton, you are not going to regain that lost bone, at least to some extent.

DH I wonder about bone-mineral density measurements: when you have these trabecula microfractures, would that interfere with a bone mineral density measurement or directly alter it?

TH You lose some. In the vertebral body you might find, at most, around 5% loss. They are so tiny, so I do not think they have any greater effect on the bone mineral. So I do not think it affects the bone mineral determination value when, as we do it here, the entire vertebrae is measured.

Stephen Cowin I have a question about the bone-mineral measurement and the athletes, or even the young girls. When you say the bone-mineral content changes, do you mean that there is more bone tissue that is mineralized, or that the bone tissue in a given volume becomes more highly mineralized than normal?

TH Both might be true.

SC It is mineralized? Is there more tissue that is mineralized?

TH Yes, there must be more tissue mineralized. In these top power lifters, the world champions, I assume their L3 vertebrae were more or less compact, at least when they performed this lift. I assume that they had formed so much bone during their practicing, by taking steroids, or the combination of both. I have not heard of anyone who has a found such a high level of bone mineral in a living human being as the results shown here. They are the highest and should then be the strongest bones we have ever seen—the strongest vertebrae we have ever seen—in a human being.

SC The whole bone structure is increased, not the per-unit volume mass?

TH Yes, the whole bone structure increases. It is the total structure that I am talking about.

SC Then I have another comment on this long-term effect of long-term immobilization. There is a study in this country done during World War II, where they put conscientious objectors to bed for several months.

TH Even longer, I think, 1 year.

SC There is a point of no return, they said, as I remember, after which you could not regain the loss of the bone mass.

TH That is what we are discussing. I do not know if that is true. You saw it in fractured legs of people. They had not recovered when they compared the fractured side to the nonfractured side, many years after the fracture. They had regained complete function. There was a lower bone-mineral content on the fractured side.

Steven Garfin I would like to bring up one issue before we close the discussion. There is great concern when spinal surgery is performed, particularly when the issue of fixation comes up. In certain cases, you may have very weak bone or osteoporotic bone, particularly with the pedicle screw fixation system. Do you have any information about the initial fixation strength of such an implant? Constant loading is more likely to stabilize the fusion or to create a good situation for the spinal arthrodesis. Would you have any information about the importance of the bone mineral content?

TH Yes, that is something that has already been discussed. I mean that osteoporosis might be a contraindication for this type surgery. There again is a threshold where these screws, because of the small amount of bone, are not gripping any longer. That is an arbitrary feature, but you can say that below a certain amount of bone, there is a great risk that the implant will come loose because there is not enough bone to drive the screw into.

May I mention one thing? We are planning to measure the bone-mineral content around the implants we are discussing here; I mean in amputees, for example. That is possible to do, although I believe there are a lot of errors in doing it. People use double x-ray measurements for measuring around a prosthesis, and I believe this is a relatively insensitive way of doing it because of the properties of this technique. It is hard to measure when you have metal somewhere in the beam. So, we don't know, but that is something we want to do again, to see what happens just between the implant and the bone.

Per-Ingvar Brånemark In the lower amputee femur or tibia, it would be unloaded for a very long time, so of course the bone is not in excellent shape. Would it be possible, since now we have access to the bone directly, to apply vibration to improve it? Would it be possible also to redesign the prosthesis, including the vibrator, whatever consequence that would have? Seriously, I think that is something we should talk about.

SG I think the femur, for example, or the tibia, is ideal for applying vibrations to the skeleton. There is a lot more work to do. Perhaps we should spray something on the implant. If we believe that we are stimulating the bone in a way that it starts remodeling, in a positive way, then of course we should try it.

PIB It might be done collaboratively eventually in Gothenburg, since we have the patients to do it there.

DH Coming back to these weight lifters and the bone marrow density: Can we actually extrapolate from the correlation in the normal individuals between strength and bone-mineral density to these weight lifters?

TH No. This is where the two are different. I have tried only to compare the curves. It can't be a linear relation any longer. On the relation between strength and bone mineral content: when tested on hundreds of spines, we have had a linear relation, but here we came up to another level. I mean, another magnitude of bone mineral content. We saw that we had a nonlinear relation between bone mineral content and strength.

References

1. Wolff J, Maquet P, Furlong R, tran-ed. The Law of Bone Remodeling. Berlin: Springer; 1986.
2. Hansson T, Roos B, Nachemson A. The bone mineral content and ultimate compressive strength in lumbar vertebrae. Spine 1980;1:46–55.
3. Brinckmann P, Biggemann M, Hilweg D. Prediction of the compressive strength of human lumbar vertebrae. Clin Biomech 1989;4 (suppl 2) 1–22.
4. Eriksson SA, Isberg BO, Lindgren JU. Prediction of vertebral strength by dual photon absorptiometry and quantitative computed tomography. Calcif Tissue Int 1989;44:243–250.
5. Granhed H, Jonson R, Hansson T. The loads on the lumbar spine during extreme weight lifting. Spine 1987;12:146–149.
6. Mosekilde L, Bentzen SM, Ortoft G, Jorgensen J. The predictive value of quantitative computed tomography for vertebral body compressive strength and ash density. Bone 1989;10:465–470.
7. Edmondston SJ, Singer KP, Day RE, Breidahl PD, Price R. In vitro relationships between vertebral body density, size and compressive strength in the elderly thoracolumbar spine. Clin Biomech 1994;9:180–186.
8. Singer K, Edmondston S, Day R, Breidahl P, Price R. Prediction of thoracic and lumbar vertebral body compressive strength: Correlations with bone mineral density and vertebral region. Bone 1995;2: 167–174.
9. Hansson T, Keller T, Spengler D. Mechanical behavior of the human lumbar spine. II. Fatigue strength during dynamic compressive loading. J Orthop Res 1987;5:479–487.
10. Brinckmann P, Biggemann M, Hilweg D. Fatigue fracture of human lumbar vertebrae. Clin Biomech 1988;(suppl 1):1–23.
11. Keller T, Spengler D, Hansson T. Mechanical behavior of the human lumbar spine: Creep analysis during static compressive loading. J Orthop Res 1987;5:467–478.
12. Neumann P, Keller T, Ekström L, Perry L, Hansson T, Spengler D. Mechanical properties of the human lumbar anterior longitudinal ligament. J Biomech 1992;25:1185–1194.
13. Hansson T, Roos B, Nachemson A. Development of osteopenia in the fourth lumbar vertebra during prolonged bed rest after operation for scoliosis. Acta Orthop Scand 1975;46:621–630.
14. Hansson T, Roos B, Nachemson A. The ability of the operated scoliosis patient to regain bone mineral lost from the axial skeleton. Orthop Trans 1983;7:1:27.
15. Hansson T, Sandström J, Roos B, Jonson R, Andersson G. The bone mineral content of the lumbar spine in patients with chronic low back pain. Spine 1985;2:158–160.
16. Carter DR. Mechanical loading history and skeletal biology. J Biomech 1987;20:1095–1109.

17. Carter DR, Orr TE, Fyrhie DP. Relationships between loading history and femoral cancellous bone architecture. J Biomech 1989;22:231–244.
18. Hart RT, Davy DT. Theories of bone modeling and remodeling. In Cowin SC (ed): Bone Mechanics. Boca Raton, FL: CRC Press; 1989:253.
19. Fyhrie DP, Carter DR. A unifying principle relating stress to trabecular bone morphology. J Orthop Res 1986;4:304–317.
20. Frost HM. Suggested fundamental concepts in skeletal physiology. Calcif Tissue Int 1993;52:1–4.
21. Jee WSS, Li XJ. Adaptation of cancellous bone to overloading in the adult rat: A single photon absorptiometry and histomorphometry study. Anat Res 1990;227:418–426.
22. Burr DB, Martin RB. Errors in bone remodeling: Toward a unified theory of metabolic bone disease. Am J Anat 1989;186:1–31.
23. Frost HM. Vital biomechanics: Proposed general concepts for skeletal adaptations to mechanical usage. Calcif Tissue Int 1988;42:145–155.
24. Hansson T. Bone mineral content and biomechanical properties of lumbar vertebrae [thesis]. Gothenburg, Sweden: Gothenburg University; 1977.
25. Freeman MAR, Todd RC, Pirie CJ. The role of fatigue in the pathogenesis of senile femoral neck fracture. J Bone Joint Surg 1974;56B:898–905.
26. Vernon-Roberts B, Pirie CJ. Healing trabecular microfractures in the bodies of the lumbar vertebrae. Ann Rheum Dis 1973;32:406–412.
27. Frost HM. Presence of microscopic cracks in vivo in bone. Henry Ford Hosp Med Bull 1960;8:27–35.
28. Hansson T, Roos B. Microcalluses of the trabeculae in lumbar vertebrae and their relation to the bone mineral content. Spine 1981;4:375–380.
29. Rubin CT, Lanyon LE. Osteoregulatory nature of mechanical stimuli: Function as a determinant for adaptive remodeling in bone. J Orthop Res 1987;5:300–310.
30. McLeod KJ, Rubin CT. The effect of low-frequency electrical field on osteogenesis. J Bone Joint Surg 1992;74A:920–929.
31. Cameron JR, Sorenson J. Measurement of bone mineral in vivo. Science 1963;142:230.
32. Nilsson BER. Posttraumatic osteopenia: A quantitative study of the bone mineral mass in the femur following fracture of the tibia in man using americium-241 as a photon source. Acta Orthop Scand 1966;suppl 91.
33. Roos BO, Sköldborn H. Dual photon absorptiometry in lumbar vertebrae: Theory and method. Acta Radiol 1974;3:266–280.
34. Hvid IN, Bentzen SM, Jörgensen J. Remodeling of the tibial plateau after knee replacement. Acta Orthop Scand 1988;59:567–573.
35. Steinberg GG, McCarthy CK, Baran DT. Quantifying bone loss from the proximal femur after total hip arthroplasty. Proceed Orthop Res Soc 1991:221.

Chapter 8

Osseointegration and Mechanical Capacity in Bone Tissue of Varying Quality

Rickard Brånemark

Osseointegration is a clinical reality in oral rehabilitation of patients suffering from total edentulousness as well as loss of a single tooth.[1] Outside the oral cavity, osseointegration has been used in the craniofacial region. Applications such as a bone-anchored hearing aid, and prostheses for the ear, nose, or eye are well documented by Tjellström[2] and Worthington et al.[3] In hand surgery, osseointegration has been used in the treatment of rheumatoid arthritis in the metacarpophalangeal joints.[4]

Osseointegration has proven to be a useful rehabilitation alternative. The anchorage between a titanium fixture and the bone tissue is characterized by the absence of intervening soft tissue.[5] But what do we know about the mechanical capacity in the interface—the region between living bone tissue and the titanium fixture? What do we know about the mechanical capacity for different modes of loading? What is the stress pattern in the surrounding tissue? Is it possible to say anything about the mechanical capacity related to bone density? What about remodeling of bone, and how will this influence the mechanical capacity of osseointegration?

Method

The setup is designed for the proximal ventral aspect of the tibia.[6] Here, it is possible to insert three fixtures in a parallel row. The fixtures can be tested in torsion, with pull-out tests, and in some instances, with lateral loading tests in different directions, transverse or parallel to the long axis of the bone.

The setup consists of an ordinary testing machine for mechanical testing (Instron, Canton, MA) and a specially designed machine for torque tests. Both machines are connected via an analog digital converter to an ordinary computer. On the computer, it is possible to register torque vs angle of deformation, and for pull-out or lateral loading tests, the applied load vs the deflection or extension. The testing fixtures used for the beagle dog experiments have a diameter of 3.7 mm, corresponding to clinical intraoral fixtures (Nobel Biocare, Gothenburg, Sweden). The two lateral fixtures have specially designed heads that can be attached to the mechanical testing machine, and the central fixture has a square head for torsional testing (Fig 8-1).

The fixtures are installed with a gentle surgical technique. Great care is taken to secure parallelism between the three fixtures via a series of guides. Efforts have also been made to avoid unnecessary loading of the fixtures when applying the mechanical tests. After in-

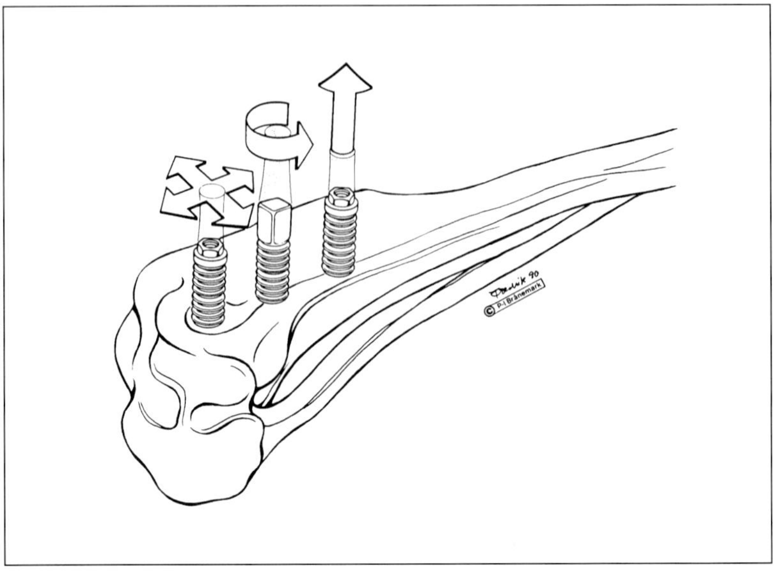

Fig 8-1 Three fixtures are installed in the proximal tibia. They are tested in pull-out, torsion, and lateral loading. (Figs 8-1 to 8-3 from Brånemark et al.[6] Use with permission.)

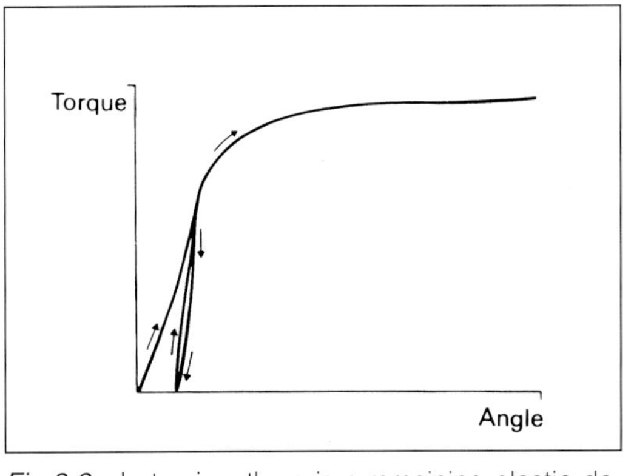

Fig 8-2 In torsion, there is a remaining plastic deformation when unloading.

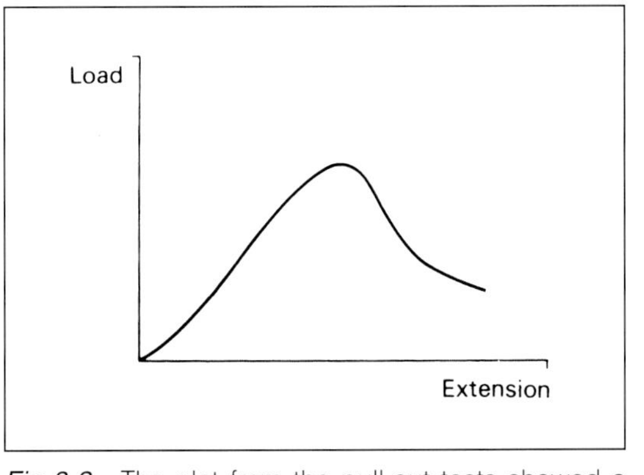

Fig 8-3 The plot from the pull-out tests showed a clear break.

stallation, the fixtures are left unloaded for, in this experiment, a healing time of 3 to 6 months. The fixtures are then exposed, and the mechanical testing is performed in vivo.

Results

All data presented are preliminary. Numerical values from the beagle dog experiment are shown in Table 8-1. The torsion test revealed an unexpected feature. When unloading, there was a remaining plastic deformation, even at an early phase (Fig 8-2), which indicates that there is no strong bond between the titanium surface and the bone tissue. The pull-out experiment showed the bone tissue to have a remarkable strength, since the pull-out force in some cases was about 2 kN. The plot shows a clear break (Fig 8-3).

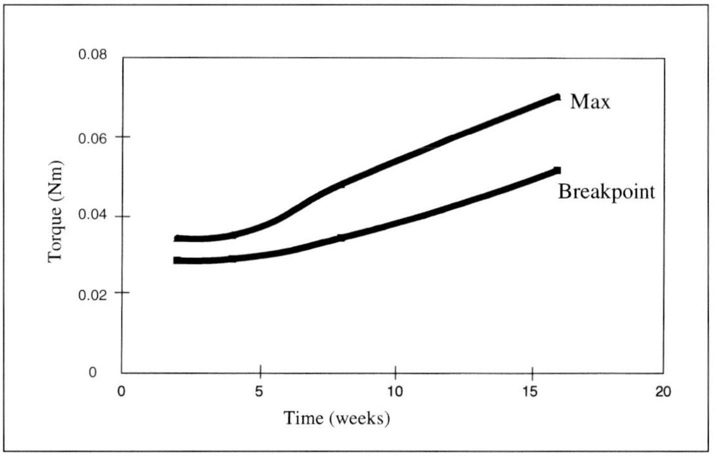

Fig 8-4 Breakpoint torques and maximal torques in the rat experiment at 2, 4, 8, and 16 weeks.

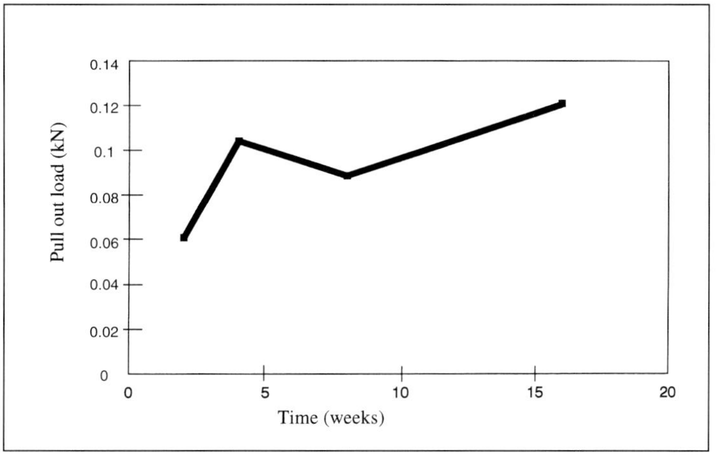

Fig 8-5 Pull-out loads from the rat experiment at 2, 4, 8, and 16 weeks.

Table 8-1 Calculations of Shear Stresses in Pull-out and in Torsion

Animal No.	*Pull-out (MPa)*	*Torsion (MPa)*
1	39.8	7.6
2	40.1	5.1
3	38.0	10.2
4	42.4	8.6
5	37.5	7.6
6	46.8	9.0
7	45.0	8.4
8	49.0	8.2
9	45.2	7.8
10	53.4	8.3
11	36.3	—
12	35.3	—
Mean	42.4	8.1
SD	5.6	1.3

The beagle dog experiment setup was modified and adapted for other animal studies. This fixture is mainly designed for testing on smaller animals, such as the rat. Mechanical tests were performed on normal rats after healing intervals of 2, 4, 8, and 16 weeks (Figs 8-4 and 8-5). There was a marked increase in torsion between 8 and 16 weeks. Pull-out tests indicated a limited influence by healing time.

The effect of irradiation was evaluated. One hind leg of a series of rats was irradiated with 10, 20, and 30 Gy. The rats were then left for 3 months. Then, three fixtures were installed as previously described. The fixtures were left for a healing time of 8 weeks. The fixtures were then exposed and mechanically

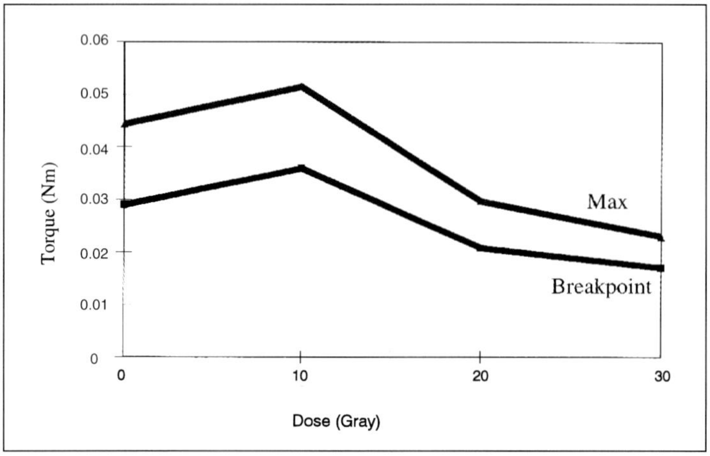

Fig 8-6 The effect of different doses of radiation on torsion.

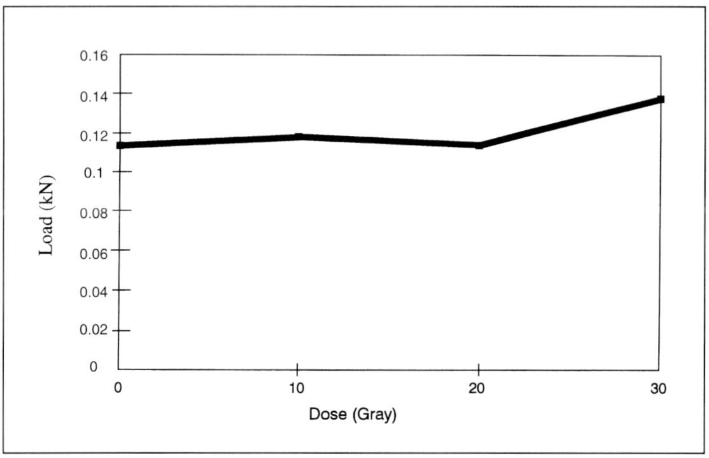

Fig 8-7 No major effect of radiation on pull-out load was observed.

tested in torsion and with pull-out tests. The torsion test, at 8 weeks for different doses of irradiation, illustrates a decreased mechanical capacity for doses above 20 Gy (Fig 8-6). The pull-out test results, on the other hand, were not negatively affected by irradiation (Fig 8-7). As with the findings in the beagle dog, the torsion plot indicated an early plastic deformation. The pull-out plot was also similar to that in the beagle dog experiments.

We have also studied the effects of arthritis on osseointegration. We have used an experimental arthritic rabbit knee model where antigens are injected in the rabbit knee joint.[7] Hereafter, we installed three so-called rat fixtures on the lateral femoral condyle of the diseased joint. The torsion tests and the pull-out tests indicate a decreased mechanical capacity (Figs 8-8 and 8-9).

In patients suffering from rheumatoid arthritis or osteoarthritis, Linder et al[8] installed titanium fixtures just below the knee joint in the tibia. At the time of arthroplasty, these fixtures were exposed and tested in torsion. The torsion plot was consistent with the previous findings of an early plastic deformation. There was a marked difference in the mechanical capacity between different patients. A typical finding was that those suffering from a rheumatoid arthritis had a much lower mechanical capacity than those suffering from osteoarthritis (Fig 8-10).

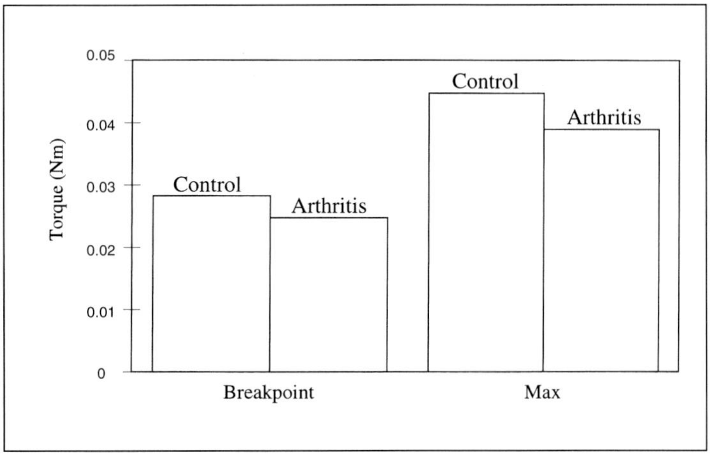

Fig 8-8 The effect of experimental arthritis on torsion.

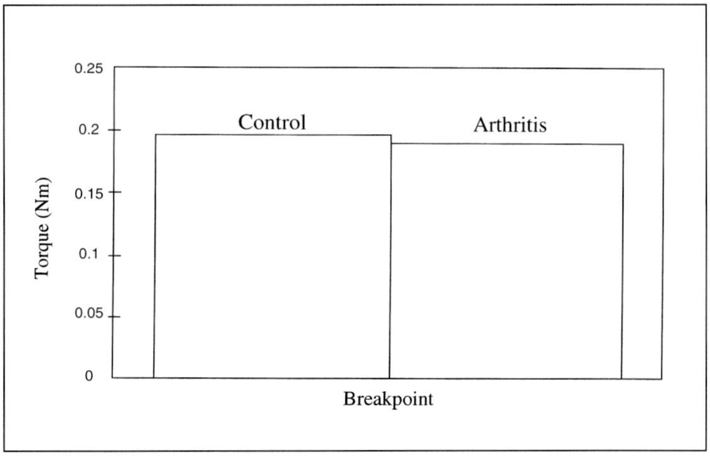

Fig 8-9 The effect of experimental arthritis on pull-out load was marginal.

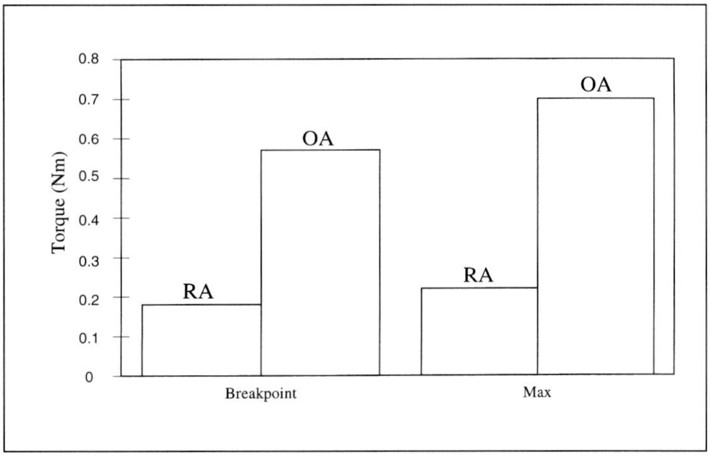

Fig 8-10 The mechanical capacity in torsion was reduced in patients with rheumatoid arthritis (RA) as compared to those with osteoarthritis (OA).

The data presented here are all macroscopic mechanical data. Thorough histologic examination of the interface region for each individual implant is desirable. With histomorphometry, it is possible to estimate the surface area that has a direct bone contact. With this information, it is possible to calculate shear stresses or shear strength in the true interface for the torsion tests, and in a theoretical interface in the pull-out tests.

We have collected some data from the dog experiments, as illustrated in Table 8-1. The pull-out shear strength is approximately 40 MPa, and the torsion shear strength is approximately 8 MPa. These findings clearly indicate that it is important to design the anchorage element so the load will be transferred out in the bone tissue, which has much greater shear strength than the interface (40 compared to 8MPa).

The mechanical capacity of osseointegration in bone tissue of varying quality must be taken into consideration when using osseointegration as a rehabilitation technique. Irradiation, arthritis, or osteoarthritis can decrease the mechanical capacity. Perhaps modification of anchorage element design and healing time can reduce the negative mechanical effects in these situations. If not, another option is optimizing the remodeling of the bone tissue. From a therapeutic point, this might be achieved with a controlled increased loading of the anchorage element with time.

The level of loading must then be individually set for each patient related to the mechanical capacity in that patient's bone tissue.

DISCUSSION

Göran Lundborg Well, I noticed with great interest your data on torsion tests on the rheumatoid patients, and especially the difference between rheumatoid arthritis and osteoarthritis. The rheumatoid patients are often on heavy drug medication, anti-inflammatory drugs, steroids, or even methotrexate and other drugs. I guess that in your cases, this may have influenced the results. Have you taken these things into account? I wonder, are there any experimental data on the effects of steroid treatment or cytostatic agents of any kind? These are very relevant questions from a clinical point of view.

Rickard Brånemark The data presented here don't take into consideration any medication. There are also quite different healing periods for the different clinical patients due to practical reasons. So I cannot actually comment on that. To my knowledge, there are no studies presenting data regarding the effect of steroids on the mechanical capacity. There is no study in which you actually perform a mechanical test. But otherwise, I would say it is common knowledge that steroids will influence the mechanical capacity in a negative way.

Tommy Hansson These are very interesting results you have here. My question is, as I see it, you have these screws in the bone at 90 degrees to the grain of the long space axis. Do you think there are differences in the properties when you do something parallel versus something 90 degrees to the parallel to the long axes? Have you seen anything on this? Of course, what we are talking about here are implants. This is not considering the mandible, which is an interesting issue.

RB That should be tested. It is slightly more difficult because we cannot test humans. The other way to test it is on animals. Perhaps you could just modify the model. I think that for our purposes, I would actually like to amputate an animal and then try to develop the best setup.

Lars Magnus Bjursten I was intrigued by the curve that you showed on the angular displacement. I wonder, what kind of displacements do you actually have at the interface at the breaking point there? Do you have any idea if it is in micrometers or millimeters?

RB I have tried to calculate it, but I don't remember the figures. The interesting thing is that even if you have 0.1 degree of angle deformation, you still see this very early plastic or remaining deformation. This actually indicates that there is no ideal elastic deformation going on in the interface at a level of the resolution, at least at a resolution that this equipment allows. That's a somewhat peculiar finding, and we did not expect that.

LMB So you have even been able to go down to 0.1 degree?

RB Yes, I have tested that in some situations. I have been talking to Dr Skalak about this phenomenon, and his idea is that the micro-irregularities in the interface will in some way be responsible for this behavior.

Per-Ingvar Brånemark I have one piece of interesting information. You said you tried to unscrew fixtures in the lower jaw. Clinically, that test failed, even in the hands of the most ambitious oral surgeon. Unscrewing involves the risk of fracture. A more serious question is in the radiated cases in the maxillofacial region. Of course the results are not as good as those in the ordinary nonirradiated cases. One could question, of course, if it would be possible to improve the mechanical capacity of carrying this load in those cases by your procedures. On the other hand, we have 10-year cases that have had 60 to 70 rads and there was no problem at all. So, it seems to me that there are quite some variations in the individual situations. Hyperbaric oxygen seems to be very beneficial for this part of the body when it comes to achieving and maintaining osseointegration. Do you have any idea about the effects, beside the vascularity, on the mineralization and remodeling of bone?

TH I know nothing about this.

PIB But it is fact that hyperbaric oxygen has a strong influence. It is a dramatic effect. This is critical now in the large number of cases of radiation treatment of the maxillofacial region.

RB We will study this as a continuation of the irradiated rat experiment and try to add hyperbaric oxygen treatment to see if it could reverse the negative effects on the irradiation. Of course we will have to try things like growth hormones.

Stephen Cowin I also find your results very interesting. I wanted to tie them to something that we found—when we assumed there was a very rough surface on the interface, we did not get very much ingrowth. This would suggest that if you did your torque test, it might turn out to be high value. But then if you do your pull-out test, it would be lower. I don't know whether the reverse of that would be true. But if you really polish the surfaces of the threads, it would enhance the ingrowth and increase what you get in a pull-out test, even though your torque test results would go down.

RB If I could just comment. Your indication is that there would be sort of low ingrowth in the threads. But in the clinical situation, you would have full ingrowth.

SC If we put something on the surface, such as hydroxyapatite coating, it would make the bone bond very well to the implant, and that may inhibit ingrowth. We would not get much ingrowth, but that would give good resistance to twisting. But it would reduce the pull-out strength. The reverse would be true if you made a very smooth screw thread. You would make it much easier for the bone to penetrate the screw threads, so then when you try to pull out, you would have a larger volume of the bone resisting. So I think it is important that the faces of the screw threads are low friction, and that may even help with installation.

RB Yes, fairly smooth may be beneficial. And that is actually what I would expect to happen.

Richard Coutts This is somewhat related to the dilemma that we face in orthopedics. The goal is obviously to achieve bonding and fixation, but the axiom in orthopedic surgery is never put in something that you can't get out. Maybe a polished surface would make it easier to get it out if you had to.

TH I see the problem—a fixture coming out even if it is not supposed to come out.

RC It is a fine line that you have to walk there.

RB If I may comment, the objective is to transfer stress out into the bone. I think this is important to consider when you design an anchorage. For the axial loading, for instance, I don't think that a smooth cylinder implant is the perfect choice.

SC There is a difference in what you consider conditions to be at the interface. If you have a smooth interface, your comment was right, but when you have a screw thread, the interface is effectively very rough, although the surfaces of the screw thread are smooth.

RB Yes. It also depends on the loading, of course. In torsion, a smooth screw can slip out easily.

LMB I looked at the data you presented from the arthritic patients, and some of them are very osteoporotic. Are these mechanical data you present related to the amount of bone you have, or is it some kind of interfacial feature that controls the results?

RB The data presented here are just raw mechanical data, but Åke Carlsson and co-workers have performed histologic examinations of these fixtures, and we will try to correlate those findings to see if that could explain the situation.

LMB Have you been able to correlate the data, histologic data, on animals to the mechanical data? Is the implant removal torque, for instance, related to the amount of contact area that you have at the implant surface?

RB The only careful histologic examination we have done so far is in the normal beagle, so I cannot answer that question yet.

Åke Carlsson I cannot recall all details since the paper was published some years ago. However, there were great variations in bone quality, and some rheumatic patients had very poor bone structure.

RB Yes, there is also a very nonuniform distribution of patients with the rheumatoid arthritis. For example, those who have had arthritis for 30 years compared to those who have had symptoms for only a few years. There are also the different healing times and so on. These different indications give different capacities.

PIB Do you have any comments, for example, on the experience of Robert Ling and his group, who are exploring surface characteristics, polished versus nonpolished hip joint prosthesis, and the results?

RB Well, I don't think it applies to the questions here, but perhaps to questions of mechanical stability. It is quite another situation in the cemented femur where the roughness sinks a bit into the cement. The rough surface produces wear debris in a greater amount, especially when there are some defects in the cement mantle. It has affected the bone in the form of gamma globulin formation and so on. I do not think this really applies to the discussion here about mechanical stability.

Dick Heinegård There are animal models being used to develop therapy for the inflammatory processes.

RB Actually, I am not that familiar with the model that Professor Thomsen in Gothenburg is using, but I think that Professor Bjursten might have some comment.

LMB Yes, we developed that model several years ago just by injecting antigen into the joint. It basically produces a local arthritis in one joint. I do not know if you have this kind of severe arthritis, but it is localized, and I do not know the effects on the adjacent bone tissue. It is an antigen that is injected directly into the joint.

References

1. Brånemark P-I. Introduction to osseointegration. In Brånemark P-I, Zarb G, Albrektsson T (eds): Tissue-Integrated Prostheses: Osseointegration in Clinical Dentistry. Chicago: Quintessence; 1985:11–76.
2. Tjellström A. Osseointegrated systems and their applications in the head and neck. Adv Otolaryngol Head Neck Surg 1989;3:39–70.
3. Worthington P, Brånemark P-I. Advanced Osseointegration Surgery: Applications in the Maxillofacial Region. Chicago: Quintessence; 1992.
4. Lundborg G, Brånemark P-I, Carlsson I. Metacarpophalangeal joint arthroplasty based on the osseointegration concept. J Hand Surg 1993;18B: 693–703.
5. Brånemark P-I, Lausmaa J, Ericson L, Thomsen P, Brånemark R, Skalak R. Anatomy of osseointegration and the transfer of load. In Fonseca RJ, Davis HW (eds): Reconstructive Preprosthetic Oral and Maxillofacial Surgery, ed 2. Philadelphia: Saunders; 1994.
6. Brånemark P-I, Brånemark R, Skalak R. Osseointegration and rigid fixation. In Yaremchuk MJ, Grass JS, Manson PN (eds): Rigid Fixation of the Craniomaxillofacial Skeleton. London: Butterworth-Heinemann; 1992.
7. Thomsen P, Bjursten LM, Ahlstedt S, Bagge U, Ericson LE. Proliferative synovitis in rabbit knee-joints induced by antigen and preformed immune complexes. Scand J Rheumatol 1985;14:239–251.
8. Linder L, Carlsson Å, Marsal L, Bjursten LM, Brånemark P-I. Clinical aspects of osseointegration in joint replacement: A histological study of titanium implants. J Bone Joint Surg 1988;70B, 550–555.

Chapter 9

Tumor Resection and Skeletal Reconstruction in the Extremities

Björn Gunterberg

Surgery remains the key to successful treatment of sarcomas of the extremities, although the development of adjuvant therapies has influenced the surgical procedures in the direction of less disabling operations. Limb salvage is the term used most often in this connection, indicating that because there may be no or small differences in oncologic outcome between amputation and resection followed by reconstruction, either method could be used depending on the particular situation.

Skeletal reconstruction can be achieved in different ways. Autologous bone grafts were used for segmental defects early, and allografts were used already in the 1960s, also for joint reconstruction.[1-3] However, the history of limb salvage to a large extent parallels the history of endoprosthetic joint replacement, where Gluck is a portal figure.[4] Scales, at a symposium in Leeds in 1969, referred to Gluck's paper from 1890,[5] stating that the problems of joint replacement were still not solved, which is a valid statement even today. He recognized that tumor surgery might require extensive resection of bone involving the joint.

Scales emphasized that the implant: 1) must be made of materials that are chemically and physically stable in the biologic environment; 2) must be able to sustain the applied stresses, particularly those of fatigue, indefinitely; 3) must provide stable joints with bearing surfaces having a low friction torque, ideally approaching that of cartilage on cartilage; 4) must not produce products of wear that will cause damage; and 5) must be permanently fixed to bone. These requirements for joint replacement are not met today. However, titanium devices for other purposes may now be permanently anchored to bone.[6]

To establish a forum for sharing knowledge and promoting development in skeletal reconstruction following tumor resection, the First International Symposium on Limb-Salvage (ISOLS) was organized in 1981. Since then, this symposium has become an every-other-year event, with participation of experts from most parts of the world. The development in this field is progressing, but no major breakthrough has occurred. Different techniques, materials, and designs have been presented with varying acceptance. Nevertheless, in a few decades a change from approximately 20% limb salvage to approximately 80% limb salvage in the treatment of malignant tumors of the extremities has taken place. Is this a major progress? Most of us would answer yes.

However, the issue may be debated. The reasons are: 1) the most common techniques for limb salvage using massive allografts or metallic megaprostheses are tainted with major complications in approximately 50% of the cases; 2) the true long-term results are

Table 9-1 Types of Reconstruction after Bone Tumor Resection

Type	*Segmental defect*	*Joint replacement*
Autograft	Graft	Graft (arthrodesis, exceptionally mobile joint)
Allograft	Graft	Graft (mobile joint, arthrodesis in exceptional cases)
Prosthesis	Prosthesis	Prosthesis
Composite (allograft + prosthesis)	Graft	Prosthesis

still to be reported; 3) the patients have to restrain their physical activities, often more than an amputee has to, not to risk damage or wear of their reconstructions; 4) the patients will need service operations at intervals to replace worn or broken parts, to repair graft failures, etc. Patients with reconstructed extremities in many cases have "glass limbs."

Reconstruction Methods

Table 9-1 presents a summary of the methods commonly used for extremity reconstruction. Allografts are favored in some centers although most centers use megaprostheses. Also, some anatomic locations appear more suited for special reconstructions, eg, composite type for the upper humerus.[7] Many variants of reconstruction have been presented. Autoclaved tumor-bearing bone has been reimplanted. Segmental bone defects have been restored with the help of external fixators used to transport bone from one end of the defect to the other during an extended period, leading to bone formation in the wake. Vascularized fibula autografts have been used to create a more reliable and accelerated union with host bone as well as to promote hypertrophy of the graft. Combinations of allografts and autografts have been used. Expandable prostheses have been tested in growing children. Ceramic implants have also been used.

Gothenburg Experiences

Reconstruction of the Proximal Femur with a Long-Stem Prosthesis and Acrylic Cement as a Spacer

This method was devised in our department in 1974 and was first presented at the ISOLS in 1981.[8] From 1974 to 1992, 20 patients underwent surgery. Complications, oncologic results, and function according to Enneking et al[9] were recorded. Age ranged from 18 to 73 years, with an average of 50 years. Indications for surgery were metastasis, 11; chondrosarcoma, 4; giant cell tumor of bone, 2; and osteosarcoma, myeloma, and fibrous dysplasia, 1 each. The prostheses used were Moore (16) and Bateman (4). The follow-up was 1 to 20 years, with an average of 4 years.

Two revisions were done, 1 and 7 years postoperatively because of pain, one possibly due to soft-tissue impingement and one because of acetabular wear. One patient had a hemipelvectomy because of chronic osteomyelitis of the pelvis, and one with the only local recurrence (metastasis) also had a hemipelvectomy. All 11 patients with metastases are now dead (average survival, 14 months). All four patients with chondrosarcoma are alive with an average survival time of more than 8 years, and both patients with giant cell tumor have been alive 17 and 18 years postoperatively. The average functional score remained at a constant and high level of 80% to 85% (normal function = 100%).

Table 9-2 Massive Autograft Reconstructions After Bone Tumor Resection

Case	*Resection*	*Reconstruction*	*Diagnosis*	*Complications*	*Functional Score*	*Follow-up (yr)*
1	Pelvis, 9 cm gap	Iliofemoral fusion, Pelvic reconstruction	Ewing's sarcoma	---	27/30 – 90%	6
2	Pelvis, 11 cm gap	Ischiofemoral fusion, Pelvic reconstruction	Chondrosarcoma	---	23/30 – 77%	6
3	Prox. tibia-fibula, 11 cm	Femorotibial fusion	Chondrosarcoma	Fatigue fracture	27/30 – 90%	8
4	Prox. humerus, 13 cm	Humeroscapular fusion	Chondrosarcoma	Traumatic fracture	27/30 – 90%	12
5	Prox. humerus, 8 cm	Humeroscapular fusion	Giant cell tumor	---	28/30 – 93%	10
6	Prox. humerus, 14 cm	Humeroscapular fusion	Aneurysmal bone cyst	---	27/30 – 90%	7
7	Distal humerus, 15 cm	Humero-ulnar fusion	Osteosarcoma	Infection, pseudo-arthrosis	19/30 – 63%	16
8	Distal radius, 6 cm	Radius reconstruction mobile joint	Giant cell tumor	---	21/30 – 70%	18

The patient who has the longest follow-up, 20 years, has a functional score of 93% (Fig 9-1).

Tumors of the proximal part of the femur can be resected and reconstructed successfully with a long-stem prosthesis supported with a spacer of acrylic cement. This is a simple and inexpensive method. The procedure is particularly suitable for metastases but has also proven successful in locally benign aggressive lesions and low-grade malignant tumors.

Massive Autografts for Bone Tumor Reconstruction: An Old Method Revisited

The complication rate following bone tumor reconstruction with massive allografts or megaprostheses is fairly high, and the durability of the reconstructions may be limited. This led us to investigate our patients who had reconstructions with massive autografts and had long follow-up times. Eight consecutive patients with a minimum follow-up of 6 years were studied. They all had reconstructions bridging a major bone defect, using ilium and/or a nonvascularized fibula. Two were internal hemipelvectomies (Fig 9-2), and the others were combined shaft and joint resections (Fig 9-3). No pure segmental defects were included. Complications and functional outcomes were evaluated.[9]

All complications, except the pseudarthrosis, were resolved (Table 9-2). No patient needed amputation. Some results were negatively affected by necessary sacrifice of nerves and/or muscles.

Massive autografts in bone tumor reconstruction can give lasting, functional results with an acceptable complication rate.

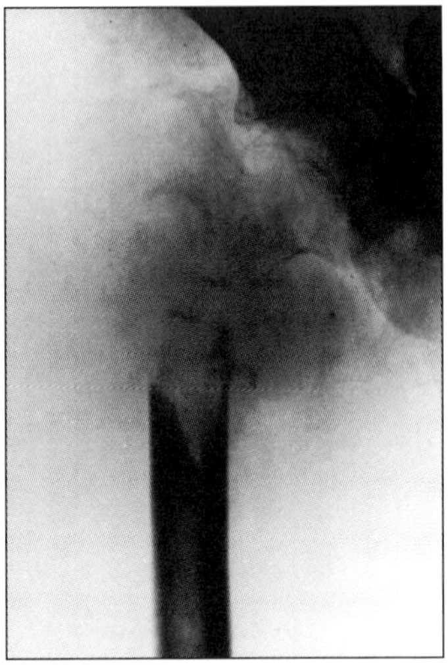

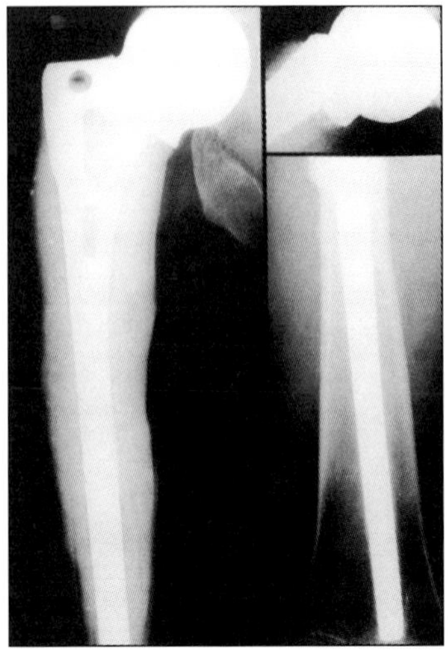

Fig 9-1 Aggressive fibrous dysplasia of the proximal femur in a 19-year-old woman.

Fig 9-1a (far left) Preoperative radiograph.

Fig 9-1b Radiographs 19 years after resection and reconstruction with a long-stem prosthesis and acrylic cement as a spacer. Excellent function is retained.

Fig 9-2 Chondrosarcoma of the ilium and acetabulum in a 30-year-old man.

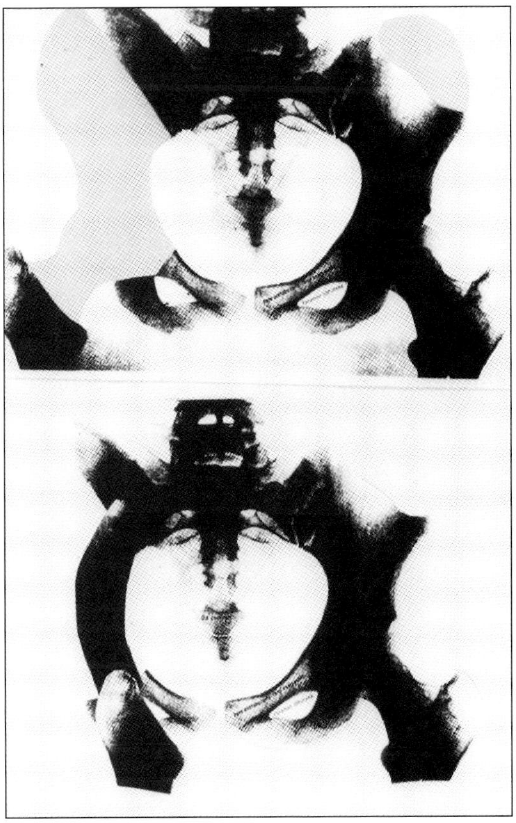

Fig 9-2a Schematic illustration of the resection and reconstruction. Light-shaded parts were removed. Part of the left ilium was used for the reconstruction.

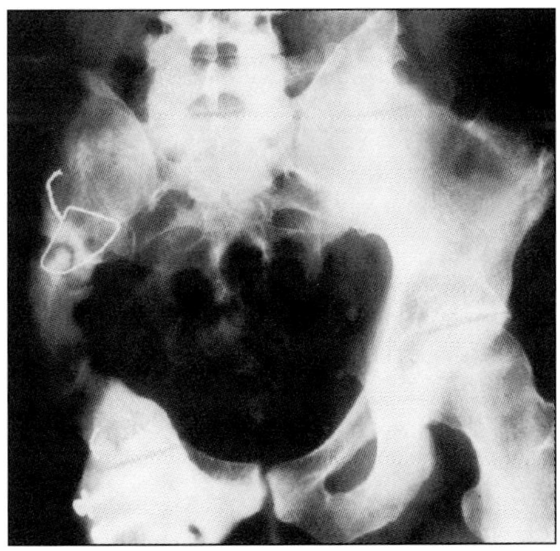

Fig 9-2b Radiograph 4 years postoperatively.

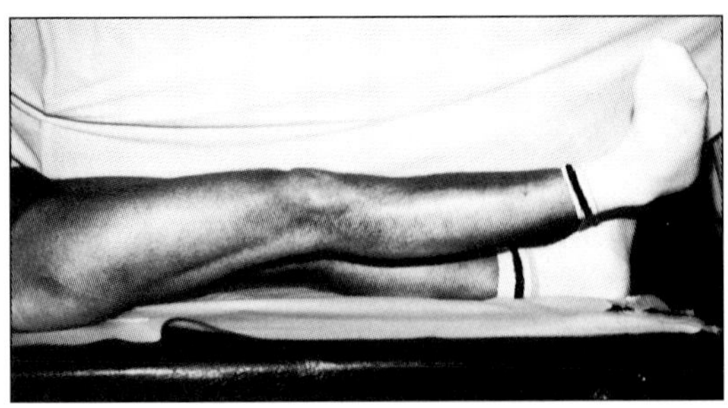

Fig 9-2c The patient is pain-free and has excellent function 6 years postoperatively.

Fig 9-3 Aneurysmal bone cyst of the proximal humerus in a 25-year-old man.

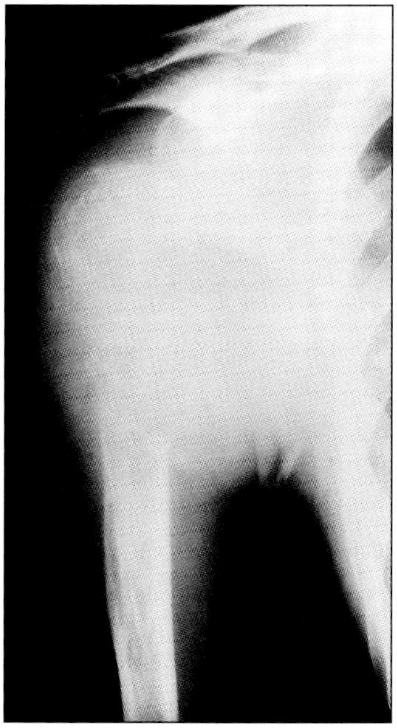

Fig 9-3a Preoperative radiograph.

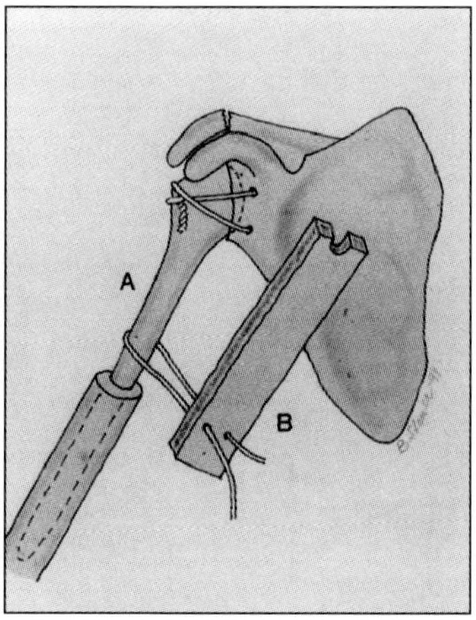

Fig 9-3b Schematic illustration of reconstruction with autologous nonvascularized fibula and ilium.

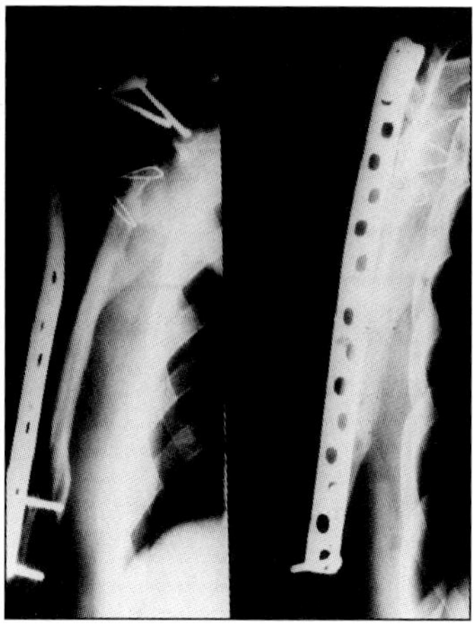

Fig 9-3c Radiographs 6 years postoperatively. An AO-plate was used to enhance stability. Excellent function is retained.

Modular Megaprosthesis in Bone Tumor Reconstruction

We used the Kotz modular femur and tibia (KMFT) reconstruction system in 10 patients from 1987 to 1993. Five patients had reconstruction of the proximal femur, two of the distal femur, and three of the proximal tibia (Fig 9-4). The diagnoses were osteosarcoma in five; chondrosarcoma, four; and giant cell tumor, one. Ages ranged from 16 to 72 years, with an average of 31 years. Complications and oncologic and functional outcomes[9] were recorded. Only seven patients have had a follow-up exceeding 2 years.

No early complications occurred. Two proximal tibia reconstructions had to be reoperated after 5 and 6 years because of fractured prostheses (1) and worn-out bushings (1). One patient died of metastases; all the others are alive with no evidence of tumors. The average functional score was 66% (range, 33% to 83%). The knee reconstructions scored better than proximal femur reconstructions in this small series.

Our results seem to support previous reports on late mechanical failures in knee reconstructions, although function was good as long as the system was intact. The short-term results thus seem good, but in the long run several of these patients may require amputation because of unanticipated problems. Possibly an early but lasting bone-graft operation with arthrodesis may be a better alternative.

Fig 9-4 Osteosarcoma of the proximal tibia in a 17-year-old girl.

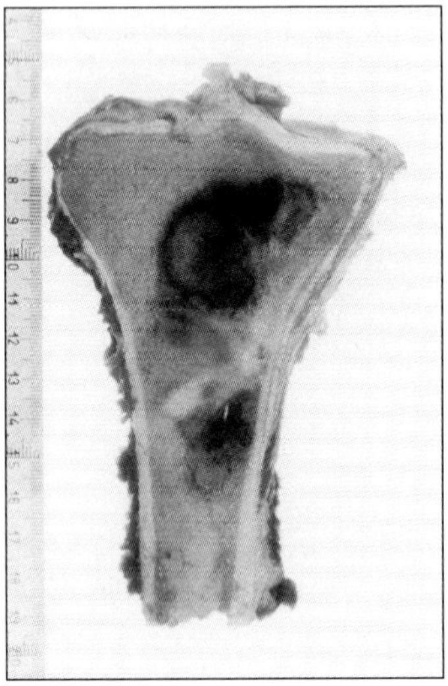

Fig 9-4a The removed specimen with partly necrotic tumor, as an effect of preoperative chemotherapy.

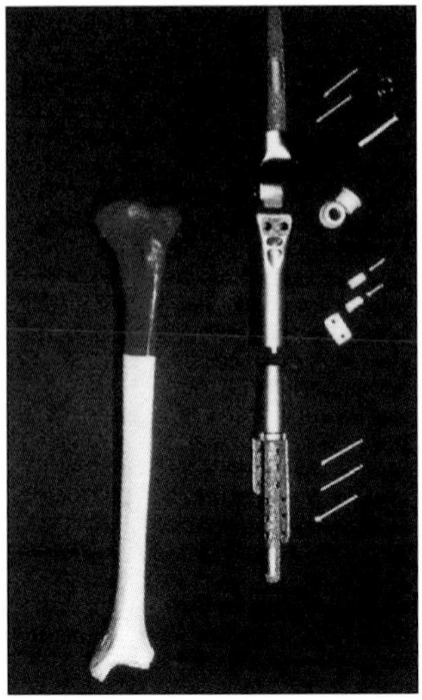

Fig 9-4b Illustration of resection and implants for reconstruction (KMFT reconstruction system).

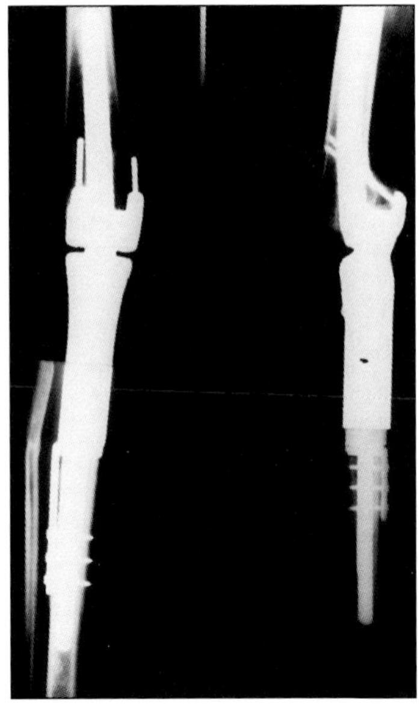

Fig 9-4c Postoperative radiographs.

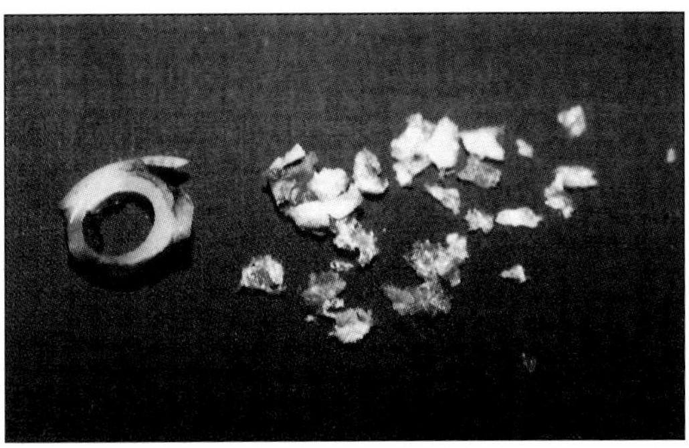

Fig 9-4d Worn and damaged plastic bushings exchanged 6 years postoperatively.

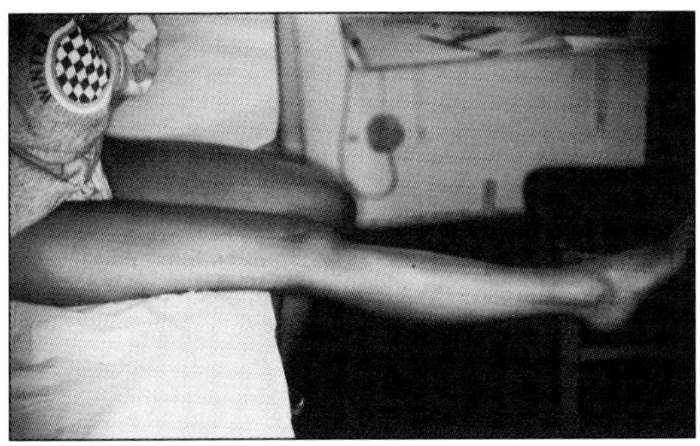

Fig 9-4e Good function remains 7 years after resection and reconstruction.

Summary

Much of the background in the development of limb-salvage surgery for malignant bone tumors was based on the idea that prognosis was usually poor and the goal should be to provide as good a quality of life as possible for the remaining limited lifetime. However, oncologic treatment, mainly based on cytotoxic drugs, has become more and more effective, especially in osteosarcoma and Ewing's sarcoma treatment. Also, the principles for surgical tumor removal with adequate margins of healthy tissue have become more firmly established. Thus, the survival after bone sarcoma treatment has improved considerably. Approximately 50% to 70% of patients with osteosarcoma of the extremities or Ewing's sarcoma are expected to survive today[10,11] compared to 10% to 20% three decades ago.

The methods for limb salvage have not kept pace with this development. The best-functioning devices obviously have limited durability. Wear and fatigue are the main problems. Graft reconstructions are also associated with problems such as infections, fractures, deformity, and joint pain. However, there are several reports on good long-term function after allograft reconstruction. Those grafts that survive seem to do well.[12]

Nevertheless, limb-salvage surgery is an unresolved dilemma in patients who are expected to have long survival. Which method is the best in the individual case? What are the demands of the patient? The situation is quite different for the young athlete and the sedentary pensioner. The former would need a strong limb with durability, which probably implies lesser quality of function, in contrast to the latter, who may have good function although limited strength and durability of the reconstruction. Whichever reconstruction is chosen, there is a risk of complications, mechanical, oncologic, infectious, or other, which in approximately 15% of all cases lead to amputation.[13] Bone quality and bone stock as well as soft tissues may be compromised in these patients, and poor prosthetic function may be the ultimate result. Also, most extremity reconstructions require restrictions of activity to avoid failure. This is the "glass limb" problem.

What can we expect in the future? Technology and biologic sciences may provide more sophisticated devices or transplants, but today the final solution is not in sight. Prophylactic measures or curative pharmacologic tumor treatment seem to be decades ahead. The first priority is to provide optimal prosthetic function, which might be achieved with an osseointegrated prosthesis, for those who have no alternative but amputation. When this concept is fully developed, it may be an option in certain cases where limb salvage could be done but the performance of a conventional reconstruction is not compatible with the patient's requirements.

DISCUSSION

Shu Chien When would you start using osseointegrated prostheses in the type of cases that you showed us?

Björn Gunterberg I think the main problems are the patients who are expected to have a long life ahead of them and expect to be active. A good question I heard recently in a meeting is: Why did you amputate this case? The patient wanted to go skiing, which could not be allowed with reconstruction.

SCh You showed several cases for which you used cemented prostheses, for instance, in the upper tibia position and in the shoulder. Do you think that we could use an osseointegrated prosthesis, eg, a titanium prosthesis of a similar shape, that would not need to be cemented?

BG I think that would be a very good development because it is not a good solution to cement a prosthesis in the humerus. It is done, but survival of this prosthesis is not expected to be a very long one. I did not show any cemented tibia prostheses.

Göran Lundborg Extensive bone grafting is sometimes required, for instance, when you transport one part of the iliac crest to the other side to replace parts of the pelvis. I wonder, with such a huge bone grafting procedure with the considerable requirement of stability, what are the problems here in terms of fracture of the graft? What do you think about using vascularized bone grafts as opposed to free grafts? Has this been discussed regarding the fibula?

BG We do use vascularized fibula grafts in many cases but in the pelvis. I have usually not done this because that would unfortunately add maybe another 4 hours to the operation, which now takes 8 to 10 hours. That is not the main reason, but we need a very strong and heavy block and it is in any case very difficult to achieve primary stability. So you have to keep these patients in casts for up to 3 or 4 months and in that time these grafts do revascularize in a way that serves its purpose. The main problem is that when they are to be revascularized we have to protect them for a long time.

Åke Carlsson Sometimes you end up with a very short stump, which makes it difficult to fit the patient with a prosthesis. Have you any experience of bone transportation using this method to elongate the stump and to make it easier to apply a conventional prosthesis or an osseointegrated device?

BG To start with, bone transportation is done in limb-salvage cases where you have a big defect and you can use the proximal part of the bone. It is sectioned and slowly pulled down to the remaining lower part of the bone, forming bone in its wake. In an amputated case, the amputation implies that you have also removed a lot of soft tissues because, when you amputate, it is because the tumor is so extensive that you have to sacrifice the soft tissues also. In that case, there is no way of elongating the stump with bone transportation because you do not have enough soft tissue for coverage. But I think these cases that you are referring to are the cases where we can hope for the best results with an osseointegrated prosthesis.

Richard Coutts I would be interested in your thoughts as to whether these pelvic reconstructions would be an area in which you want to consider an osseointegrated prosthesis. Then you could restore the contour of the pelvis. For men, that would make it easier for them to hold their pants up.

BG There are many possibilities. I showed only a few types of reconstructions. There are very many different ways of reconstructing, for instance, the pelvis, and there are very large metallic prostheses that resemble the normal pelvis. But these have to be anchored in the remaining pelvic parts that are not very thick and not very strong, to make them reliable. They do function in the short term, but in the long term they do not function, and I think that even with osseointegration the load may be too much in the pelvis. I do not feel that the pelvis is a very good place to start. I think autologous bone, which heals with the normal remaining bone, will give the best and most reliable function. I strive to reconstruct each case in the pelvis so that the reconstruction will last the rest of the patient's life, and I do not want to risk the reconstruction with foreign material too much.

RC You could actually use a combination?

BG Could possibly—there are also patients reconstructed this way.

RC It would be intriguing to actually get a metallic prosthesis in contact with bone to which bone would adhere. I think the loads would not be too great to break this.

BG I think this is an area where we can expect further development, and I am not prepared to give an answer now.

References

1. Albee FH. Fundamentals in bone transplantation: Experiences in three thousand bone graft operations. JAMA 1923;81:1429–1432.
2. Ottolenghi CE. Massive osteo and osteo-articular bone grafts: Technique and results of 62 cases. Clin Orthop 1972;87:156–164.
3. Enneking WF, Eady JL, Burchardt H. Autogenous cortical bone grafts in the reconstruction of segmental skeletal defects. J Bone Joint Surg 1980;62A: 1039–1058.
4. Gluck T. Auto plastic-transplantation-implantation von Fremdkörpern. Klin Wochenschr 1890; 27:421.
5. Scales J, Kelly P, Goddard D. Friction torque studies of the total joint replacement: the use of a simulator. In Lubrication and Wear in Joints. London: Sector Publishing; 1969:88–102.
6. Brånemark P-I, Zarb GA, Albrektsson T (eds). Tissue-Integrated Prostheses: Osseointegration in Clinical Dentistry. Chicago: Quintessence; 1985.
7. Rock MG. Intercalary allograft and custom Neer prosthesis after en bloc resection of the proximal humerus. In Enneking WF (ed): Limb-Salvage in Musculoskeletal Oncology. New York: Churchill Livingstone; 1987:586–597.
8. Stener B, Gunterberg B. Reconstruction of the proximal femur with a long-stem Moore prosthesis and acrylic cement after resection for tumor. In Chao EYS, Ivins JC (eds): Tumor Prostheses for Bone and Joint Reconstruction. New York: Thieme-Stratton 1983: 117–125.
9. Enneking WF, Dunham W, Gebhardt MC, Malawar M, Pritchard DJ. A system for the functional evaluation of reconstructive procedures after surgical treatment of tumors of the musculoskeletal system. Clin Orthop 1993;286:241–246.
10. Bacci G, Toni A, Avella M, et al. Long-term results in 144 localized Ewing's sarcoma patients treated with combined chemotherapy. Cancer 1989;63: 1477–1486.
11. Saeter G, Alvegård TA, Elomaa I, Stenwig AE, Holmström T, Solheim Ö. Treatment of osteosarcoma of the extremities with the T-10 protocol, with emphasis on the effects of pre-operative chemotherapy with single-agent high-dose methotrexate: A Scandinavian Sarcoma Group study. J Clin Oncol 1991;9:1766–1775.
12. Musculo DL, Ayerza MA, Calabrese ME. Major complications and final outcome of knee osteoarticular bone allografts. In Tan SK (ed): Limb-Salvage: Current Trends. Singapore: ISOLS; 1993:197–200.
13. Brown KLB (ed). Complications of Limb-Salvage: Prevention, Management and Outcome. Montreal: ISOLS; 1991.

Chapter 10

Tumors of the Spine

James N. Weinstein and Stefano Boriani

The connection of tumors of the spine and osseointegration may not be immediately apparent, but there are two relevant correlations that may be drawn from two viewpoints. The first correlation is clinical. As new surgical techniques and imaging modalities improve and more precise surgical planning is possible, it may be useful to consider restorations and stabilizations using osseointegrated components to achieve long-term stability and protection of the spinal column.

The second connection is more tenuous and philosophical. It is that osseointegration, like tumors, allows for growth of existing tissues, usually long after normal development is completed. In osseointegration, the hope is that growth of the bone will proceed unchecked until all the space around a prosthesis is filled with new bone. In a tumor, the new growth may be unchecked with no natural barrier. But in both cases, the understanding of normal vs therapeutic means of inducing and/or limiting growth is central to treatment. Any information gained in one arena may be useful in the other.

The diagnosis and treatment of spinal tumors have evolved dramatically over the past three decades and continue to evolve rapidly. There appear to be significant improvements in both the short- and long-term outcomes as a result of obtaining better surgical margins, applying the oncologic staging system of the extremities, and careful surgical planning with adjuvant treatments to the spine. With the tremendous progress in radiologic imaging with magnetic resonance imaging (MRI) and computed tomography (CT), careful preoperative planning now makes it possible—using new surgical techniques—to achieve adequate margins with less morbidity and mortality in treating primary spinal tumors.

Unfortunately, until recently, there has been no uniform approach applied to the diagnosis and treatment of these patients and little uniformity in reporting outcome. This lack of uniformity makes comparison of treatment protocols difficult and conclusions regarding definitive management somewhat tenuous. Today, in patients afflicted with primary bone tumors of the spine, surgical treatment offers a reasonable likelihood of functional improvement and significant improvement in pain relief, and in many cases of benign tumors, cure of the disease can be expected. Some conditions like aneurysmal bone cyst rarely require surgery. Such benign spinal tumors must be submitted to a careful cost-benefit evaluation to assess the difficulty and possible risk of a surgical procedure vs an effective nonsurgical option.

It is more imperative that the treating physician appreciate the specific symptoms and characteristics of spinal neoplasias. The principles applied to limb tumor staging and management need to be closely followed when a benign spinal tumor is expected. The correct approach is: 1) to perform a complete clinical and imaging workup before biopsy; 2) to choose the simplest biopsy technique that will allow for the best yield; 3) to decide the most appropriate surgical, medical, and possible adjuvant treatments according to the tumor's oncologic stage and the patient's overall condition; 4) to plan surgery according to a standardized spinal surgical staging of the tumor's extent; 5) to make the maximum effort to preserve neurologic function; and 6) to maintain spinal column stability. Although the treatment of patients with spinal tumors must be individualized, these specific principles should be observed in managing spinal neoplasms in general.

Primary tumors of the spine are very rare, and for the most part their relative incidence reflects that of the skeleton in general. Certain tumors, like osteoblastoma, do show a predilection for the spinal column, but these still make up only a very small portion of all spinal tumors. This chapter addresses tumors of the spine. In many instances the information presented is also applicable to metastatic tumors involving the spine.

Presentation

Tumors of the spine often present to the physician after a long history of symptoms. This may have both diagnostic and prognostic significance. While tumors of the spinal column may remain asymptomatic for some time, when symptoms do develop, they usually are a consequence of one or more of the following: 1) expansion of the cortex of the vertebral body of tumor mass with fracture and invasion of the paravertebral soft tissues; 2) compression or invasion of adjacent nerve roots; 3) pathologic fracture due to vertebral destruction; 4) development of spinal instability or hypermobility; and/or 5) compression of the spinal cord. The caveat in the spine that is important to remember is that slow-growing, yet mostly aggressive tumors that might be easily managed in the extremities may prove to be unresectable and lethal when extensively established in the spine.

Patient's Age

The patient's age at diagnosis is an important prognostic indicator and is highly correlated with tumor type. In patients under 21 years of age, benign spine lesions are responsible for most spinal tumors, while in patients over 21 years of age, one should think about malignant tumors.

Location

The location of the lesion within the vertebra is another important prognosticator. The majority of benign lesions are generally located posteriorly in the spine, even when more than one level is involved, whereas malignant tumors are more likely located with the vertebral body.

Symptoms

The most consistent complaint of patients affected by tumors of the spine is pain. Although back pain is extremely common and nonspecific, the pain associated with neoplasia tends to be progressive and unrelenting. It does not have the close association with activity that mechanical low back pain or neck pain do. Pain at night is particularly worrisome. Pain symptoms may localize to a specific spinal segment and/or be reproduced by pressure or percussion over the involved segment. Radicular symptoms are less common but are still frequently seen in patients with cervical and/or lumbar involvement. Radicular pain in the thoracic region may result in pain that radiates around the chest, forming a band of dysesthesias or paresthesias circumferentially from the level of the involved vertebra.

Local tumor growth can produce expansion of the cortex of the vertebral body, resulting initially in thinning and remodeling of the cortex and later in pathologic fracture with invasion of the paravertebral soft tissues. As the cortex expands, the periosteum is distorted and stretched to stimulate associated pain receptors. Local extension of tumor tissue into the paravertebral muscles or adjacent nerve roots or spinal cord can result in radicular symptoms or myelopathy, or may simply produce pain and paresthesias. Pathologic fracture may result from extensive bone loss associated with vertebral body destruction, producing acute pain symptoms similar to those seen in a traumatic compression fracture, which may lead to development of spinal hypermobility. Radicular symptoms, similar to those seen in a herniated nucleus pulposus, may be caused by a spine tumor, leading to confusion in the diagnosis and treatment.

Spinal deformity[1] may also be associated with the onset of pain and usually results from paraspinal muscular spasm. Scoliosis is frequently associated with osteoid osteoma[2,3] and sometimes with osteoblastoma. These two benign tumors often present with localized pain at night associated with paravertebral muscle spasm and limitation of motion.[4] The onset of such a scoliosis may be rapid. The tumor is usually found on the concave side of the deformity and usually at the apex of the curve.[3] Although these deformities usually disappear after tumor excision, curves neglected for prolonged periods will become structural. The bone scan is highly sensitive in identifying a focus consistent with osteoid osteoma, and it can significantly improve the efficacy of treatment by providing an earlier diagnosis and accurate localization of the tumor.[4,5]

Neurologic deficits are fairly common in patients presenting with spinal tumors. They are most commonly seen in association with rapidly expanding malignant lesions, but any slowly progressive, benign, expansive neoplasm may produce a deficit if left alone long enough. Nonetheless, as many as 70% of patients with spinal neoplasms will manifest clinical weakness by the time the correct diagnosis is made. This should emphasize the importance of having a high index of suspicion when observing patients with persistent back or radicular pain, and particularly those with a history of known systemic malignancies.[6] Bowel and bladder dysfunction may develop prior to diagnosis in as many as half of patients with spinal cord compression. Patients with compression at the level of the conus medullaris may present with isolated sphincter dysfunction, but it is far more common to see associated lower extremity impairments. The neurologic assessment in these patients must be thorough and should include an evaluation of bladder function, including cystometric studies, if any deficits are suspected.

Diagnosis

In reviewing primary neoplasms of the spine, pain and weakness are the most common presenting complaints (Table 10-1).[7] Pain is present in nearly 85% of patients, and 20% report radicular symptoms. At the initial evaluation, objective neurologic deficits are identified in approximately 35% of patients with benign tumors and in up to 70% of patients with malignant tumors. A palpable mass is rarely found in patients.

Table 10-1 Patterns of Presenting Symptoms

Presenting symptom	No.	%
Back pain	25	30.5
Radicular pain	8	9.75
Weakness	7	8.5
Back pain and weakness	15	18.3
Radicular pain and weakness	8	9.75
Mass	4	4.9
Pain and mass	9	11.0
Pain, weakness, and bowel and bladder dysfunction	4	4.9
Asymptomatic	2	2.4
Total	82	100

Diagnostic Techniques

Radiographs

Radiographs should be obtained in any case where neoplasm is suspected. Anteroposterior (AP) and lateral views of the involved vertebra can provide considerable information about the nature and behavior of the lesion, and may be sufficient to identify some characteristic tumor types. An early vertebral lesion may be difficult to detect since radiographic evidence of bone involvement is not apparent until 30% to 50% of its regular bone has been destroyed. The most classic early sign of vertebral involvement is the "winking owl" sign seen on the AP view. The loss of the pedicle ring unilaterally results from the destruction of the cortex of the pedicle, usually by the tumor invading from the vertebral body proper.

Vertebral collapse, secondary to erosion of bone by the tumor, is another common radiographic finding. A pathologic compression fracture may be difficult to differentiate from a traumatic injury, particularly in patients with osteoporosis. But a periosteal reaction that seems too old for the acute trauma, whether the presence of a soft-tissue mass or soft-tissue calcification, should alert the physician to perform a more definitive diagnostic test. Distinguishing neoplastic vertebral destruction from pyogenic osteomyelitis may also prove difficult. The differential diagnosis is often based on the preservation of the intervertebral disk in patients with neoplasm. The disk is highly resistant to tumor invasion, and disk height is usually maintained even after extensive vertebral collapse.

Bone Scan

A bone scan is a sensitive test for neoplastic disease of the spine and can depict lesions as small as 2 cm.

Computed Tomography

Computed tomography scanning offers improved sensitivity in the detection of spinal neoplasms. Computed tomography is highly sensitive in the demonstration of ulcerations and bone, depicting neoplastic involvement far more reliably than plain radiography. For surgical staging and planning, CT often remains as the most helpful imaging modality.[7-11]

Myelography

Once the only reliable way to assess spinal cord and nerve root compression, myelography has now given way to MRI in most cases. However, in combination with CT scanning, myelography remains extremely helpful in the ability to depict spinal cord compression due to fracture and bony soft-tissue impingement. It is often less precise in the presence of stainless steel devices producing metal artifacts or scatter, thus preventing early detection of recurrences.

Magnetic Resonance Imaging

Magnetic resonance imaging has proven useful in evaluating a variety of spinal diseases and is well tolerated, noninvasive, and safe. The ability to quantify tumor growth on T2-weighted images to provide superior soft-tissue detail, and to obtain multiplanar images, has enhanced the diagnostic and treatment capabilities considerably.

Biopsy Techniques

The importance of a carefully planned biopsy, not unlike that in the extremities, must not be underemphasized in the spine. To date, the tissue of biopsy in tumor surgery of the spine has been ignored. The hazards of biopsy in the extremities have been discussed extensively in that it significantly alters treatment in one of every three patients; this fig-

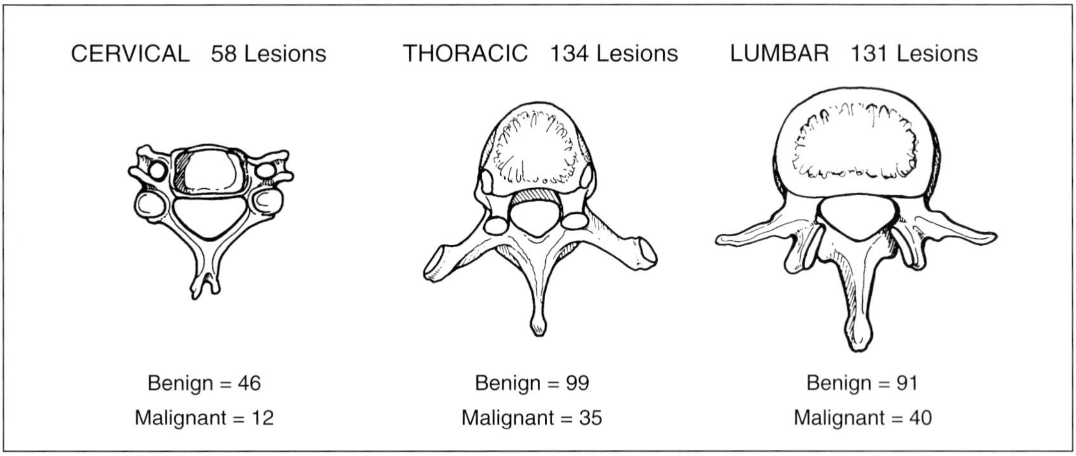

Fig 10-1 Primary spine tumors by location (Rizzoli Institute series. From Boriani S, Weinstein JN. Differential diagnosis and surgical treatment of primary, benign, and malignant neoplasms. In Frymoyer JW, Ducker TB, Hadler NM, Kostuik JP, Weinstein JN, Whitecloud TS III (eds): The Adult Spine: Principles and Practice. Philadelphia: Lippincott-Raven; 1997:951–987. Used with permission.)

ure may be even higher in spinal lesions.[12] This risk can be reduced significantly when the biopsy of the spine is performed by the treating rather than the referring physician.

Most specimens of lesions of the spinal column are obtained by incisional or needle biopsy techniques with CT guidance. There are times when biopsies are not indicated for primary spinal tumors. Radiographic and other staging procedures often allow the surgeon to perform definitive surgical procedures at one setting, for example, in the case of osteoid osteoma. In many cases, this avoids the hazards of biopsy, but at the same time enhances the opportunity for applying good tumor surgery principles.

Primary Tumors of the Spine

Primary tumors of the spine are uncommon. At the Rizzoli Institute, 323 primary spine neoplasms were seen over the past 50 years; 236 were benign. Of that 236, there were 13 benign tumor types identified (Fig 10-1). Five-year survival is reported at 86% in benign tumors vs 24% for patients with malignancies of the spine.[7]

Benign Tumors of the Spine

Osteochondromas (exostoses)

Vertebral involvement occurs in approximately 7% of patients, but neurologic compromise is only rarely seen. Sixteen cases of osteochondroma have been reported with spinal cord compression. Sixty percent of those lesions arose in the cervical spine and 19% in the thoracic spine or below. Although nearly 90% of those patients experience good neurologic recovery following decompression, two patients died of cervical cord compression and respiratory failure secondary to an exostoses of the upper cervical spine. Because these lesions are slow to progress, surgical excision of these tumors often provides excellent recovery of neurologic function (Fig 10-2).

Osteoblastoma/Osteoid Osteoma

In the spine, these benign lesions show a propensity for the posterior elements. In one review of 36 patients with osteoid osteoma,

Fig 10-2 Osteochondroma arising from the transverse process of L-5.

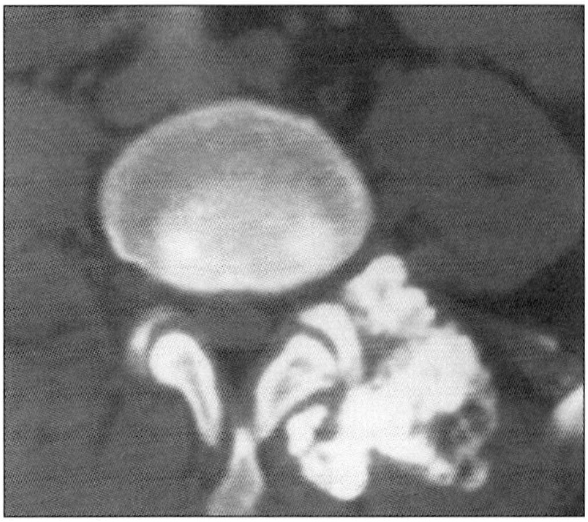

Fig 10-2a Computed tomographic (CT) scan.

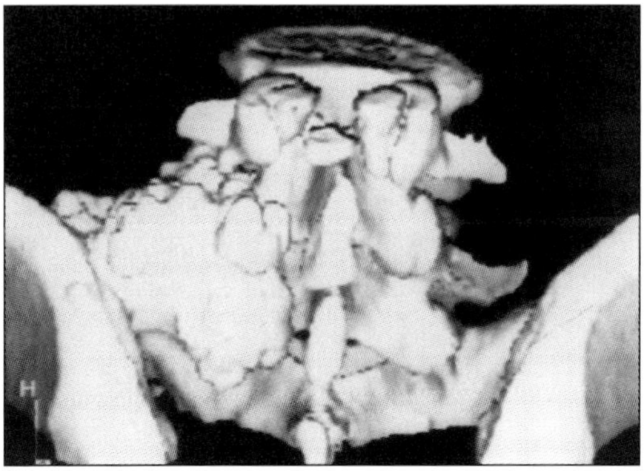

Fig 10-2b 3-D CT scan picture.

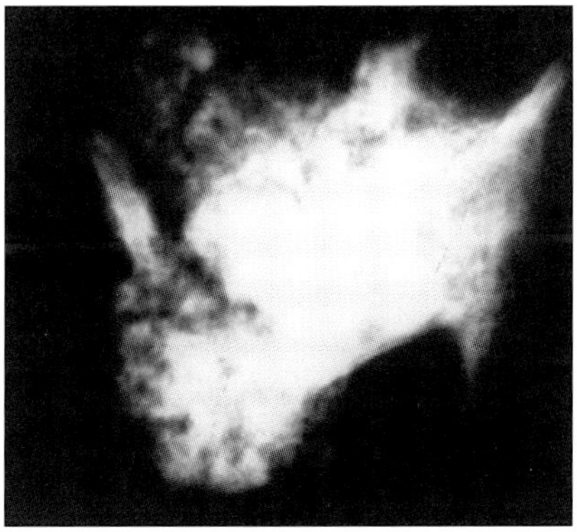

Fig 10-2c Radiograph of the resected specimen.

25% occurred in the spine located in the posterior elements.[3] Osteoblastoma shows an even greater propensity for the spine. In a review by Marsh et al,[13] 41% of the lesions in the series were located in the spine and all originated in the posterior elements. In the thoracic or lumbar spine, 80% of the cases of painful scoliosis were associated with osteoid osteoma (Fig 10-3). Scoliosis is associated with about 40% of thoracic and lumbar osteoblastomas. The remaining patients only complain of night pain. In Pettine and Klassen's series of 41 patients with osteoid osteoma or osteoblastoma of the spine, 63% had a significant scoliosis.[14] Nearly all patients in that series had improvement or resolution of their scoliosis when the tumor was removed within 15 months of the onset of symptoms, but only one patient in 11 had any improvement when symptoms had been present more than 15 months.

Radiographic demonstration of an osteoid osteoma can be difficult in the spine. The lesion, by definition, is less than 2 cm and is easily obscured by the overlapping shadows of the vertebral column. Computed tomography usually demonstrates the lesion very well (Fig 10-4). The most sensitive test for locating an osteoid osteoma is a bone scan. The treatment of osteoid osteoma is extralesional excision (Fig 10-5). It does not require removal of the intralesional sclerotic reaction, but the

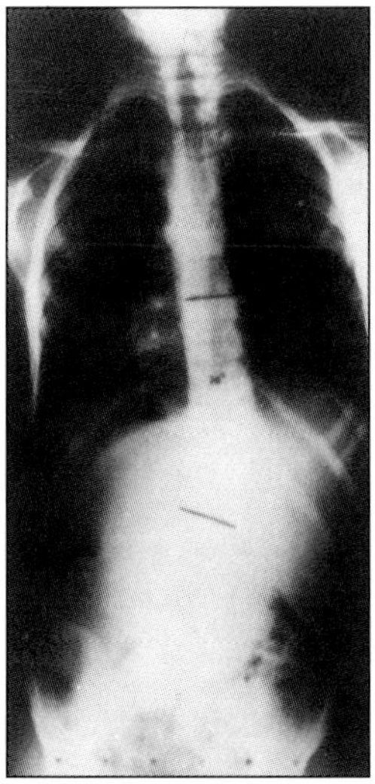

Fig 10-3a A 13-year-old boy with back pain caused by osteoid osteoma. Anteroposterior radiograph demonstrates thoracolumbar scoliosis.

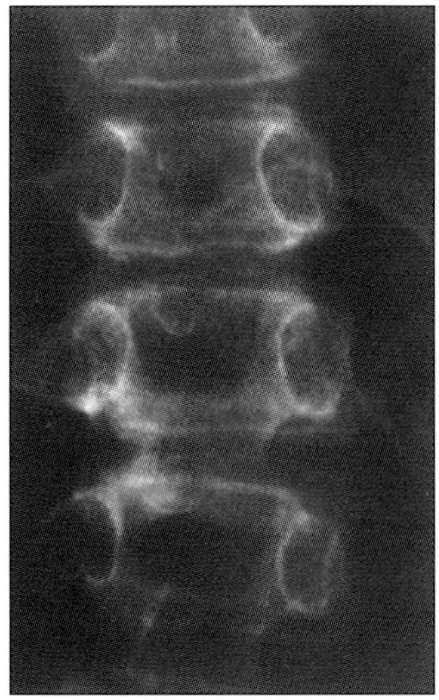

Fig 10-3b Radiograph in more focused view demonstrates a probable nidus in the posterior elements of T-12 on the left.

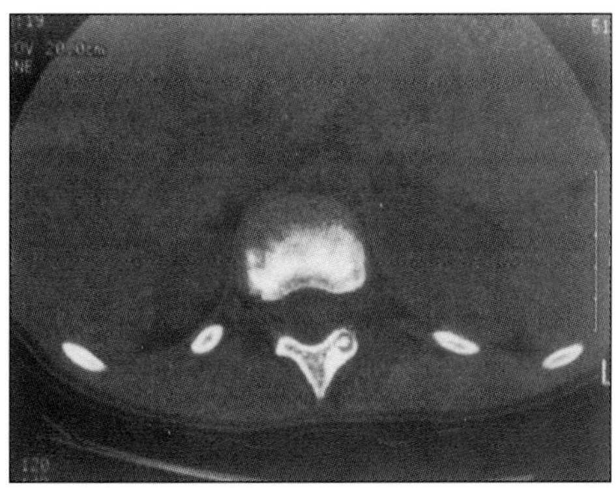

Fig 10-3c Computed tomographic scan of the patient demonstrates a nidus consistent with an osteoid osteoma in the posterior elements of T-12 on the right. The lesion was excised, resulting in complete relief of pain and resolution of the scoliosis.

tumor must be completely removed. Excision provides reliable pain relief, and the spinal deformity is improved or disappears in the majority of patients.

For osteoblastomas, curettage had been advocated in the past as the treatment of choice,[13,15] but, in our view, osteoblastomas must be treated according to their oncologic staging: intralesional excision is adequate for stage 2 lesions even if it doesn't prevent recurrence. Radiation therapy, as an adjuvant to excision, is required in stage 3 lesions, since intralesional excision is not adequate. Marginal resection, when feasible, is the best oncologic option (Fig 10-6).

Aneurysmal Bone Cyst

The spine is the preferred site of occurrence for this pseudotumoral condition,[16] found most commonly in the thoracolumbar spine and involving the posterior elements in approximately 60% of the cases. These lesions have a tendency to involve adjacent vertebra[13] and may involve three or more vertebrae in sequence (Fig 10-7a–e).

Fig 10-4 Osteoid osteoma.

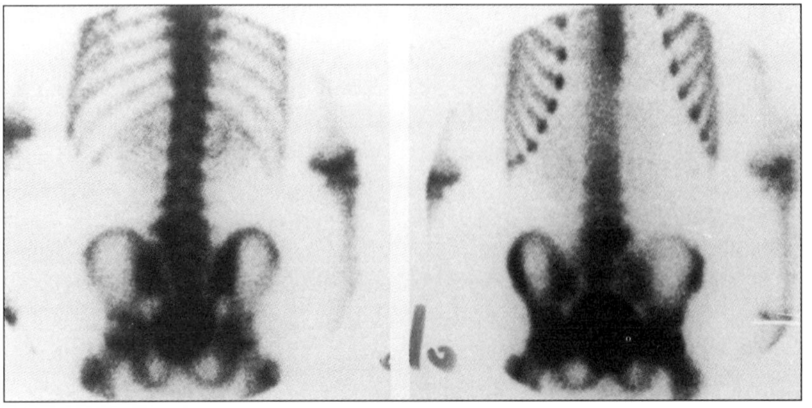

Fig 10-4a Posterior (*left*) and anterior (*right*) prone technetium bone scan of an 8-year-old boy who had night pain relieved by aspirin for approximately 6 months. The pain, however, interfered with daily activities. He constantly complained of pain during school, but more often at night. Increased uptake is evident at L-5 (*left*).

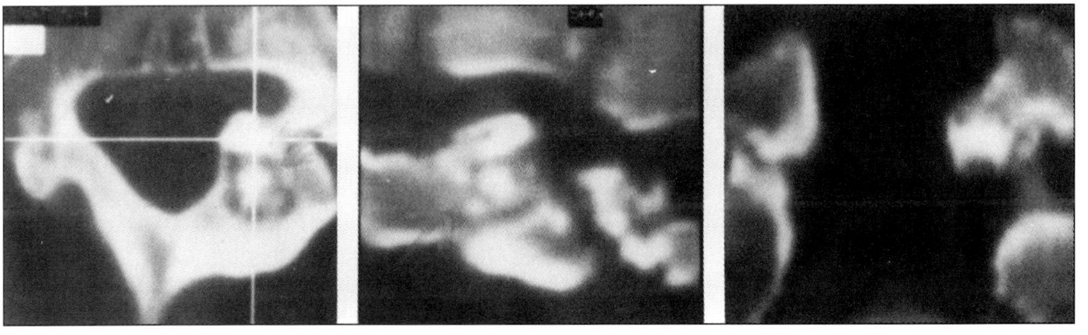

Fig 10-4b A multiplanar CT scan of the same patient identifying the nidus of the osteoid osteoma in the posterior elements in the axial (*left*), sagittal (*center*), and coronal (*right*) views.

Fig 10-5 Osteoid osteoma.

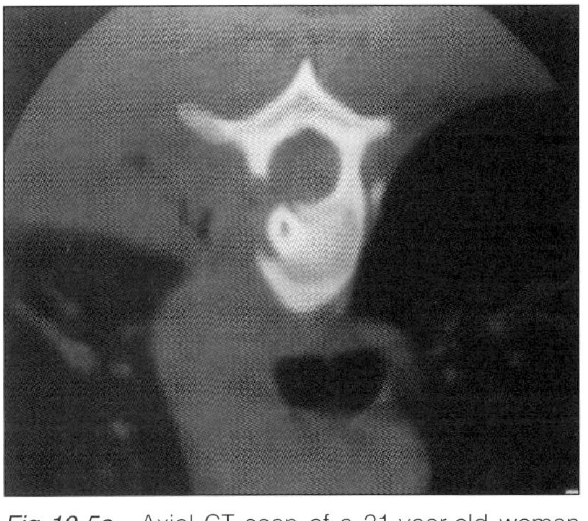

Fig 10-5a Axial CT scan of a 21-year-old woman with an osteoid osteoma in the posterior part of the left side of the vertebral body encroaching on the spinal canal.

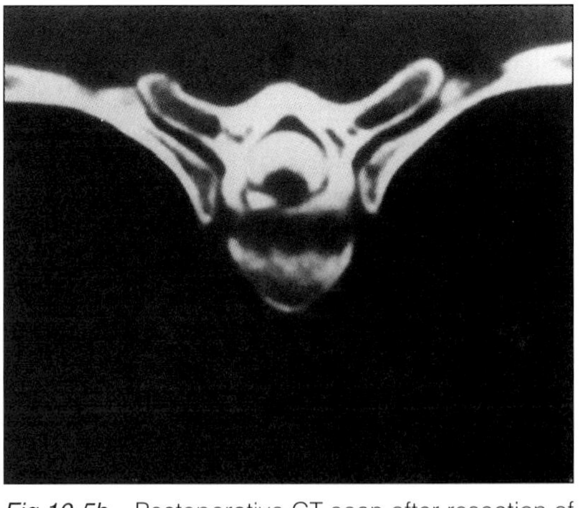

Fig 10-5b Postoperative CT scan after resection of the osteoid osteoma and placement of a rib strut graft.

Fig 10-6 A T-7 osteoblastoma.

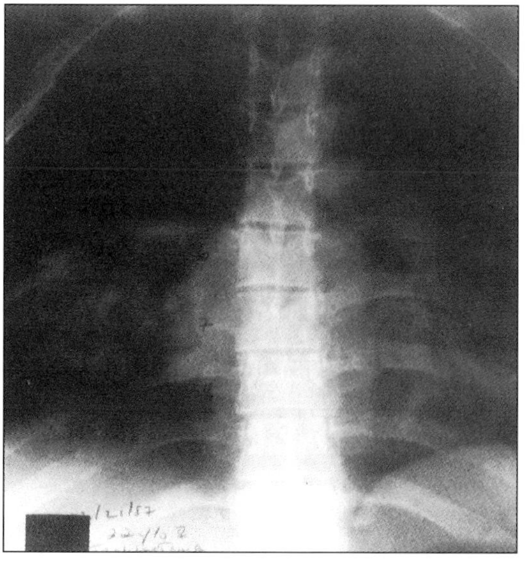

Fig 10-6a Anterior radiograph shows the expansile lesion from the body of T-7 including and up to the body of T-6; the ribs of T-7 and T-8 are clearly involved. Some calcification of the lesion is apparent.

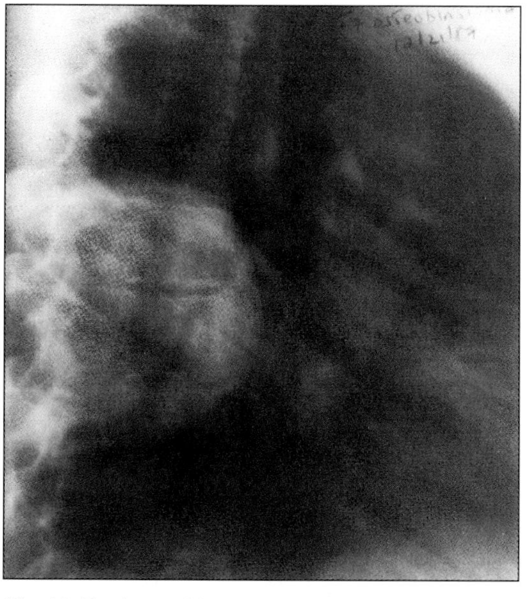

Fig 10-6b Lateral image shows a large expansile mass in the area of T-6, T-7, and T-8, with some calcification.

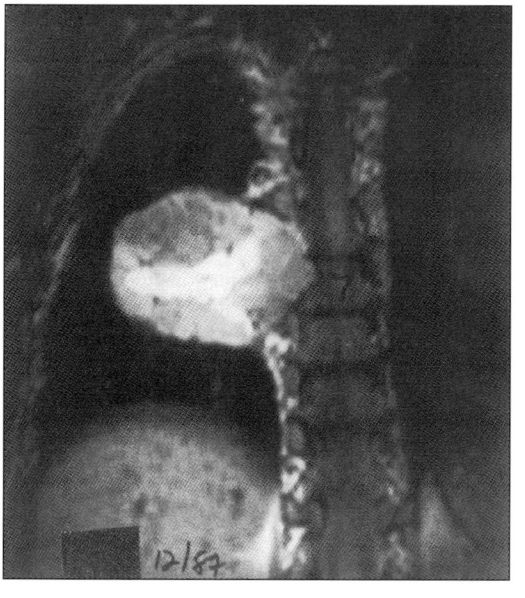

Fig 10-6c Coronal MRI shows growth of a large osteoblastoma off the vertebral bodies of T-6 and T-7 with some central calcification.

Fig 10-7 Plain anteroposterior (a) and lateral (b) radiographs of L-4 tumor involving the left L-4 transverse process and pedicle. Diagnosis: osteoblastoma/aneurysmal bone cyst/solid. Computed tomographic scan (c) and MRI (d) of the lesion preoperatively. (e) Computed tomographic scan and AP and lateral radiographs following excision of this solid aneurysmal bone cyst.

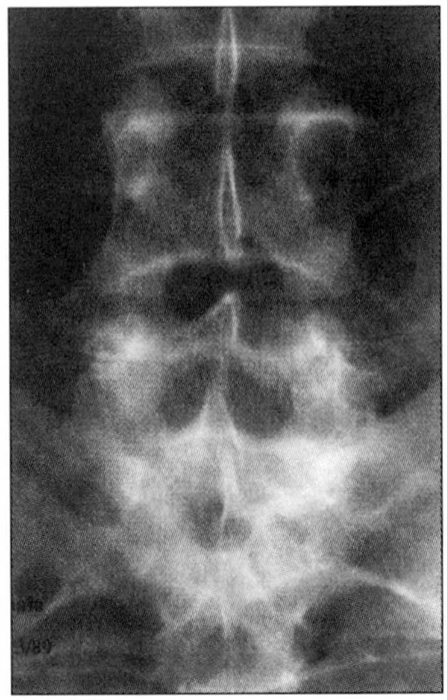

Fig 10-7a

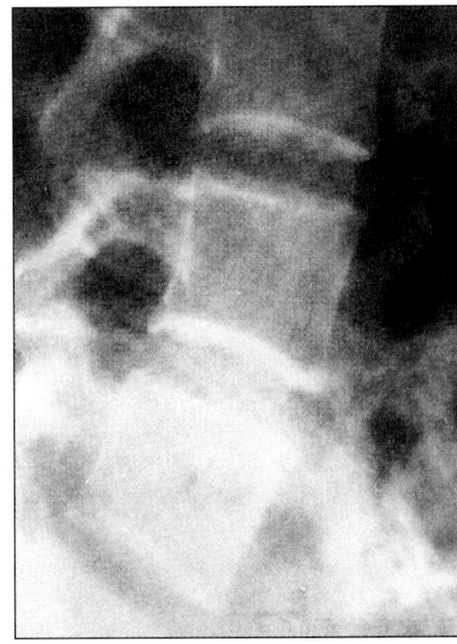

Fig 10-7b

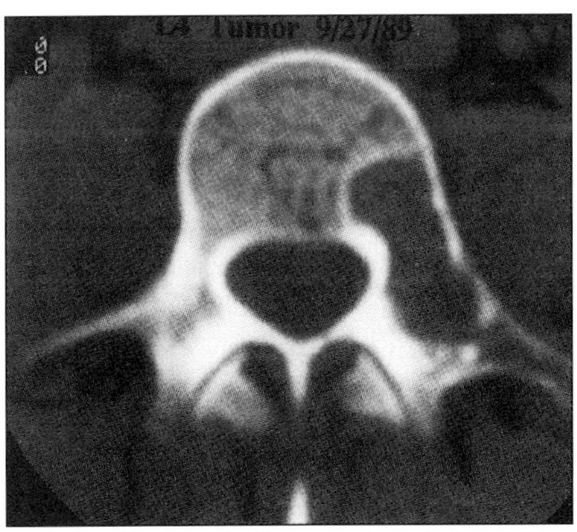

Fig 10-7c

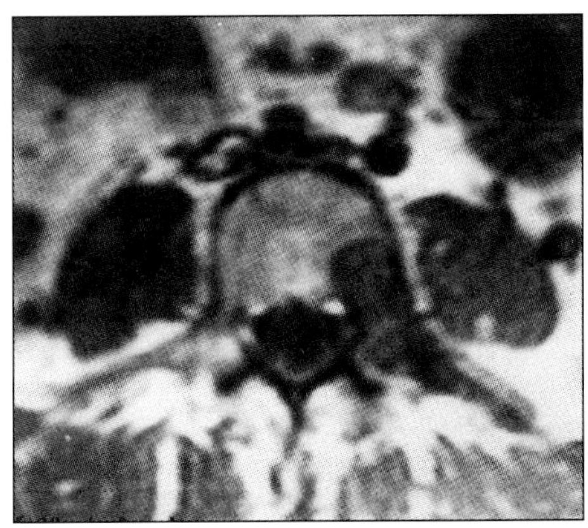

Fig 10-7d

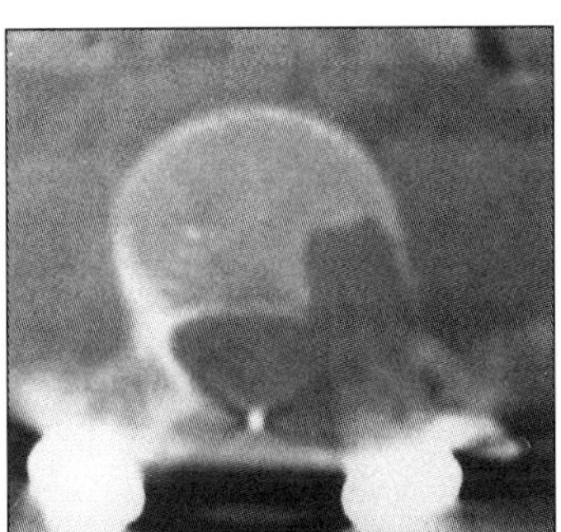

Fig 10-7e

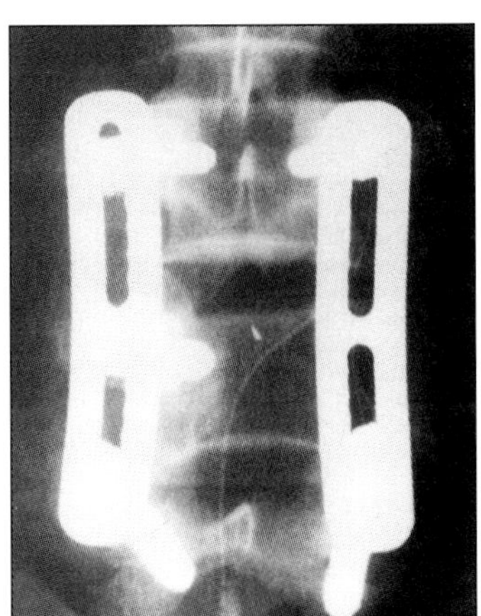

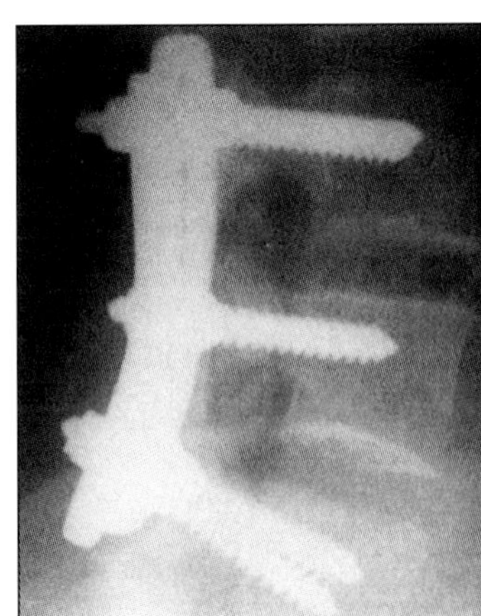

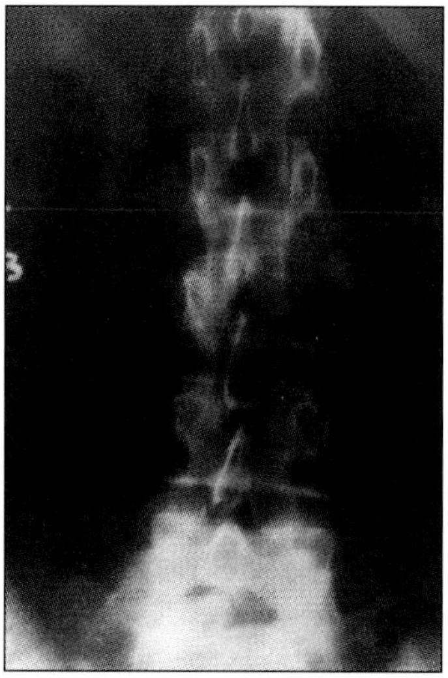

Fig 10-8a Anteroposterior radiograph of hemangioma T-8.

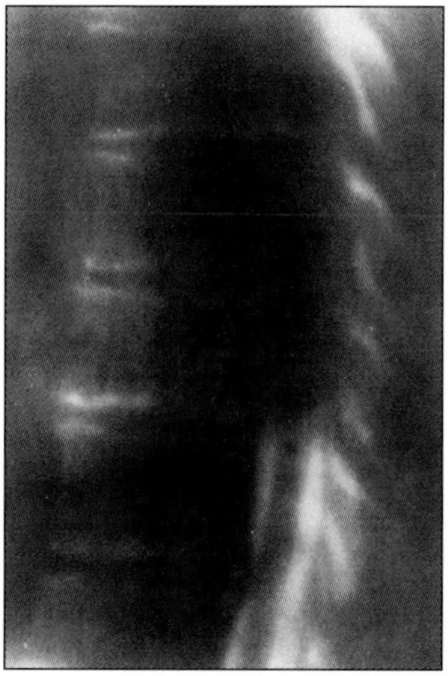

Fig 10-8b Lateral thoracic myelogram demonstrating obstruction of contrast flow. Diagnosis: hemangioma at T-8 with soft-tissue extension.

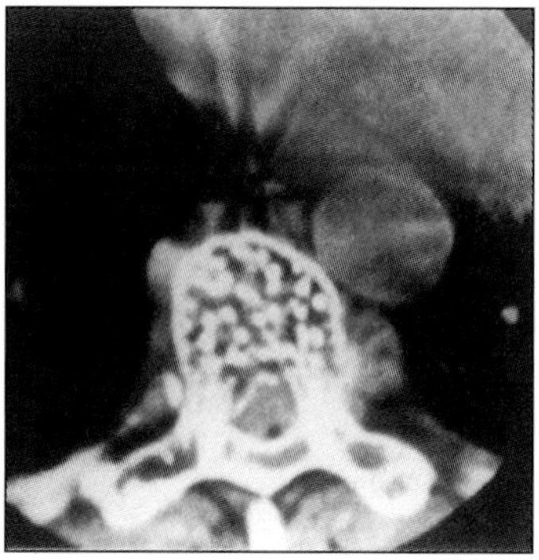

Fig 10-8c Computed tomographic scan of T-8 hemangioma with extradural extension.

Radiographs typically demonstrate an expansive osteolytic cavity and a bubbly appearance of the bone. The cortex is often eggshell thin and blown out. In our experience over the past few years, selective arterial embolization has become the first line of treatment capable of healing, or at least involuting the cyst, and at the very least reducing the intraoperative bleeding potential.

Hemangioma

Hemangiomas are found at autopsy in approximately 10% of people. These are rarely symptomatic and/or of clinical importance. Reports of deformity and/or pain associated with vertebral hemangiomas are uncommon; however, cases of soft-tissue extension with nerve root compression have been documented (Fig 10-8). These are easily diagnosed with plain radiographs, although CT or MRI will provide more information about nerve root or spinal cord impingement. Ordinary radiographs classically show prominent vertical striations produced by abnormally thickened trabeculae.

When symptomatic, these lesions frequently respond to low-dose radiation alone, but selective arterial embolization seems a safer and more effective kind of treatment. Angiography of the spinal cord is mandatory for lesions of the thoracolumbar area to identify the primary vascular supply to the cord (Fig 10-9). When spinal cord compression develops and surgical treatment is considered, embolization or operative ligation[17] should be performed before curettage.

Giant Cell Tumors

Tumors of the spine are generally benign and usually present as stage 2 or 3 lesions, but they are sometimes unpredictable. Spinal involvement is usually seen in patients in their third and fourth decades of life. Symptoms are often present for months to years before the patient sees a physician. Pain is a constant complaint. Cord compression is seen in

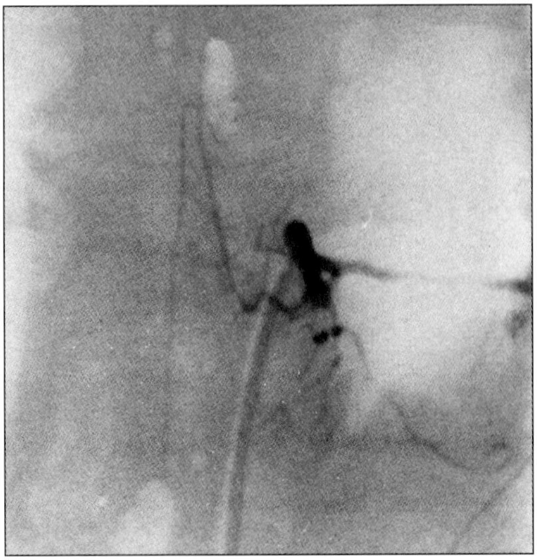

Fig 10-9 Angiogram demonstrating the normal artery of Adamkiewicz. Note the smaller ascending branch and larger descending branch forming the ASA. This gives the characteristic "hairpin" configuration (courtesy of Dr. Alfredo Lazo, Harper-Grace Hospitals, Detroit, Michigan).

about 30% of cases. Plain radiographs usually demonstrate an area of focal rarefaction and giant cell tumors are usually seen in the vertebral body, but may expand the surrounding cortical bone extensively as the tumor enlarges (Fig 10-10). Computed tomography is especially important in the evaluation of these tumors since complete resection is very important to eradicate local disease and prevent recurrence. It is also essential to use CT or MRI to identify a recurrence. Therefore, instrumentation that is radiologically compatible is recommended, ie, titanium or carbon fiber. En-bloc resection of these lesions is the treatment of choice in stage 3 lesions, completely avoiding the risk of adjuvant radiation therapy and/or chemotherapy.

Eosinophilic Granuloma

Eosinophilic granuloma often produces focal destruction of bone,[18] which is often benign and sometimes self-limiting. It is most commonly seen in children before the age of 10. Lesions of the skull are most common, but any bone may be affected. Vertebral involvement occurs in approximately 10% to 15% of the cases. The ensuing vertebral destruction is classically referred to as vertebra plana following complete collapse of the vertebral body. This is where the vertebral body collapses and settles, and the radiographs demonstrate the flattened disk of cortical bone retained between the two intact adjacent intervertebral disks.[16]

On rare occasions, vertebral collapse produces pain and focal spasm, and in the cervical spine torticollis is not uncommon. Mild-to-moderate kyphosis is frequently observed. Neurologic symptoms may or may not develop with associated vertebral collapse. Diagnosis of eosinophilic granuloma is often confirmed with findings of open needle biopsy. Treatment is somewhat controversial, but it is clear that many lesions will heal without any treatment, or at least after biopsy. Local corticosteroid injection after frozen section has been advocated by some as the treatment of choice, while surgical decompression is indicated in rare cases of cord compression. Courses of chemotherapy are only indicated in disseminated forms of histiocytosis.

Malignant Tumors

Malignant tumors arose almost exclusively from the body (85%), whereas the prevalence of benign tumors from the body dropped to 70%. Of the benign tumors, 58.7% arose in the first two decades of life, vs 27.7% of malignant neoplasms.

Malignant Multiple Myeloma and Solitary Plasmacytoma

Multiple myeloma is an uncommon neoplastic process with an incidence of two or three cases per 100,000 people among the general

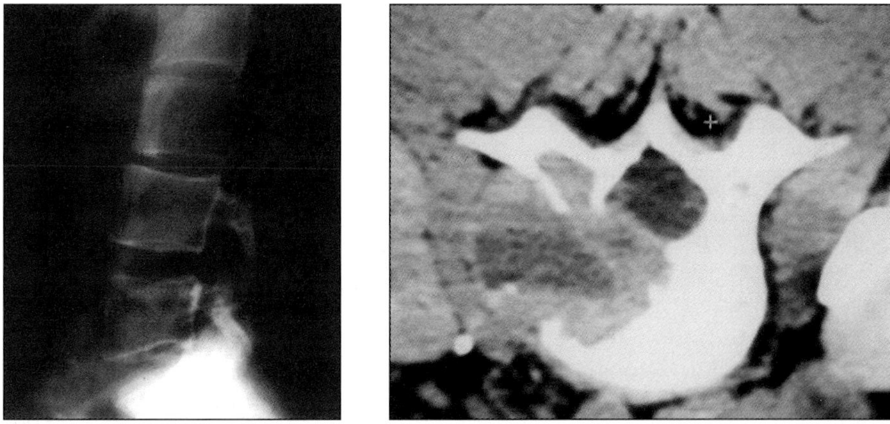

Fig 10-10 a and b Giant cell tumor at L-2. Plain radiograph (a) and axial CT (b) revealing a stage 3 lesion.

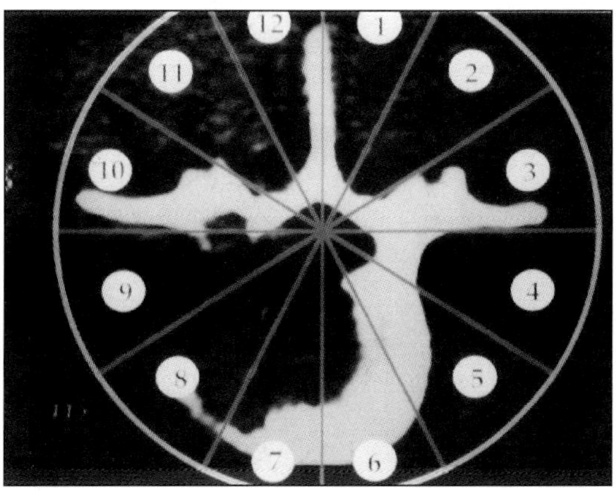

Fig 10-10c According to the WBB surgical staging, the lesion occupies the radiating zones 6 to 10, layers A to D.

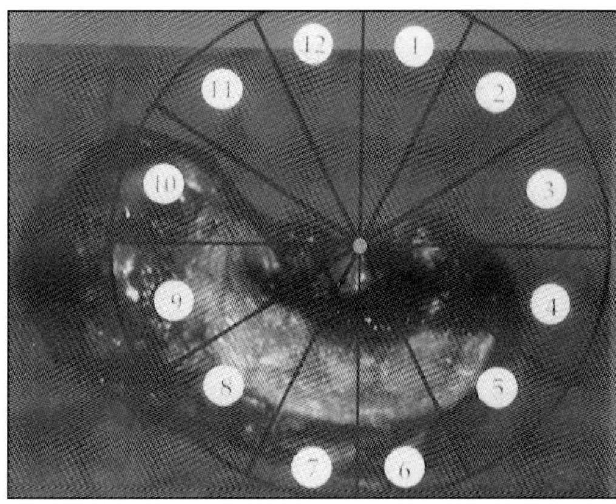

Fig 10-10d Resected specimen after en-bloc resection of the vertebral body (marginal margin) by double approach vertebrectomy. Reconstruction was achieved by posterior pedicular plates, anterior plate, and allograft.

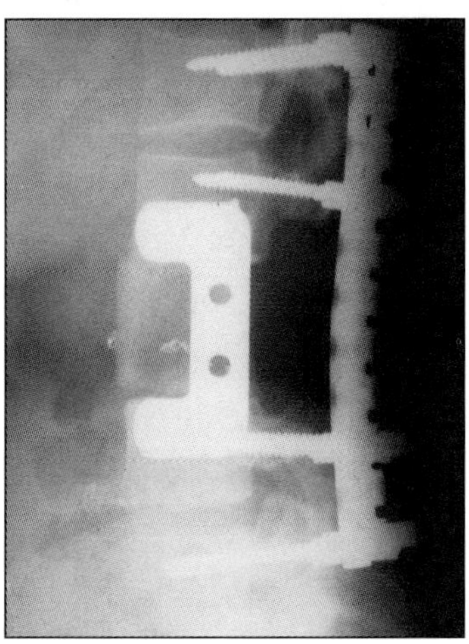

Fig 10-10e No recurrence 36 months later.

population. True solitary plasmacytoma is a rare entity, representing only 3% of all plasma cell neoplasms.[19] Patients with solitary and multiple myeloma differ in terms of age and sex distributions and survival.[20] Although the course of multiple myeloma is usually rapidly progressive and lethal, patients with solitary plasmacytoma may have prolonged survival despite eventual progression. The treatment of solitary plasmacytoma must, therefore, be considered somewhat differently than the treatment of a spinal focus of multiple myeloma.

The spinal lesion in multiple myeloma represents a metastasis in a progressive, systemic disease that commonly results in death within 2 to 3 years despite local or systemic treatment. The prognosis for survival of patients with disseminated myeloma is poor; the 5-year survival rate is 18%, and the median survival is 28 months. In cases involving the spinal column, the outcome is even worse. Valderrama and Bullough reported that 76% of such patients were dead within a year of diagnosis, and all were dead within 4 years.[21]

The treatment of choice of the affected vertebra with solitary plasmacytoma and/or in multiple myeloma is radiation therapy.

Primary Osteosarcoma

This is a high-grade tumor typical of those most active in the metaphysis of long bones (knee and shoulder) in the adolescent. Long-term survival has dramatically improved, and it is now possible to perform limb-salvage surgery owing to preoperative intra-arterial multiple-drug chemotherapy. This treatment has yet to be performed effectively in the spine, the site of occurrence of approximately 2% of all osteogenic sarcomas. An oncologically correct treatment of these lesions in the spinal column is difficult, since en-bloc resection without preoperative chemotherapy is frequently impossible; therefore, the outcome in these tumors remains very poor. In a review of 27 cases of osteosarcoma of the spine, median survival was 10 months from the time of diagnosis; only seven patients survived more than 1 year, and one patient survived more than 5 years.[22] Similar results were seen by Barwick et al,[8] who reported a median survival of 6 months for ten patients.

Although the early indication is that en-bloc resection of the tumor may result in longer survival, especially if combined with chemotherapy, longer follow-up is needed before assumptions can be made about our ability to cure these tumors.

Ewing's Sarcoma

This is a round cell malignant tumor typical of the first three decades. Approximately 3.5% of all Ewing's tumors arise in the spinal column, with the majority originating in the sacrum.[23] The thoracic spine is more frequently involved in the mobile spine (15 of 29 cases). Although prognosis is generally poorer than for those with an extremity lesion, long-term survival is reported.[24]

The treatment of choice for Ewing's sarcoma involves multi-agent chemotherapy and high-dose radiation therapy. The primary role of surgery is to perform en-bloc resections; these procedures sometimes become feasible after chemotherapy and radiation therapy. At times, a few courses of chemotherapy can reduce the mass dramatically and produce an almost complete necrosis. Even if wide resection is done, it is wise to perform radiation therapy. With current regimens of therapy, excellent local control and encouraging disease-free survival rates are obtainable in the spine, though the long-term outlook for most patients remains grim.[25]

Chordoma

Chordoma is a low-grade malignant tumor typically arising from the clivus and the spine as it originates from the remnants of the notochord[26,27] (Fig 10-11). It occurs predominantly in the fifth or sixth decade of life. Although this tumor is characterized by its slow but relentless local spread, it is a fully malignant lesion capable of distant metastases. Initial symptoms are usually mild and progress slowly as the tumor expands. Chordomas may reach a considerable size before metastasizing, and symptoms of dyspnea, constipation, urinary frequency, or nerve root compression may appear before patients present to their physician. In a review of 22 cases of chordoma of the mobile spine (4 in the cervical, 6 in the thoracic, 12 in the lumbar spine), 11 cases had cord compression as the presenting symptom.[28]

Kaiser et al[2] demonstrated that simply exposing the tumor during resection increases the recurrence rate of this tenacious lesion from 28% to 64%. The feasibility of wide resection is more frequent in sacrococcygeal chordoma[29] even if high sacral lesions may necessitate a colostomy and functionally important nerve roots will be sacrificed; in such cases significant morbidity should be expected. Surprisingly little morbidity of functional loss is referred, provided the S2 roots can be spared bilaterally.[30,31] In the mobile spine, the huge myxoid masses found at presentation and the surgically inaccessible sites (such as the high cervical vertebrae) make resection difficult or impossible to perform. In cases like these, intralesional excision as completely as possible is the only possible treatment, combined with neutron-beam radiation therapy.

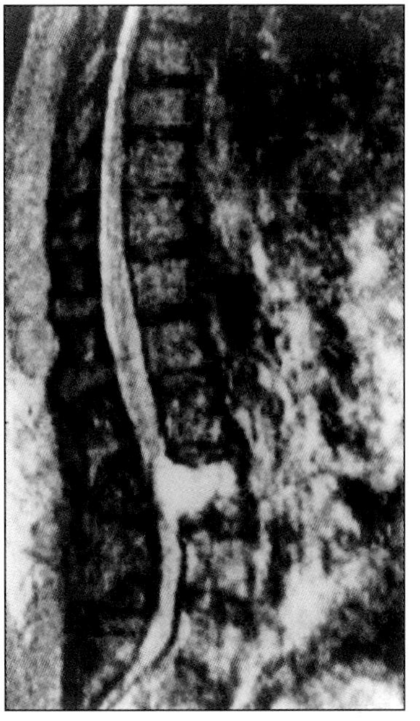

Fig 10-11 T2-weighted sagittal MRI of an L-3 chordoma.

Chondrosarcoma

Approximately 100% of chondrosarcomas arise in the spinal column. These conditions grow slowly, sometimes manifesting at the onset with signs of cord compression (9 of 19 cases observed at the Rizzoli Institute). These tumors are resistant to radiation therapy and chemotherapy, and because of the difficulties in performing wide-margin resections, lesions of the vertebral column have a poor prognosis independent of histologic grade. Long-term survival has been reported in low-grade chondrosarcoma in patients who have been treated with repeated independent excisions of recurrent disease, but few lesions can be managed this way with any success. A complete surgical excision is required to cure the patient with chondrosarcoma, although this is sometimes impossible to obtain. In some vertebral locations, this requires difficult and long procedures via a combined anterior and posterior approach (Fig 10-12).

Treatment of Spinal Primary Tumors Based on Oncologic Staging and Anatomic Extent

The tumors of the spine are oncologically staged in the same way as in the long bones, while the surgical staging is very different.

Fig 10-12 a–d Computed tomographic scan of axial and coronal images of a chondrosarcoma arising from the transverse process of L-5, invading asymmetrically L-4 and the sacroiliac joint, radiating zones 9 to 11, layers A to C. Submitted to en-bloc sagittal resection by double combined surgical approach, according to the planning shown in c and d. Wide margin achieved.

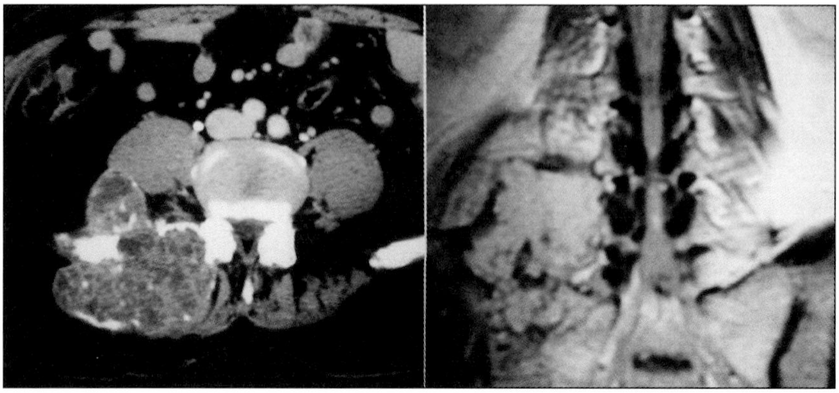

Fig 10-12 a and b

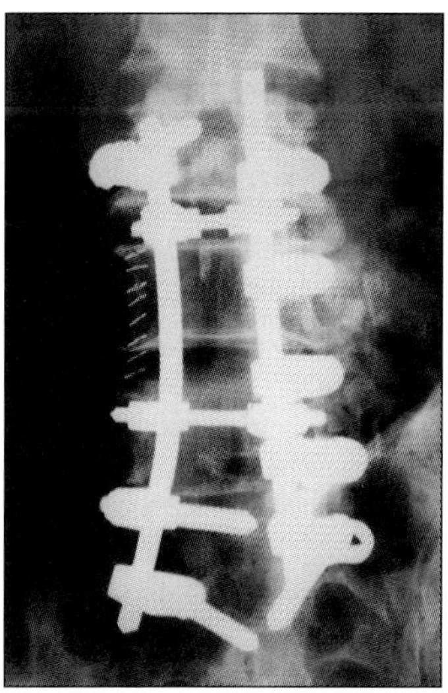

Fig 10-12c

Fig 10-12d

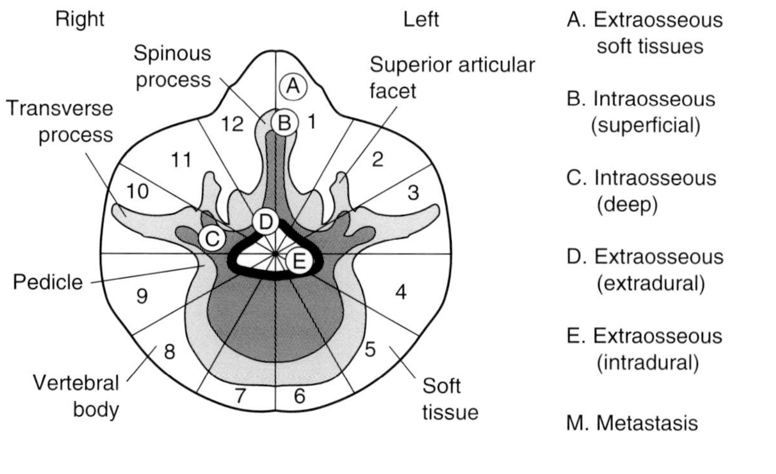

Fig 10-13 Anatomic extent of spine tumors by radiating zones (locations) and layers.

Surgical Staging

A surgical staging system describing the anatomic extent of a spinal tumor is used to provide a uniform classification to communicate data within and across institutions, and, more importantly, to plan appropriate surgical interventions aimed at attaining surgical margins devised by the Enneking staging system. Such a surgical staging system was originally introduced by Weinstein[10] and subsequently modified[32,33] and submitted to clinical evaluation.[7,8]

On the transverse plane, the vertebra is divided into 12 radiating zones (numbered 1 to 12 in a clockwise order) and into five layers (A to E from the prevertebral to the dural involvement). Also the longitudinal extent of the tumor is recorded (Fig 10-13).

Some commonly used terms: "vertebrectomy" (removal of all the elements of the vertebra), "corpectomy," and "somectomy" (removal of the vertebral body), have no significance from an oncologic viewpoint of outcome if they are not specified according to the oncologic terminology:

Excision means a piecemeal removal of the tumor.

Resection is the en-bloc removal of the tumor. This can be performed along the pseudocapsule (**marginal resection**) or outside the pseudocapsule, removing the tumor with a continuous shell of healthy tissues (**wide resection**).

Palliation is a surgical procedure aimed to a functional purpose (cord decompression, fracture stabilization) with or without parital (piecemeal) removal of the tumor.

In our experience, there are three possibilities to perform an en-bloc resection in the thoracolumbar spine: vertebrectomy, hemivertebrectomy, and resection of the posterior arch.

Vertebrectomy (marginal/wide resection of the vertebral body). A resection of the vertebral body can be performed if the tumor is confined to zones 4 to 8 or 9 to 5 (Fig 10-14), which means centrally located, and at least one pedicle free from tumor. To perform an en-bloc resection, the procedure must be performed in two stages: the posterior approach (patient supine) allows excision of

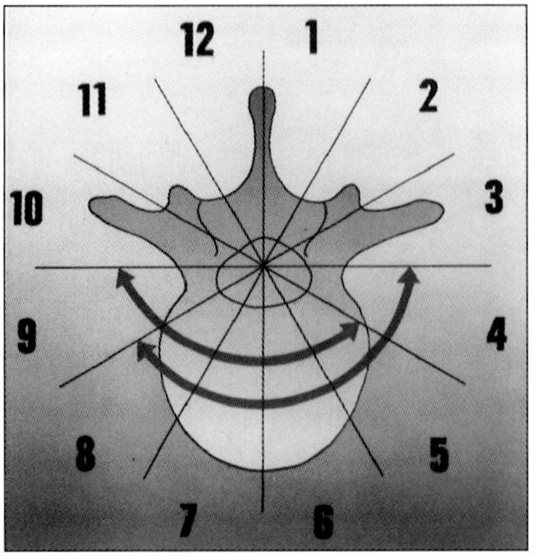

Fig 10-14 Indication to vertebral body en-bloc resection: tumor confined to locations 4 to 8 or 9 to 5: The resection is achieved by combined double approach (vertebrectomy).

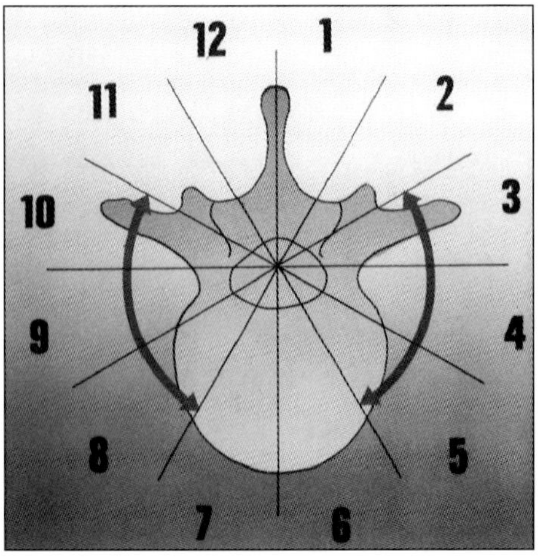

Fig 10-15 When the tumor occupies zones 3 to 5 (or 10 to 8), this means that the tumor has arisen eccentrically within the body, the pedicle or the transverse process. En-bloc resection can be performed by combined double approach (sagittal resection).

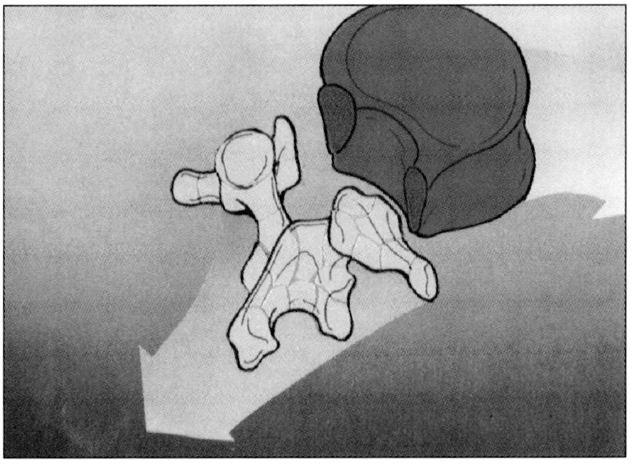

Fig 10-16 When the tumor is located between zones 10 and 3, en-bloc resection can be performed by a simple posterior approach.

posterior elements, which then enables one to section the annulus fibrosus and the posterior longitudinal ligament; and hemostasis of the epidural venous plexus. The anterior approach (transpleural thoracotomy, retroperitoneal) allows ligature of segmental vessels (at the lesional level, above and below); proximal and distal diskectomies; and en-bloc removal of the vertebral body.

Hemivertebrectomy (marginal/wide sagittal resection of the vertebra). When the tumor occupies zones 3 to 5 (or 10 to 8), this means that it arises eccentrically within the body, the pedicle, or the transverse process (Fig 10-15). In these cases, a resection of even more than one level can be performed. This resection may include, if necessary, one or more ribs, if performed in a single-stage procedure allowing a 300-degree view of a thoracolumbar vertebra.

The patient is placed in a lateral decubitus position. A "T" incision is performed; in the thoracic spine, a midline posterior incision combined with an oblique thoracic incision on the rib of the affected level is performed. In the lumbar spine, the lateral branch is the classic retroperitoneal approach. The posterior healthy structures are removed, including the pedicle (to make room for the dural displacement). The nerve root(s) of the affected segment is (are) ligated, if necessary. The vertebra is cut by chisel or osteotome far from the tumor (at least one zone free from tumor) after protecting the major vessel (isolated by the anterior approach), obtaining an en-bloc resection.

Marginal/wide resection of the posterior arch. When the tumor is located between zones 10 to 3, en-bloc resection can be performed by a simple posterior approach (Fig 10-16). To achieve this result, a wide laminectomy must expose the dural sac above and below the tumor. Lateral dissection must expose the pedicles, which are sectioned by osteotome or Gigli saw.

The approach to these rare, primary benign and malignant tumors of the spine is quite different than that in the limbs. First, one must obtain a diagnosis; second, staging of the lesion must be done to decide the adequate margins; and third, planning surgery to achieve those margins with use of these criteria must be performed. This third goal is the major difference between spinal tumor surgery and limb tumor surgery. An oncologic and surgical staging system for the spine is therefore required.

The main aim must be to treat tumors appropriately from an oncologic standpoint. To this end, techniques for surgical en-bloc resection of the mobile spine have been developed and proven effective. This has been made possible by a surgical planning system that aids the surgeon in achieving appropriate oncologic objectives. Stability of the spine and associated neurologic symptoms are solved as a consequence, but they are not the first objective of treatment. Treating these problems before achieving the correct diagnosis will prevent one from the correct treatment paradigm and exposes the patient to a potential risk of recurrence and dissemination of disease (Fig 10-17).

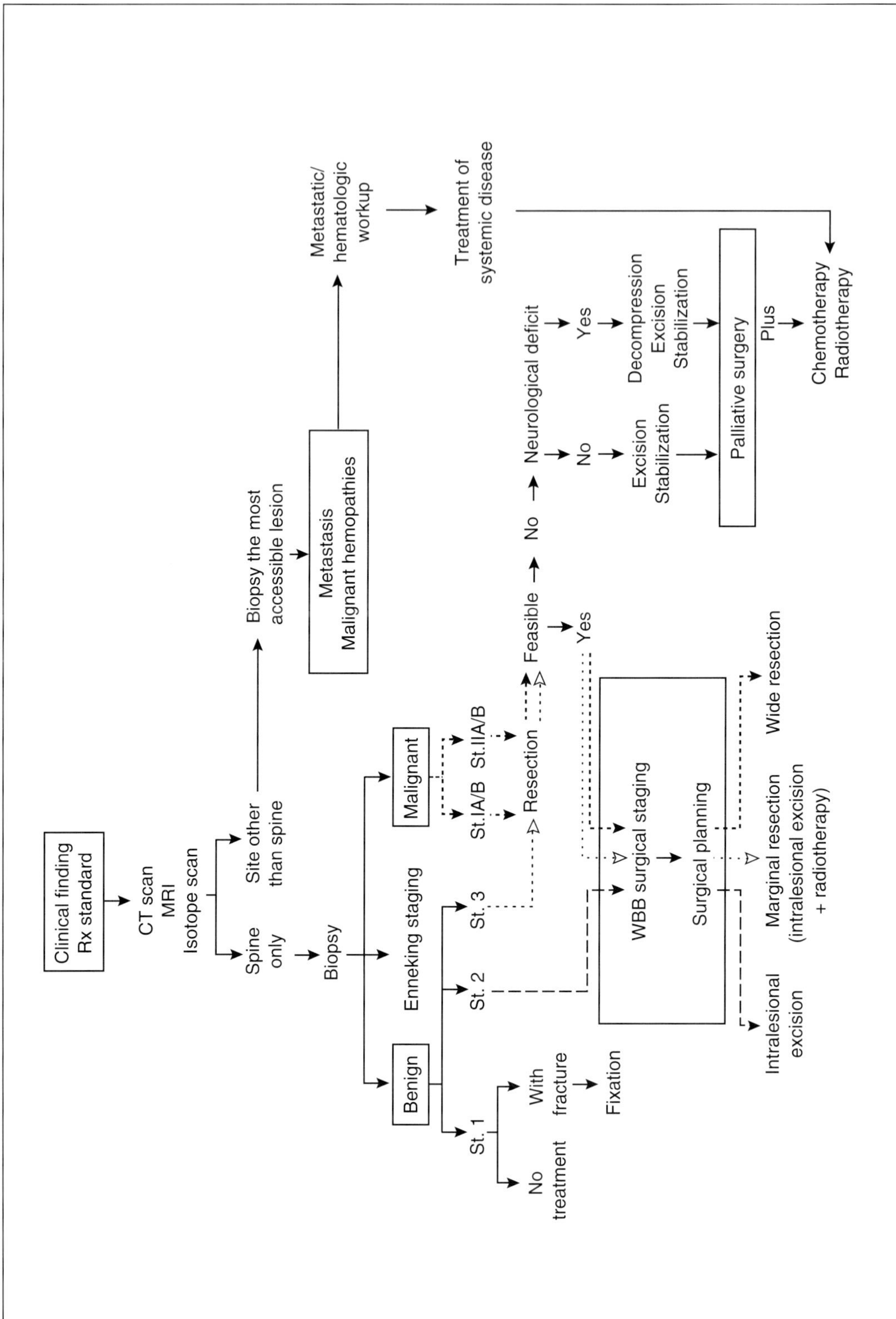

Fig 10-17 Algorithm for approach to spine tumors.

References

1. Sim FH, Dahlin DC, Stauffer RN, Laws ER. Primary bone tumors simulating lumbar disc syndrome. Spine 1977;2(1):65–74.
2. Akbarina BA, Rooholamini SA. Scoliosis caused by benign osteoblastoma of the thoracic or lumbar spine. J Bone Joint Surg 1981;63A(7):1146–1155.
3. Keim HA, Reina EG. Osteoid-osteoma as a cause of scoliosis. J Bone Joint Surg 1975;57A(2):159–163.
4. Capanna R, Boriani S, Mabit Ch, Donati D, Savini R. Osteoid osteoma of the spine. Rev Chir Orthop 1991;77:545–550.
5. Waxman D. Bone scans are of sufficient accuracy and sensitivity to be part of the routine work up prior to definitive surgical treatment of cancer. In Van Scoy-Mosher MB (ed): Medical Oncology: Current Controversies in Cancer Treatment. Boston: Hall Medical; 1981:69–76.
6. Springfield DS, Enneking WF, Neff JR, Makley JT. Principles of tumor management. Instructional Course Lect 1984;33:1–25.
7. Weinstein JN, McLain RF. Primary tumors of the spine. Spine 1987;12(9):843–851.
8. Barwick KW, Huvos AG, Smith J. Primary osteogenic sarcoma of the vertebral column: A clinicopathologic correlation of ten patients. Cancer 1980;46:595–604.
9. Campanacci M, Boriani S, Savini R. Staging, biopsy, surgical planning of primary spine tumors. Chir Org Mov 1983;75:99–103.
10. Weinstein JN. Surgical approach to spine tumors. Orthopedics 1989;12(6):897–905.
11. Weinstein JN, Hart R, Boriani S, Biagini R, Currier B. Spine tumors: Surgical staging and clinical outcome. Application to giant cell tumors of the spine. Presented at the 21st ISSLS Annual Meeting, Seattle, Washington, 1994.
12. Mankin HJ, Lange TA, Spanier SS. The hazards of biopsy in patients with malignant primary bone and soft-tissue tumors. J Bone Joint Surg 1982;64A:1121–1127.
13. Marsh BW, Bonfiglio M, Brady LP, Enneking WF. Benign osteoblastoma: Range of manifestations. J Bone Joint Surg 1975;57A(1):1–9.
14. Pettine KA, Klassen RA. Osteoid-osteoma and osteoblastoma of the spine. J Bone Joint Surg 1986;68A(3):354–361.
15. Griffin JB. Benign osteoblastoma of the thoracic spine. Case report with fifteen-year follow-up. J Bone Joint Surg 1978;60A(6):833–835.
16. Compere EL, Johnson WE, Coventry MB. Vertebra plana (Calve's disease) due to eosinophilic granuloma. J Bone Joint Surg 1954;36A(5):969–980.
17. Hay MC, Paterson D, Taylor TK. Aneurysmal bone cysts of the spine. J Bone Joint Surg 1978; 60B(3):406–411.
18. Sherk HH, Nicholson JT, Nixon NE. Vertebra plana and eosinophilic granuloma of the cervical spine in children. Spine 1978;3(2):116–121.
19. Corwin J, Lindberg RD. Solitary plasmacytoma of bone vs. extramedullary plasmacytoma and their relationship to multiple myeloma. Cancer 1979;43: 1007–1013.
20. Bergsagel DE, Rider WD. Plasma cell neoplasms in cancer. In De Vita VT, Hellman S, Rosenberg SA (eds): Principles and Practice of Oncology, ed 2. Philadelphia: Lippincott; 1985.
21. Valderrama J, Bullough PG. Solitary myeloma of the spine. J Bone Joint Surg 1968;50B:82–90.
22. Shives TC, Dahlin DC, Sim FH, Pritchard DJ, Earle JD. Osteosarcoma of the spine. J Bone Joint Surg 1986;68A:660–668.
23. Whitehouse GH, Griffiths GJ. Roentgenologic aspects of spinal involvement by primary and metastatic Ewing's tumor. Can Assoc Radiol J 1976;27:290–297.
24. Russin LA, Robinson MJ, Engle HA, Sonni A. Ewing's sarcoma of the lumbar spine: A case report of long-term survival. Clin Orthop 1982;164: 126–129.
25. Kornberg M. Primary Ewing's sarcoma of the spine: A review and case report. Spine 1986;11:54–57.
26. Kaiser TE, Pritchard DJ, Unni KK. Clinicopathologic study of sacrococcygeal chordoma. Cancer 1984;54: 2574–2578.
27. Mindell ER. Chordoma. J Bone Joint Surg 1981;63A:501–505.
28. Boriani S, Chevalley F, Weinstein JN, Campanacci L, Delure F, Biagini R. Chordoma of the spine above sacrum: Treatment and outcome of 21 cases. Spine 1995 (in press).
29. Stener B, Gunterberg B. High amputation of the sacrum for extirpation of tumors: Principles and technique. Spine 1978;3:351–366.
30. Gennari L, Azzarelli A, Quagliuolo V. A posterior approach for the excision of sacral chordoma. J Bone Joint Surg 1987;69B:565–568.
31. Rich TA, Schiller A, Suit HD, Mankin HJ. Clinical and pathologic review of 48 cases of chordoma. Cancer 1985;56:182–187.
32. Boriani S, Weinstein JN, Biagini R. A surgical staging system for therapeutic planning of primary bone tumors of the spine: A contribution to a common terminology. Spine (in press).
33. Weinstein JN. Primary tumors. In Weinstein S (ed): The Pediatric Spine. New York: Raven Press; 1994.

Chapter 11

Spine Reconstruction, Stabilization, and Fusion

Steven R. Garfin

Surgeons are increasingly able to reconstruct the spine following instabilities created by surgery, trauma, degenerative conditions, tumors, infections, or deformities (scoliosis, kyphosis, spondylolisthesis). Some of the expansion of opportunity is related to a clearer understanding of underlying disease processes; increasing familiarity with the anatomy of the spine (in all planes) and, therefore, development of safer approaches to the spine; better preoperative visualization of pathosis as afforded by computed tomography (CT) and magnetic resonance imaging (MRI); improvement in the treatment of patients (preoperatively, intraoperatively, and postoperatively) with the help of critical care specialists, intensive care specialists, and gerontologists; better intraoperative prevention of catastrophic occurrences due to electrophysiologic monitoring of the spinal cord using somatosensory-evoked potentials (SSEP) and motor-evoked potentials (MEP); and improvement in techniques to stabilize, as well as fuse, the spine.

The available techniques to reconstruct the spine vary according to anatomic location and the surgical approach—the latter depends on the location of the pathosis, spinal cord involvement, and deforming forces. Anatomically, the type of instrumentation and the surgical approach vary depending on whether the disorder occurs in the cervical, thoracic, or lumbar spine or at the lumbosacral junction. The approach is predicated on the underlying pathosis, which can be:

1. Degenerative in nature—eg, anterior compression related to a large central disk herniation, or multilevel degenerative changes.
2. Trauma—leading to bone or disk retropulsion into the canal, or instability due to ligamentous disruption in subluxations-dislocations of a motion segment.
3. Infections—which erode the bone and lead to pathologic fractures and epidural involvement (by abscess, or bony/soft-tissue debris).
4. Tumors—which can involve the anterior or posterior elements, and lead to vertebral body collapse, or direct invasion through bone into the canal.
5. Deformities, such as scoliosis, kyphosis, and spondylolisthesis—which lead to instabilities that can be surgically treated, frequently with combined anterior and posterior surgery.

In some disorders (eg, infections of the anterior vertebral bodies/elements), surgery may be staged to debride the infected site and to insert a bone graft (reconstruction) followed by posterior instrumentation with rigid fixation and realignment of the spine, which holds the anterior graft in place by stabilizing the spine (Fig 11-1).

Fig 11-1 A 60-year-old man with L-3 osteomyelitis. He was treated with anterior corpectomy and bone graft strut fusion L2-4. Ten days later he was returned to the operating room for posterior segmental fixation.

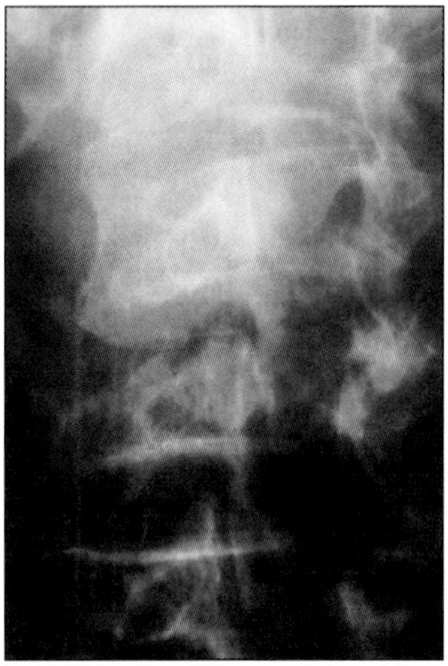

Fig 11-1a Osteomyelitis of the bone is seen in the lower portion of the figure in the compressed irregular L-3 vertebral body.

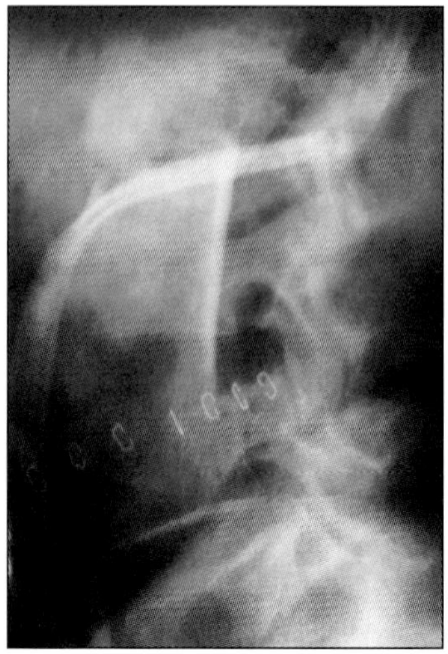

Fig 11-1b Lateral view showing the bone graft in place from L2-4 spanning the L-3 corpectomy site.

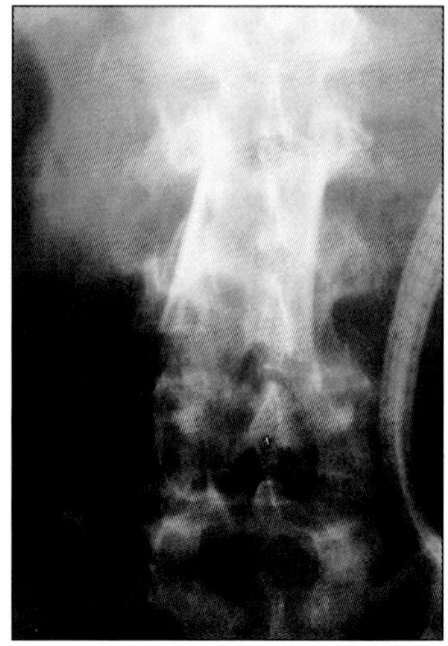

Fig 11-1c Anteroposterior radiograph demonstrating the strut graft in position.

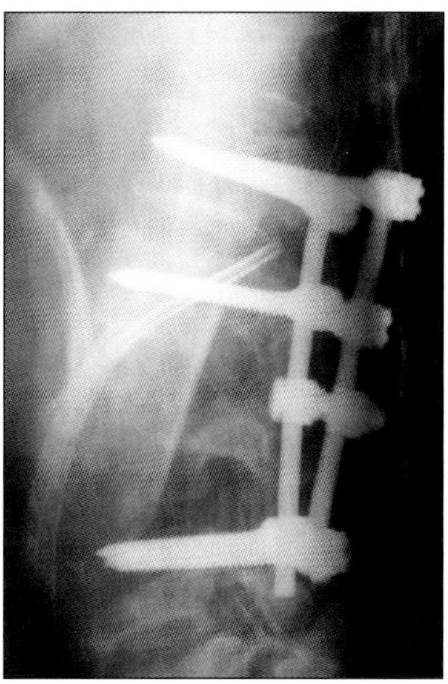

Fig 11-1d Lateral radiograph following posterior segmental fixation with pedicle screw fixation L2-4. A-O universal spinal system is made of titanium. This was supplemental with a posteriolateral lateral fusion.

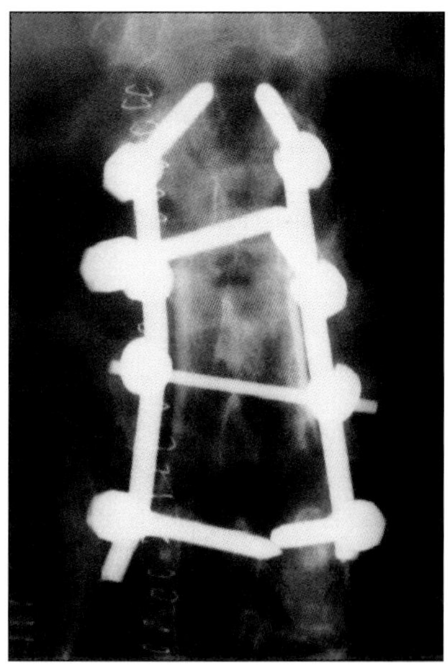

Fig 11-1e Anteroposterior radiograph with the posterior instrumentation in place.

Bone Grafts

In general, nearly all reconstructions depend on obtaining a solid bone mass (fusion) to reconstruct the spine when vertebral elements are removed. The lone exception is for tumors, where the patient has a short life expectancy (less than 3 years), and/or where radiation therapy and extensive chemotherapy may be used. In these cases, methyl-methacrylate may be inserted, rather than trying to obtain a bony union in the presence of re-growing cancer cells; and/or radiation therapy and chemotherapy, which may significantly retard new bone growth and the ability to obtain a fusion.

Posteriorly, an autograft is the preferred long-term spine-stabilizing factor. This is usually obtained from the posterior iliac crest. The autograft from this location provides a significant amount of cancellous bone that functions in osteoconduction and osteoinduction, and provides osteoprogenitor cells. Solid bone healing occurs in approximately 80% to 90% of the cases over 3 to 9 months.

An allograft, applied posteriorly, does not work well and is a poor choice, although it is occasionally used by necessity (eg, lack of adequate bone stock from previously harvested iliac crests). The processing of the bone for an allograft may affect its ability to function in any more than an osteoconductive manner. Fresh-frozen bone has potential osteoinduction capabilities. However, it also runs the risk of transmitting diseases (there are only one or two reported cases of acquired immunodeficiency syndrome that have been proven to be transmitted from fresh-frozen bank bone).

Anterior Reconstruction

Anterior reconstructions offer more options. Anteriorly, choices include either allograft or autograft struts, which work equally well; methyl-methacrylate, which may be supplemented with mesh as a bridgework or hydroxyapatite blocks; and metal cages, which help create fusion/stability across a disk space or have been used to reconstruct the spine after a corpectomy. Anterior fusion with autograft is frequently successful in obtaining a healed fusion. In the process of healing, it provides structural support for the vertebral body or bodies it is used to replace. Limitations in this technique relate to the amount of bone available and the location where it is inserted. The anterior iliac crest, for anterior spinal recontruction, is the most frequently used, but spanning more than two to three cervical vertebral segments, or one to two thoracic or lumbar segments, is difficult.

If longer bony reconstructions are necessary, many physicians will choose the fibula, as it is structurally sound and the removal of the middle two thirds of the fibula is not associated with knee or ankle joint morbidity. However, in the cervical spine, placing this straight bone in the normally lordotic spine may create difficulty in seating the strut. Using the fibula in the thoracic or lumbar spine may create difficulty in adequately filling the defects since the fibula is narrow. Additionally, it is primarily a cortical shell with a thin cancellous center, which leads to a long (2-year) healing process. Iliac crest bone has a significant amount of cancellous bone, whereas the fibula is more cortical, providing more osteoconduction (with some osteoinduction) than the iliac crest, which offers osteoconductive, osteoinductive, and osteoprogenitor cell capabilities.

Methyl-methacrylate, as noted, should primarily be reserved for reconstruction following removal of a vertebra (or multiple vertebrae) in treating cancer or metastases.

Infections should be treated with bone, not supported directly by metal implants, because of the possibility of prolonging the infection because of the metal implant.

Hydroxyapatite is available in multiple synthetic formulations. At this time, the Food and Drug Administration (FDA) has prohib-

Fig 11-2 A 17-year-old girl had a chordoma at L-5. It was progressing radiographically. The treatment was complete extirpation of L-5 (anterior and posterior) with anterior and posterior reconstruction. Because the tumor was primarily in the vertebral body, the postoperative plan included focused high-dose radiation to L-5. Because of this, polymethyl-methacrylate (PMMA) was used anteriorly since it would not be affected by the radiation therapy, which was started immediately after surgery. Stabilization was created posteriorly with posterior instrumentation and autograft iliac fusion L-4 to the sacrum following an L-5 laminectomy, facetectomy, pediculectomy, and removal of transverse processes.

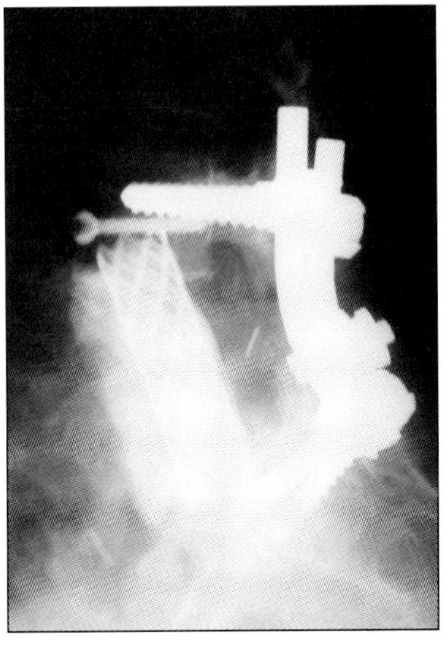

Fig 11-2a Lateral radiograph demonstrating a titanium Harm's cage anteriorly and TSRH instrumentation posteriorly in place from L4-S1, spanning the corpectomy site. The cage was filled with PMMA. The screw in L-4 was passed through the anterior portion of the L-4 body into the top of the cage to help fix it in position.

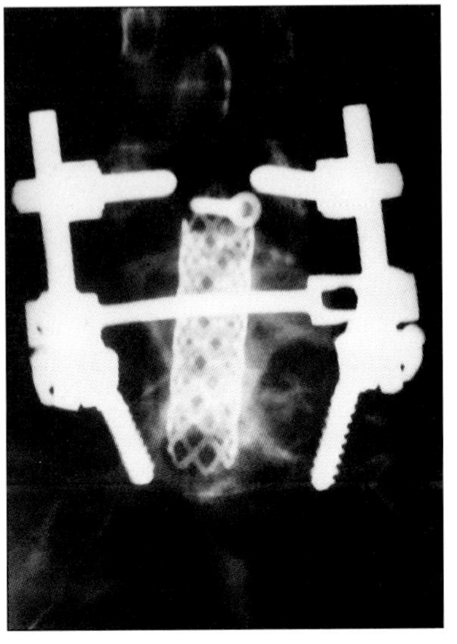

Fig 11-2b Anteroposterior radiograph of the reconstruction. The cage can be seen centrally; the posterior pedicle screw instrumentation (stainless steel) is lateral.

ited the use of hydroxyapatite in the spine (except in controlled settings) because of a high rate of failure—including shattering of implants and a number of reported cases of retropulsion of fragments into the canal, collapse across the replaced vertebral element, and a high rate of nonunion with increased patient complaints necessitating repeat surgery. Although many new devices are not approved by the FDA for use, hydroxyapatite blocks are specifically prohibited, except for controlled use in a few centers.

A number of cages are available to help structurally support and rigidly stabilize the spine (Fig 11-2). Cages are cylinders, fabricated from either carbon fibers or metal (stainless steel or titanium) that may be threaded, designed to be inserted into a disk space and fixed to the vertebral end-plates. The majority of the cages that are available have holes in the outer portion of the cylinder walls, as well as a hollow core, that can be filled with cancellous bone. Theoretically (and hopefully), bone grows from the adjacent vertebrae through the holes into the cancellous bone that fills the center of the cage. Additionally, with all currently available devices, it is recommended that the bone graft is placed between and lateral to the cylinders. This acts as a bridge, assuming that a fusion occurs in these areas. Some cages that are currently under study are BAK and Ray cages.

For vertebral body replacement, there are metal cages that can be inserted from vertebra to vertebra (across a vertebrectomy site). These also are hollow in the center, have holes through their walls, and are designed to be filled with cancellous bone graft. They provide immediate structural support, allowing bone to grow across and through the cages from vertebra to vertebra. These devices may or may not be supplemented with other anterior instrumentation, or following insertion anteriorly, posterior rigid fixation across the laminae or through the pedicles.

Posterior Instrumentation

Correction of spinal deformities, stabilization of the spine, and reconstruction from a posterior approach generally involve a midline incision with subperiosteal exposure of the posterior vertebral elements. In the cervical spine, this includes the lateral aspect of the facet joints. In the thoracic spine, the exposure includes the transverse processes, which are posterior to the head and lateral curve of the ribs. In the lumbar spine, to obtain a fusion, the tips of the transverse processes, deep (anterior) to the facet joints, must be exposed, which may include the sacral ala if the fusion extends to the sacrum.

Cervical Spine

For stabilization of C1-2 using advanced techniques, screws can be placed from the inferior aspect of the C-2 facet joint, directing them upward at a steep angle to cross the C1-2 facet joint, fixing to the anterior portion of the lateral mass of C-1 (Fig 11-3). This screw is placed lateral to the spinal cord and odontoid process and medial to the vertebral artery. If the screw is placed too anteriorly, it could enter the oropharynx. This fixation is used for C1-2 instabilities created by odontoid fractures, failed previous posterior C1-2 fusions, resection of C-2 for tumor or infection, transverse ligament disruption from trauma, and/or for rheumatoid instabilities. This is then supplemented with a posterior bone graft to obtain fusion. Rigid fixation to C-2, without fixation to C-1, can be obtained through a C-2 pedicle screw that enters the lateral aspect of the facet, directed medially through the pedicle.

Fixation of screws in the lower cervical spine involves primarily insertion of screws into the lateral mass.[1-3] These screws generally start in the mid-portion of the facet and are directed 10 to 20 degrees lateral, and then either straight ahead or up to 25 to 45 degrees cephalad. If these screws are directed too medially, the spinal canal can be entered (Fig 11-4). If they are too straight, without any lateral projection, the vertebral artery can be injured and, depending on their caudal-cephalad direction, nerve roots can be injured.

Thoracic and Lumbar Spine

Thoracic and upper lumbar posterior fixation can be obtained through hooks fixed to rods or screws secured to plates or rods[3-10] (Fig 11-5). In the thoracic spine, the most conventional fixation is through hooks, although pedicle screws are used outside of North America (and occasionally in North America). In the lumbar spine, particularly the lower lumbar spine, the use of pedicle fixation is increasing, although these devices are not cleared by the FDA for routine use in all disorders in this manner.

Indications for posterior fixation are degeneration (to help achieve a solid fusion following a laminectomy or for severe pain), trauma, tumor, or to stabilize bone grafts that are placed anteriorly in reconstructions for trauma, tumors, or anterior vertebral

Fig 11-3 Example of a C2-1 posterior cervical screw fixation. This supplements posterior C1-2 fusion with autograft iliac crest bone.

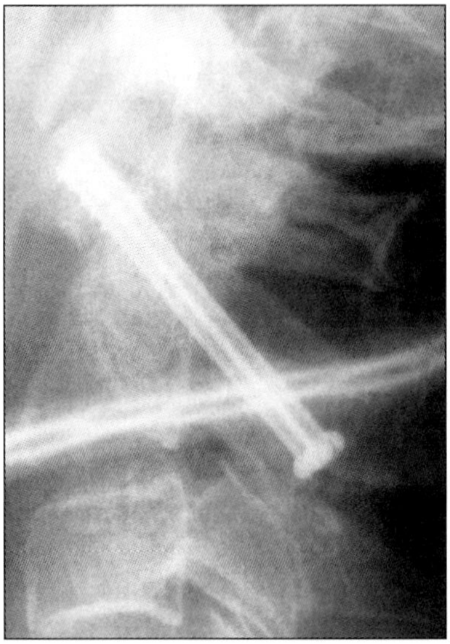

Fig 11-3a Lateral radiograph showing Reunion screws passed from the posterior inferior edge of the C-2 facet, angling steeply anteriorly across the C1-2 facet joint, entering C-1 laterally.

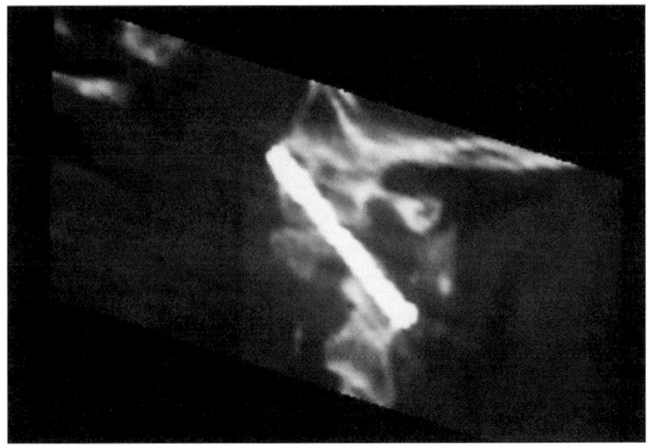

Fig 11-3b Lateral CT scan reconstruction showing the course of the screw from the C-2 facet, across the C1-2 facet joint directed anteriorly.

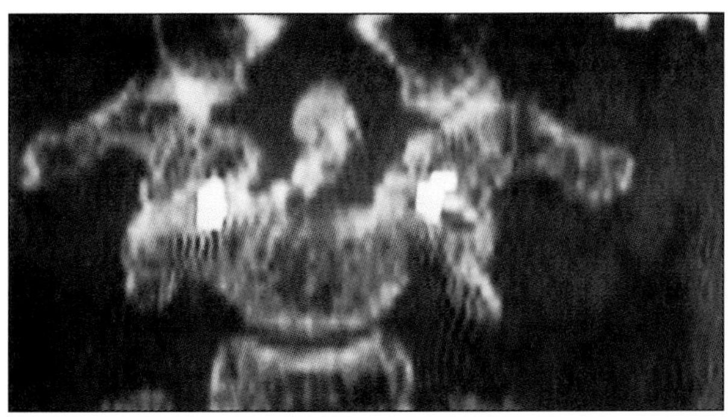

Fig 11-3c Coronal reconstruction showing the screw crossing the C1-2 facet joints bilaterally.

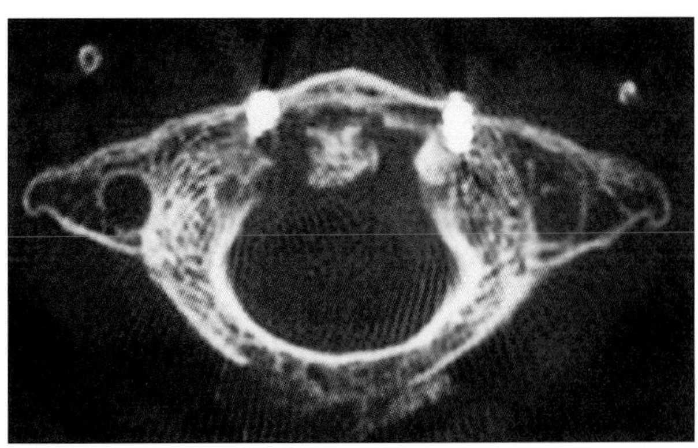

Fig 11-3d Cross section of C-1 showing the screws lateral to the spinal cord, medial to the vertebral artery foramen, engaging the anterior cortex of C-1.

Fig 11-4 Example of lateral mass fixation (axis). The distal screw on each side is placed in the T-1 pedicle.

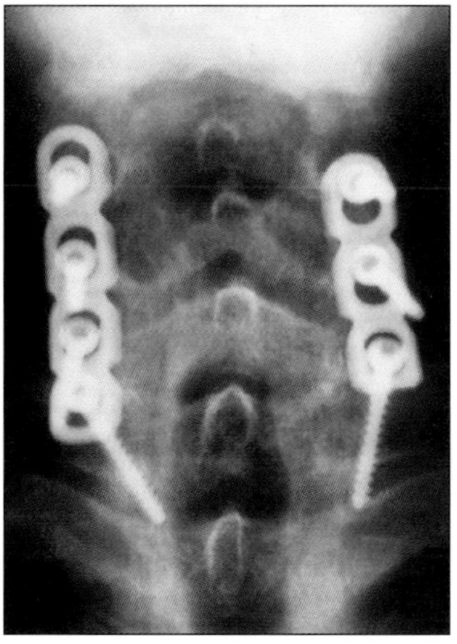

Fig 11-4a Anteroposterior radiographs demonstrating lateral mass fixation with the Axis system. The distal screws are medially based as they are in the first thoracic pedicle and the more proximal screws are in the lateral mass.

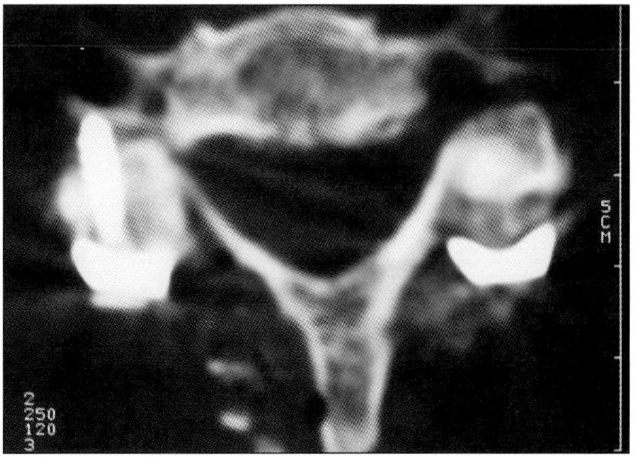

Fig 11-4b Computed tomography scan demonstrating position of the screws in the lateral mass.

Fig 11-5 An example of Edward's instrumentation used in a compression mode to treat a T12-L1 fracture subluxation.

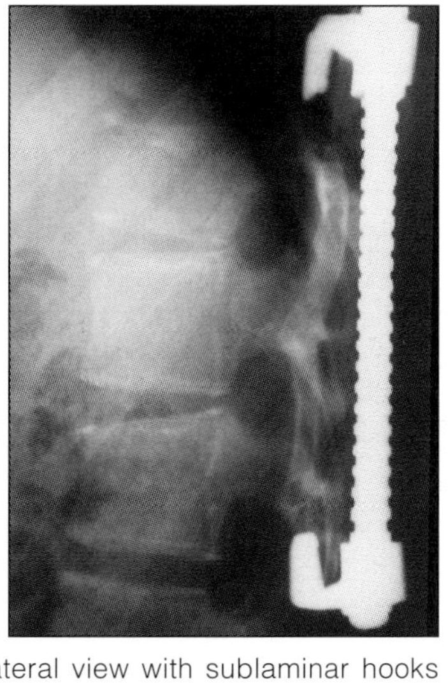

Fig 11-5a Lateral view with sublaminar hooks and a Universal rod that can be ratcheted for more compression.

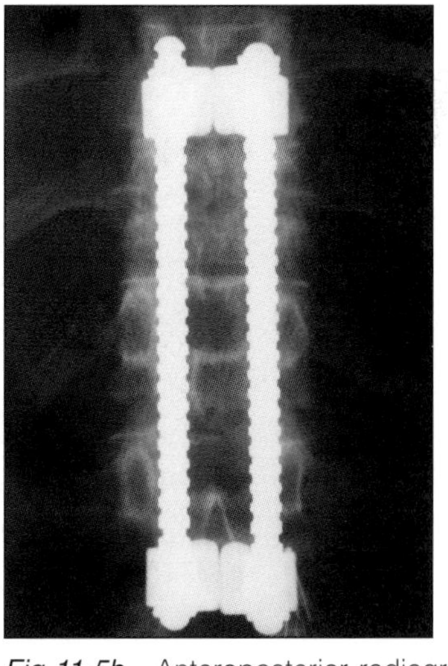

Fig 11-5b Anteroposterior radiograph demonstrating the Edward's instrumentation in place. Washers are placed proximal to the cephalad hooks and distal to the caudal hooks to prevent loss of compression. The instrumentation supplements a posterior iliac crest autograft to obtain a solid fusion.

body/disk space infections. Posterior instrumentation is also used in the treatment of scoliosis and/or kyphosis, frequently preceded by anterior diskectomy (releases) and fusions. Low lumbar and sacral fixation is best performed with screws placed in the pedicles, affixed to plates or rods. Hook fixation in the lower lumbar spine is associated with a 5% to 20% failure rate, due primarily to loss of fixation and the creation of a flat back deformity in the lumbar spine (loss of lumbosacral lordosis by pure distraction instrumentation posteriorly). Complications with these devices include infections, breakage of instrumentation, loss of fixation, violation of the canal, injury to spinal nerve roots and/or the dura, nonunions, and the necessity to re-operate. Successful results are as dependent on choosing the patient properly (patient and disease process) as on performing the surgery correctly and inserting the instruments properly.

Anterior Approaches/Fixation

Anterior surgery has more options and approaches. The upper cervical spine can be approached transorally or through a high retropharyngeal approach. This, primarily, allows access to the C1-2 facet joints and the C1-2 bodies. Through these approaches, vertebrectomies can be performed and bone graft inserted. This high in the neck, it is difficult to insert metal with any degree of stability and, particularly through the transoral approach, the risk of infection is increased if metal is inserted. In general, these approaches are reserved for tumors, upper cervical infections, or rheumatoid deformities with significant encroachment on the spinal canal by the odontoid process and associated pannus. The more conventional Smith-Robinson approach can be used to approach the lower portion of C-2 through T-2.

Odontoid fractures are now frequently treated with anterior screw fixation. The risks include screws backing out of the bone and lack of union, as well as violation of the canal, either posteriorly or proximally through the tip of the odontoid. Reconstruction of C-3 through C-7, primarily for tumors, infections, or myelopathy associated with burst fractures, bone retropulsion, disk herniation, or cervical stenosis from degenerative conditions, can be accomplished with bone grafts supplemented with plates and screws. Anteriorly, autograft or allograft can be used. The insertion technique including fashioning the graft to fit into the vertebrae is important in helping fix the graft. If there is not an associated infection, the graft can be held in place and the spine further stabilized with plates and screws.

The Caspar system is the oldest available and requires bicortical fixation to obtain a rigid construct.[11] The A.O. Moscher and Danek Orion Plates are more frequently used in North America, as these require only unicortical fixation. The bone screw is locked to the plate with a small, secondary screw, which increases the mechanical stability and success of these devices. Metal implants should not be used to supplement a fusion in the face of an infection, though second-stage posterior stabilizing procedures can be used in these settings.

Anterior reconstruction across the thoracic and lumbar spine is also achieved with bone, but can be supplemented with screws and rods (Edward's, TSRH) or screw and plate fixation devices (Z-Plate and Kaneda device)[12-14] (Fig 11-6). These instrumentations allow some degree of distraction prior to inserting the bone graft strut, followed by compression, to stabilize the spine and create a solid interface between the intact bone and the bone graft to increase the likelihood of obtaining a solid fusion.

Additionally, these anterior devices can be used to supplement methyl-methacrylate or vertically oriented cages. The screws are placed from one anterolateral aspect of a ver-

Fig 11-6 A 24-year-old woman with increasing midthoracic pain and symptoms of thoracic radiculopathy and early myelopathy related to an expanding T-11 hemangioma. Because of the symptoms, surgery was elected. An anterior corpectomy of T-11 was performed through a thoractomy approach. Iliac crest autograft was placed from T10-12 and fixed in position with a Z-plate made of titanium.

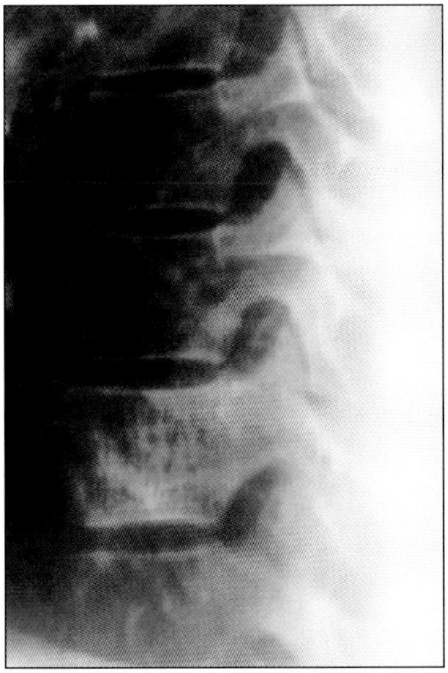

Fig 11-6a The lateral radiograph demonstrating striations in the bone with slight ballooning of the anterior and posterior cortex consistent with hemangioma.

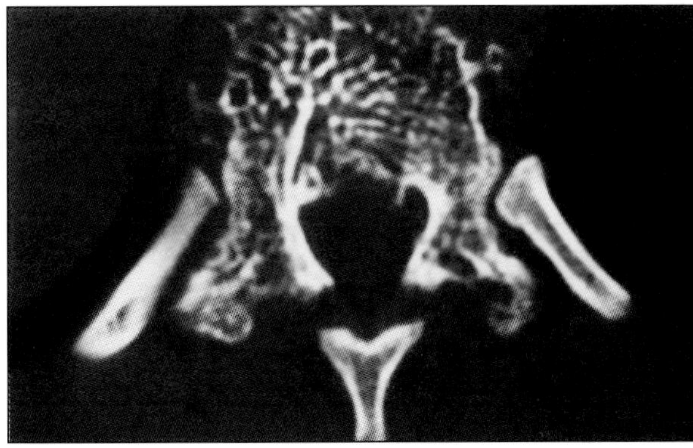

Fig 11-6b Computed tomography scan cross section through T-11 demonstrating the hemangioma appearance of the bone.

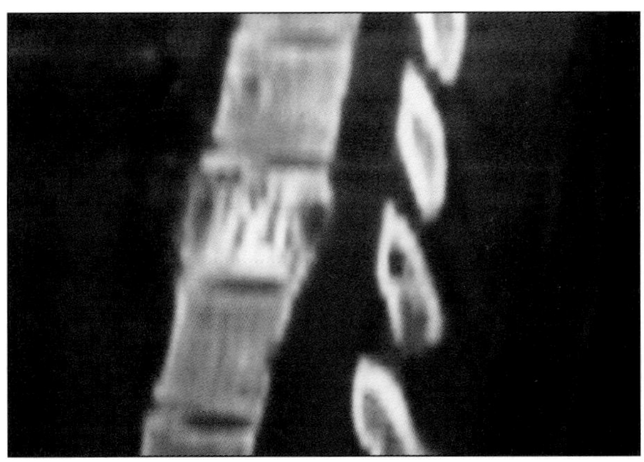

Fig 11-6c Sagittal CT reconstruction demonstrating the lesion in L-1.

tebral body (usually the right side) through the opposite lateral cortex. They help provide rigid fixation and stabilization of the spine, allowing earlier rehabilitation and mobilization of the patient, and theoretically increase the union rates. Indications for this approach relate primarily to trauma and tumor, although bone graft, as in the cervical spine, is used in the face of infection, although not supplemented with metallic implants.

Fixation across the lumbosacral junction is difficult to obtain anteriorly. In general, a bone graft is placed into the sacrum, extending as cephalad as necessary, supplemented with posterior pedicular instrumentation devices.

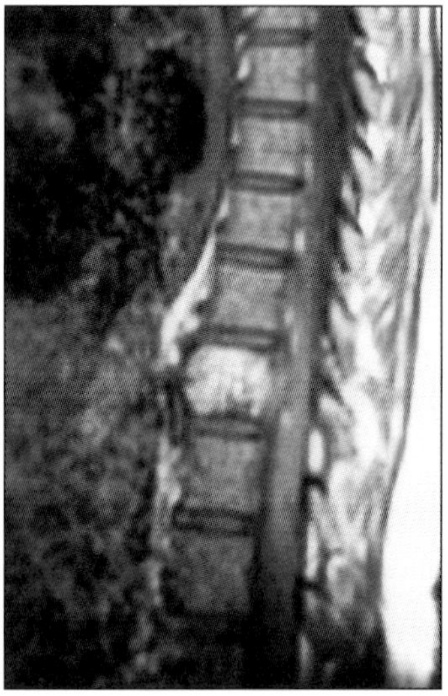

Fig 11-6d Magnetic resonance image demonstrating signal changes in T-11 (consistent with hemangioma).

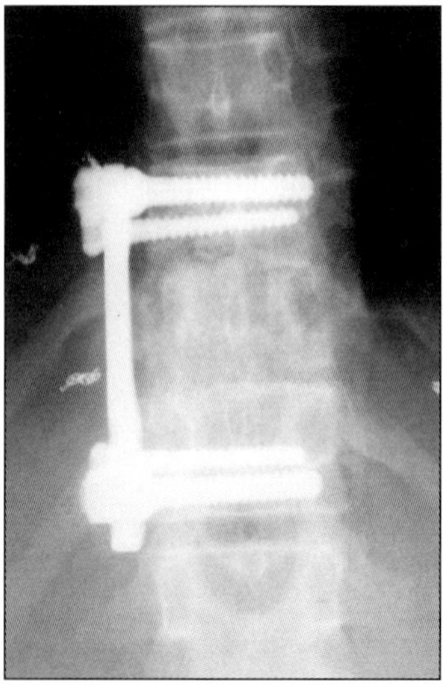

Fig 11-6e Postoperative AP radiograph showing the Z-plate in place with two vertebral body screws in T-10 and two in T-2, fixing the plate, which is placed laterally.

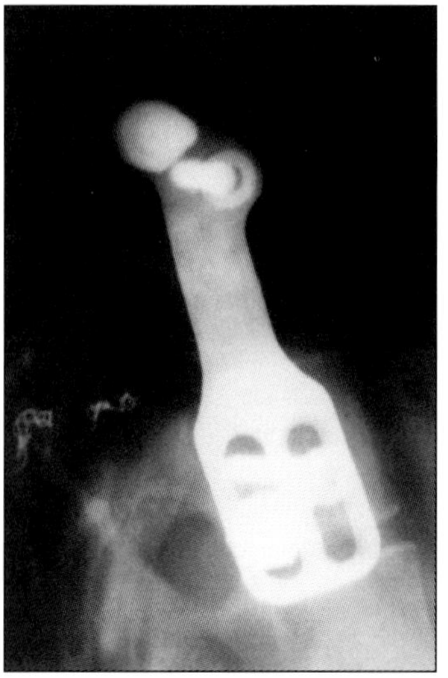

Fig 11-6f Lateral radiographs of the Z-plate in position. An autograft strut spans the T-11 corpectomy site. It is inserted into the caudal aspect of T-10 and the superior aspect of T-12.

A long incision is required to approach the thoracic or lumbar spine anteriorly. The thoracic spine is approached through a thoracotomy incision. A lumbar retroperitoneal approach is used from L-1 through L-4. A thoraco-abdominal approach, taking down the ninth or tenth rib and the diaphragm, may be necessary to approach T-11 through L-1. The L4-5 and L5-S1 disk spaces, L-5 vertebra, and the sacrum anteriorly can be approached retroperitoneally or transperitoneally. This gives direct access to the L4-5 through S-1 elements, anteriorly, for diskectomies, L-5 corpectomies, and fusion.

Complications

The complications associated with anterior approaches and instrumentation primarily relate to the approaches themselves. In the cervical spine, the esophagus, trachea, and carotid arteries, as well as the recurrent laryngeal nerves, are mobilized to access the vertebral elements. If screws/plates are used, they, as well as grafts, can migrate anteriorly, leading to esophageal perforation. In the thoracic spine, a rib is removed and the approach involves mobilizing a lung, and on the left side the thoracic aorta and on the right side the vena cava. Additionally, segmental vessels must be mobilized.

Instrumentation placed in these positions can erode the aorta on the left side, if placed too anteriorly, and the vena cava on the right side. On the lower right side, the thoracic duct also can be injured. Additionally, if the screws are placed too posteriorly, the spinal cord may be violated, and if they are too long, the contralateral lung can be punctured, leading to a pneumothorax or hemothorax. In the lumbar spine, the iliac vessels bifurcate at approximately L-4, and these can be injured by screws. Anteriorly, if a graft is placed at L-5, the bone graft can back out and injure the iliac vessels.

Re-operations

Anterior re-operations are difficult to perform because of the surgical risks involved with re-operating through the anterior soft-tissue structures. Staged posterior surgeries may be useful in deformity surgery to help avoid rapid correction of deformities, stretching of the spinal cord, and iatrogenically induced spinal cord injuries related to overstretching and cord/cauda equina mechanical injury.

Considerations

It is now possible to obtain stable reconstruction of all vertebral elements. However, the anatomic approaches involve risk, as well as detailed familiarity with the anatomy. The larger the reconstruction, the more difficult the surgery and the intraoperative and postoperative management. Risk from the approach, as well as insertion of metal and grafts, include injury to significant anatomic structures. These can occur early or late, particularly if the bone graft does not heal, if it backs out or fractures, or if the metal fails and/or migrates.

Rigid internal fixation devices are now available for use either posteriorly or anteriorly. Posterior instrumentation is usually fixed either to the lamina or pedicle in the thoracic or lumbar spine or to the lateral masses in the cervical spine. Anteriorly, bicortical fixation may be necessary. In the cervical spine, the screws are directed into the canal. In the thoracic and lumbar spine, instrumentation is from the lateral aspect of the vertebra and crosses the vertebral body to engage the contralateral portion of the vertebral body cortex. Insertions are done "blind," by feel, and followed up with radiographic confirmation in the anteroposterior (AP) and lateral planes (in certain instances fluoroscopy may be used).

At this time, there are no bio-absorbable substances used in the spine for fixation. With posterior instrumentation, frequently the metal implants are removed. This occurs in up to 50% of procedures in the lumbar spine when pedicular screw instrumentation is used. Anteriorly, assuming the fusion is solid, the metal frequently does not fail or back out (the fusion anteriorly is more likely to succeed than posteriorly), and the metal is usually not removed because the approach is difficult, and the metal implants frequently do not lead to pain as they are not as "bulky" as posterior instrumentation.

Future considerations may be toward stronger, smaller implants, more rigid fixation to the bony elements, and the introduction of bioresorbable materials.

DISCUSSION

Shu Chien What is the role of osseointegration here?

Steven Garfin I honestly do not know, which is why I presented this. Dr Brånemark always asks me, and I always hesitate. I did not have these slides when I was in Sweden. I thought I would show these slides here and hope that somebody could come up with something. It might be that our goals are wrong. I mean, all we are trying to do is fuse the spine. Maybe if we can figure out where pain is, or where instability really is, or how to re-create ligaments, then maybe in some of these patients, when we are fusing multiple levels, there would be some way to maintain some motion. Everything I have seen so far relates to motion. Fixing something in a bone at a joint is done so one can move it and use it. Our whole goal is to stabilize, and maybe that needs to be re-thought. It would be useful if we have a way in fixing something in small, weak vertebrae.

Richard Coutts I would be interested in your thoughts with regard to bone grafts because we do have available to us an array of different bone products. Not only can you get allografts, but you can get them fresh-frozen, you can get them freeze-dried, you can get them partially demineralized, you can get a demineralized form. You can get hydroxyapatite (HA) from various manufacturers now, you can get HA mixed with collagen, you can get granules of demineralized bone. Do you have any information on what works best once you get out of the autograft range? There is no question that an autograft is the best bone replacement.

SG Most of the studies have been on extremities, and there are very few in the spine. Again, I have to bring in the factor of posterior vs anterior. Almost no allografts work via the posterior approach. You have to remember that we have all moved to a posterolateral fusion, which is transverse process to transverse process, so the graft is lying on a bed of fascia and the muscle comes back over the top. It is not bone on bone except for small contact areas. We do not need structure there. We need something that is osteo-inductive. Osteoconductive does not help quite enough. The scaffolding does not work. So most of this is scaffolding so far. I do not know how it might work in these transverse processes to fill this big gap with a bunch of dead bone. It is very clear that allograft, no matter how you prepare it, does not work well posteriorly. It is not reliable enough to do. Some studies (we just had the North America Spine Society meeting here) now show various preparations used posterolaterally. So far, although it works in the extremities, it does not appear to work in the spine.

Anteriorly, most materials work because what you need is structure. You need materials, so HA is useful. It gives you structure. It can be manufactured to be stronger but it needs structure. We need to hold the spine up. Most of the other approaches, which inject natural substances, do not work. So far, it is just autograft or allograft that works, but that will change, I hope.

Björn Rydevik You said that the method of choice today is really to try to achieve fusion for many conditions of the spine. What is, in your mind, the actual role of internal fixation: to get the best possible or the highest rate of fusion in various cases? How about the influence of age in these situations, or on the healing rate of these fusions?

SG The latter may be an interesting question to raise, but let me deal with the former first because the second is only a hypothesis. We have been involved in a national study for pedicle screws in a role of internal fixation. Looking at those data, it is pretty clear that the fusion rate in situ is not 90%, even for a one-level fusion. It is probably 60% or 70%. Adding internal fixation, which rigidly stabilizes the spine, increases the fusion rate to 85% or 90%. I think rigid fixation does a couple of things: 1) it stabilizes the spine, which increases the fusion rate; and 2) it allows deformity correction, as it does elsewhere in the body. We can correct them in the hand or other long bones if the bone is maligned. In the spine, we classically have just left vertebrae out of position and hoped for a fusion. Now with segmental fixation, we can realign the elements, make more room for the nerves, and open the neuro-foramen and correct deformities. I think that is an advantage of the segmental type of internal fixation. Additionally, it allows patients to leave the hospital quicker because they feel a little better, although I do not have any statistics on this.

The second question you asked is about osteoporosis. Internal fixation seems to work. My experience is that some of these older people with osteoporosis heal surprisingly well compared to young athletes. I wonder if the bone turnover related to osteoporosis might be conducive to healing, since things are happening, not necessarily happening in balance, but happening fast. The more I see some of these people who appear to have limited potential to heal, but do heal, I find that they tend to be osteoporotic and tend to be a little less active than many others. The metal implant seems to also help in these individuals.

RC I am ignorant on exactly how you get those screws to go where they are supposed to go. You see what bad things happen if your aim is off. So how do you ensure that your aim is on target? I think it's like screwing the odontoid. That looks tricky to me.

SG One factor is that you have to pick your patient carefully. Dr Rydevik and I have talked about it. In this country you may pick someone who cannot speak English and is not insured, as in Sweden. It means dealing with an anatomy that we do not know very much about, and we have to learn the anatomy. I mean, most orthopedic and neurosurgeons do not know enough perivertebral anatomy. So, for me, it means going to the lab, the cadaver lab, and using radiographs so I learn the anatomy thoroughly.

RC Do you use radiologic guidance when you are putting these screws in?

SG Not for pedicle screws. I don't use fluoroscopic guidance for pedicle screws because I have done so many of them that I can feel where I am going. For the other locations—for C2-1 nerves and the odontoid, I do use fluoroscopic guidance. It is usually biplanar; in fact, for the odontoid screws, I have two views with two fluoroscopes that are fixed in position all the time. Biplanar views are used not so much for the posterior as for the anterior procedures.

BR Are you aware of any developments in terms of using growth factors, or bone morphogenic protein or other growth-stimulating factors that could be used topically, locally at the site of fusion?

SG Yes, bone morphogenic protein (BMP) is being used more often in the spine. The same studies that showed that demineralized bone does not uniformly work show that BMP works spectacularly. Bone morphogenic protein is not approved for use in this country, but there are two companies in North America that are close. One of them is ready for a clinical trial, and one of them is close to a clinical trial for the spine. I am excited about it. Wayne Akeson, MD, whom everyone here knows, has a vision of the future, and strongly feels that BMP is going to change the face of orthopedics in 5 years.

References

1. Anderson PA, Henley MB, Grady MS, et al. Posterior cervical arthrodesis with AO reconstruction plates and bone graft. Spine 1991;16:72–79.
2. Montesano PX, Juach EC, Anderson PA, et al. Biomechanics of cervical spine international fixation. Spine 1991;16 (suppl):10–16.
3. Sutterlin CE, McAfee PC, Warden KE, et al. A biomechanical evaluation of cervical spine stabilization method in human cadaver model. Spine 1988; 13:795–802.
4. An HS, Vaccaro A, Cotler JM, Lin S. Low lumbar burst fractures: Comparison among body cast, Harrington rod, Luque rod and Steffee plate. Spine 1991;16:8 (suppl):440–444.
5. McAfee PC, Farey ID, Sutterlin CE, Gurr KR, Warden KE, Cunningham BW. Device-related osteoporosis with spinal instrumentation. Spine 1989; 14:9.
6. Panjabi MM. Biomechanical evaluation of spinal fixation devices: Part I. A conceptual framework. Spine 1988;13:1129–1134.
7. Panjabi MM, Abumi K, Duranceau J, Crisco JJ. Biomechanical evaluation of spinal fixation devices: Part II. Stability provided by eight internal fixation devices. Spine 1988;13:1135.
8. Roy-Camille R, Saillant G, Berteaux D, Salgado V. Osteosynthesis of thoracolumbar spine fractures with metal plates screwed through the vertebral pedicles. Reconstr Surg Trauma 1976;15:2–16.
9. Steffee AD, Brantigan JW. The variable screw placement spinal fixation system: Report of a prospective study of 250 patients enrolled in Food and Drug Administration clinical trials. Spine 18;9:1160–1172.
10. Zdeblick TA. A prospective, randomized study of lumbar fusion: Preliminary results. Spine 1993; 18(8):983–991.
11. Tippets RH, Apfelbaum RI. Anterior cervical fusion with the Caspar instrumentation system. Neurosurgery 1988;22:1008–1013.
12. Kaneda K, Abumi K, Fujiya M. Burst fractures with neurologic deficits of the thoracolumbar-lumbar spine: Results of anterior decompression and stabilization with anterior instrumentation. Spine 1984;9:788–795.
13. Kostuik JP. Anterior fixation for burst fractures of the thoracic and lumbar spine with or without neurological involvement. Spine 1988;13:286–293.
14. Zdeblick TA, Shirado O, McAfee PC, et al. Anterior spinal fixation after lumbar corpectomy. J Bone Joint Surg 1991;73A:527–534.

Chapter 12

The Interfaces Between Humans and Limb Replacement Components

Dudley S. Childress

Limb prostheses traditionally have been interconnected with the human user through an "interface," which because of its shape and the way it fits over the residual limb is also known as a "socket." The mechanical components of the limb replacement are firmly attached to the socket. The socket and the attached parts (artificial feet, hands, etc) constitute the prosthesis. The socket must be fitted carefully to the limb and shaped to provide the interface qualities desired. These qualities are:

1. To effectively transmit loads between the body and the prosthesis in such a way as not to cause trauma to the tissue, or pain in the limb during load-bearing or other usage
2. To achieve stability on the limb so that forces and moments on the prosthesis are resisted, keeping the socket fixed on the limb in a reasonably secure manner during loading, and enabling the user to successfully manipulate and place the prosthesis
3. To suspend the prosthesis against gravity and centripetal forces during arm or leg swing

Prosthetic sockets are a little like bicycle seats in that the seat (interface) is shaped to transmit loads from the torso (gravitational and mechanical accelerations) to the bicycle (prosthesis) through soft tissues of the lower pelvis. Nevertheless, a bicycle seat is considerably simpler to design because suspension is not required, loads may be redistributed easily through postural changes of the body (including standing), control of the bicycle is accomplished through the hands and feet, and reasonable stability can be achieved even with generic seat contours.

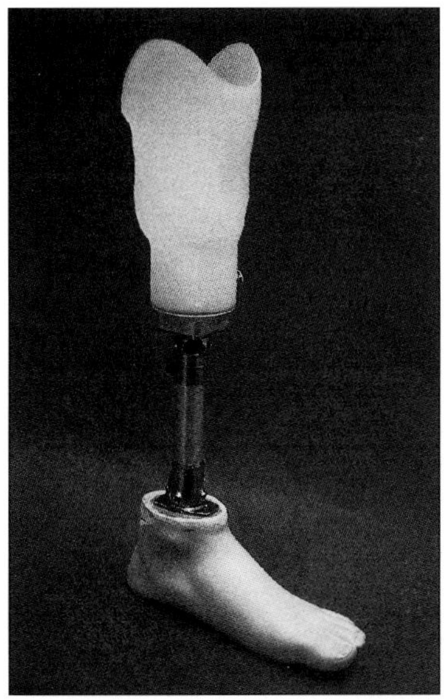

Fig 12-1 Below-knee prosthesis showing socket, pylon, and foot.

A modern transtibial (below-knee) prosthesis consisting of a computer-manufactured socket, pylon, and artificial foot is shown in Figure 12-1. The prosthesis is suspended by suction (atmospheric pressure) using an inner silicone sleeve that attaches to the distal end of the polypropylene socket. The pylon is commonly covered with a plastic foam that is shaped like the user's shank.

Determination of the shape of a customized prosthesis socket that is comfortable and functional is one of the most important aspects of limb replacement design. The shape cannot simply be an intimate fit, because some areas of the limb are more sensitive to loading than others. Load needs to be applied to tissues that can best accept it (eg, tendinous tissues) and avoided in other locations, like the ends of cut bones or the crest of the shin. The fact that effective prosthesis interfaces are often achieved is a tribute to the knowledge and skill of the people who design and fit these structures to persons who have limb loss. Nevertheless, it is often difficult to design sockets that achieve all these goals, particularly for amputees who are highly active and who place large complex loads on their prostheses, or for those people who have poor circulation in their limbs or whose limbs are sensitive and possibly covered with minimal soft tissue. Modern approaches to prosthesis design as well as more detail about prosthetics problems are covered elsewhere.[1,2]

Socket design, which has largely been empirical in nature, is now under engineering and scientific scrutiny by persons interested in tissue modeling (theoretical and computational), computer-assisted engineering, and computer-aided manufacturing. At present, it is too early to predict the impact of these new approaches on socket design and manufacture, but the potential for significant influence is great indeed.

Despite steady improvement in socket designs and socket materials over the past half century, there remains a need for improved prosthesis interfaces. One continuing hope of the field has been to someday attach the prosthesis parts directly to the skeleton, avoiding the need for a "socket" and the problems that accrue from load-bearing through soft tissues of the residual limb. Instead, load would be applied directly from the prosthesis components to the skeleton. This technique is called *direct skeletal attachment*. A connection of this kind would satisfy all the requirements of a good interface and appears to have much potential for intimate integration of the human user with the prosthesis.

Direct Skeletal Attachment

Bones form the superstructure that gives the body its form. They form the frame for muscle action, providing the ability to stand, walk, run, and jump. Loads on the body during walking can reach nearly two times body weight and three or four times body weight during running and jumping. Although there are soft tissues between the bones and between the bones of the feet and the support surface, these tissues often can accept high mechanical loads. Cartilaginous materials, for example, have low metabolic rates and few pain receptors. However, in leg prostheses that are attached to the body with sockets, loading between the prosthesis and the limb must be transmitted through soft tissues (skin, muscle, nerves, etc) that generally are not well suited for load-induced pressure. Nerves, for example, are particularly sensitive to pressure because of their need for metabolites and oxygen. Muscles and skin may be traumatized under the kinds of loading (normal and shear) that exist during running and jumping. Skin may chafe when it slips back and forth under load over frictional surfaces. Also, loading of the soft tissues of elderly persons who have peripheral vascular disease may further compromise circulation in the lower limb.

It easily can be seen that carrying loads through soft tissues has many disadvantages and why it would be advantageous to avoid it if possible, particularly for walking and running. For this reason, support of prosthesis loads directly through the skeleton appears to have considerable merit in limb prosthetics. Likewise, the direct connection provides suspension, can provide prosthesis stability, and makes for an intimate relation between skeletal movement and prosthesis movement.

As an aside, it should perhaps be mentioned that people in western cultures spend considerable time sitting on soft tissues of the body, particularly the buttocks and posterior aspect of the thighs. One might reasonably ask why loads through sockets shouldn't be as benign to soft tissues as loads on soft tissues during sitting. Answers to this question are that people avoid tissue trauma during sitting by frequent movements to ensure that loads on the soft tissues are regularly relieved; that sitting mostly involves only static loads, and loads that do not involve the whole weight of the body provided the feet touch the floor; and that the area of support in sitting is substantially larger than that for a prosthetics socket. We must also consider that just because some humans sit in chairs for extensive periods doesn't necessarily mean that sitting is a healthy endeavor. Examination of the shape of the ischial tuberosities seems to suggest that these structures are not well designed for sitting.

Although long considered a desirable goal, direct skeletal attachment of limb prostheses has been elusive. Groups in the United States, Canada, and Germany made a number of early attempts. In the United States, the Veterans Administration supported efforts during the 1970s and early 1980s to demonstrate direct skeletal attachment in animal models. Hall reported on this work, partly in the form of progress reports.[3-15] Murphy[16] has written about the history and philosophy of attachment of prostheses to bone through the skin. Hall[15] reported that the only groups that had attempted clinical trials for permanently affixing artificial limbs to the skeleton were Dümmer in Germany and Mooney et al[17] in the United States. References are limited to American work connected directly with limb prosthetics.

As often happens in science and engineering, researchers and developers working in other fields and with other objectives are the ones who find solutions that elude more direct approaches. It appears that investigators working with dental implant osseointegration and with implants in the maxillofacial region may have made the headway necessary for the clinical application of direct skeletal attachment of limb prostheses. Other papers in this workshop provide references and discuss research and clinical progress concerning direct attachment of prostheses to bone.

Hall[5,6] suggested interfacing the achilles tendon of a goat to a skeletally attached articulating joint and rubber hoof. This may be the first reference to the concept of incorporating direct muscle attachments with direct skeletal attachments. Forms of direct muscle attachment have been used in limb prosthetics, mostly in upper-limb prosthetics, since World War I. The procedure for attachment is to place a skin-lined hole or tunnel through a muscle that has been released at its insertion. The approach, called tunnel cineplasty because muscle contraction mobilizes the tunnel, can be used to activate a prosthesis via a pin through the tunnel that is attached to a cable. The cable is usually connected to a prehensile device (eg, hand). Tunnel cineplasty surgeries more or less vanished in America during the 1970s, but the method still has ardent advocates, particularly among certain German surgeons (eg, Baumgartner, Brückner, and Marquardt). The tunnel cineplasty approach or variants of it may have applications in conjunction with direct skeletal attachment of limb prostheses.

Direct Muscle Attachment

Efforts to attach devices outside the body directly or semidirectly to human muscles originated in Italy around 1900. Sauerbruch, a well-known German surgeon, independently developed similar ideas around 1916,[18] and it

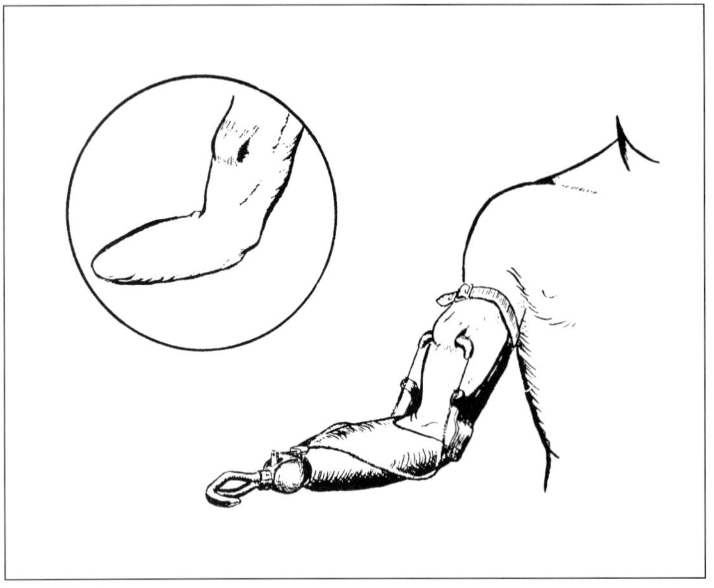

Fig 12-2 Biceps tunnel cineplasty. (Modified from Klopsteg PE, Wilson PD [eds]: Human Limbs and Their Substitutes. New York: McGraw-Hill; 1954.)

was he who reduced the concept to clinical practice, largely in connection with amputees of World War I. Kessler, a surgeon in the United States, learned the technique from Sauerbruch and transferred it to the United States during the 1930s. He wrote a book about cineplasty,[19] which is a good starting point for interested persons whose main language is English; other sources are available.[20]

In 1946, after World War II, American surgeons studied the principles of German prosthetics practice. One of these surgeons, Alldredge, studied the cineplasty techniques of Professor Lebsche in Munich, and Lebsche's concepts have been the main ones espoused in the United States since that time. Figure 12-2 illustrates the Lebsche concept for control of a below-elbow prosthesis. Lebsche generally used one muscle for activation, whereas Sauerbruch preferred to use two muscles as an agonist-antagonist pair. The drawing shows a biceps tunnel cineplasty in which the muscle pulls on a cable and opens the gripping device against an internal spring that normally holds the device closed.

Muscle cineplasty control in prosthetics has the advantage of intimacy of control. The muscle provides proprioception to the user so that he or she is aware of position, velocity, and force. The method has the disadvantage of requiring surgery and extensive training. The appearance of the limb is compromised by the surgery, and the tunnels require care to avoid infection and other problems. Sometimes the excursion or force of the muscle is not adequate. Sauerbruch put tunnels into small muscles of the forearm, but this practice did not catch on in the United States because the force and excursion available from these muscles was considered to be insufficient for good gripping force or wide opening of the hand. Some amputees in Germany still use this kind of prosthesis to good advantage.

Modern electric prostheses, controlled by position control systems that link the output of the prosthesis with the input—similar to the way the power steering of an automobile links the driver with the position of the front wheels—enables powerful prosthesis action to be controlled by small muscle forces and/or small excursions. Consequently, modern prostheses appear to open the way for alterations in the way surgeons can develop direct muscle attachment.

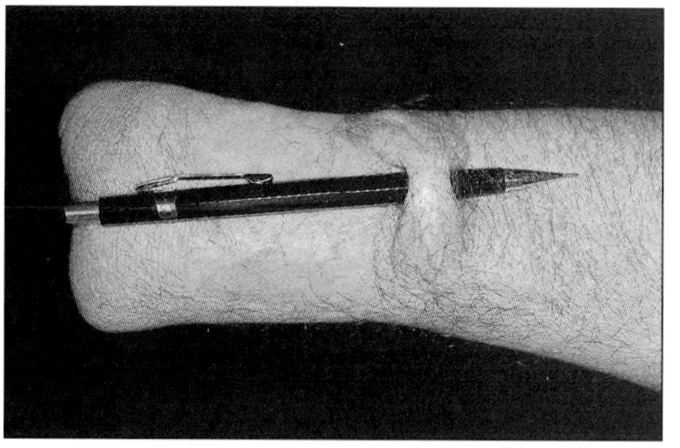

Fig 12-3 A tendon exteriorization. A pencil is placed through the exteriorization to show it well.

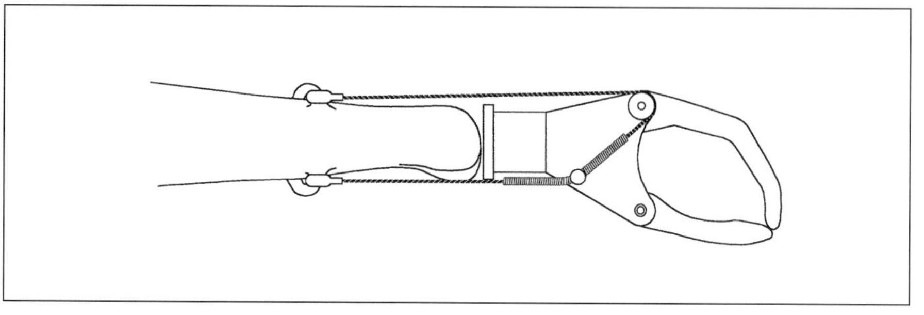

Fig 12-4 Connection of direct muscle attachments to an electric hand. The electronic position servomechanism is not shown.

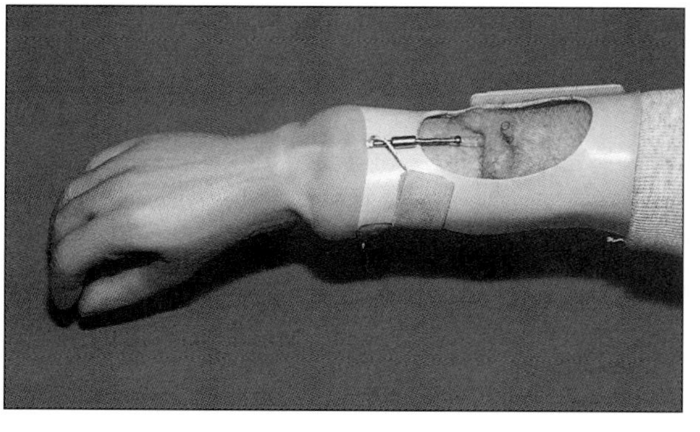

Fig 12-5 Prosthesis as fitted to amputee with tendon exteriorizations, electric hand, and cable control through a position control system.

Beasley[21] introduced the concept of tendon exteriorization as an alternative to tunnel cineplasty. Other kinds of surgical techniques might be developed to connect muscle movement with the outside world. Surgical techniques have advanced tremendously since 1916, when Sauerbruch introduced the tunnel cineplasty. Minicineplasties or microcineplasties may be desirable to improve appearance and to enable multiple muscles to be attached to multiple movements of a prosthesis. Figure 12-3 shows a tendon exteriorization as performed by Beasley. Figure 12-4 illustrates schematically how the author's laboratory connected this limb with an electric hand via a position servomechanism (not shown). Figure 12-5 shows the prosthesis, as developed by Childress et al,[22] attached to the arm shown in Figure 12-3.

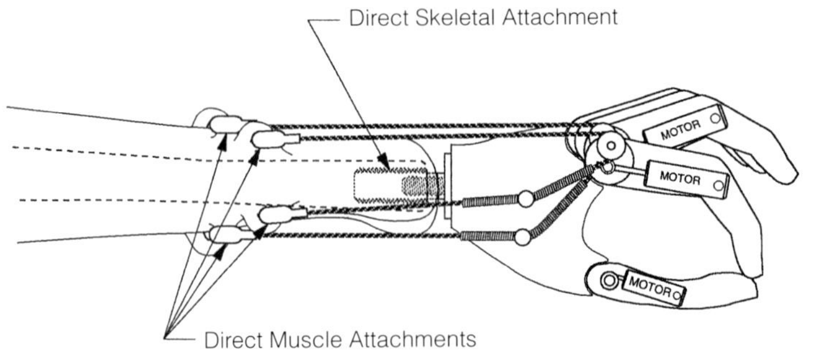

Fig 12-6 Possible multifunctional prosthesis with direct skeletal attachment and with direct muscle attachments.

Physiologic Prostheses

It seems natural to think of using direct muscle attachments in conjunction with direct skeletal attachments. A human using such a prosthesis would know precisely where the artificial device was located because it would be a direct extension of the skeleton of the body. Similarly, the user would know what the device itself was doing because the muscles directly connected with it would provide this information. Prostheses of this kind would appear to more closely approach physiologic limbs than most constructed hitherto. Figure 12-6 illustrates how direct skeletal attachments might be used in conjunction with direct muscle attachments to control individual fingers of a hand prosthesis. Progress along this line of research has much potential, but success will require close cooperation and collaboration between surgeons, engineers, scientists, and others, as the author has suggested in a recent editorial.[23]

Potential Problems

Although endoprostheses have been highly successful at the hip and knee, it does not necessarily follow that the direct attachment of limb prostheses to bones of the lower limb will follow without problems, even if infection at the interface boundary and possible thermal conduction issues are eliminated. Endoprostheses are to some measure protected by the neuromuscular system. We know that the forces of walking are far from being impulsive in healthy walking. We also know the foot and ankle can effectively adapt the limb to rough walking surfaces. There is a kind of compensatory adjustment of limb, ankle, and foot compliance that reduces loading during locomotion, compared with what it would be without these shock- and torque-reducing mechanisms.

It has been hypothesized that muscles may work actively to reduce bending moments on bones as they are produced by external loading. Such action might be difficult to achieve in a prosthesis. Artificial legs that attach directly to the skeleton will likely need to have characteristics that similarly are able to keep loading manageable during locomotion. One positive aspect of the traditional socket (interface) in prosthetics is that the soft tissue contained in the socket acts as a kind of shock absorber and load attenuator between the prosthesis and the amputee's body. Many modern leg prostheses are much more compliant than former designs, and it appears that the direction of development is toward artificial limbs that are more compliant. Designs of this kind are likely to benefit clinical application of direct skeletal attachment in lower-limb prosthetics.

The point to be remembered is that design of the leg components may be as important to the success of the bone interface as the design of the interface itself. Gunterberg, in his presentation on tumor resection and skeletal reconstruction in the extremities, has described the "glass limb" problem, which is the problem of having a limb that may be easily broken during typical human activities. Glass limbs will have to be avoided if direct skeletal attachment is to achieve its full potential in lower-limb applications.

Despite potential problems like those just discussed, as well as other problems that have not been mentioned or that remain unknown at this time, direct skeletal attachment appears to have great potential for the limb prosthetics field, a potential that is too significant to ignore.

DISCUSSION

Göran Lundborg That was a most fascinating lecture. Some things you mentioned are new to me: for instance, the exteriorized tendon concept. I think that what you have shown demonstrates two things. First, some kind of feedback system is necessary. Either you need sensibility in the hand to provide some kind of feedback for the motor control, or at least you need to watch what you are doing. Either use sensibility or the eyes or perhaps some kind of proprioceptive recognition in the muscles. Second, and most of all, I think the brain is, of course, the ultimate organ because it can adapt to all these new situations. For all these procedures you have described, I think there is an enormous capacity for functional reorganization in the brain to adapt to these things. When there are new demands put on a specific part of the body, the projection on this body part in the brain expands considerably over the brain cortex.

Dudley Childress The brain might be retrained, but our objective was to not do that. If I pick up a tennis racket and if I have any motor skills at all, I can hit the ball fairly well right away with very little reorganization. Now, to be really good I have to practice. But the point is that it's not a completely new idea and I don't want to add sensors here because it makes the prosthesis complex and it may break down. I don't want to bring sensory information into the body in modalities that are not natural to the body. That is why I want direct skeletal attachments so that you feel the way a blind person feels holding a long cane. It all comes back through the system. I want to use the body's natural functional abilities. I want to use the body's proprioception that comes from its muscles. I don't want to use a lot of technical devices. I want to have the surgeons work together with us to make simple systems that are naturally controlled.

GL You must have used some kind of afferent input with the tennis racket. I am suggesting that maybe we are going to use the body's own sensors and call on the brain to reorganize.

DC What you provide with direct skeletal attachment, I think, will be suitable.

GL It is like using chopsticks for eating rice, you feel everything. Of course there is no sensibility in the sticks, but the sticks transmit the pressure to the tissue media innervated. A very smart idea.

DC Exactly.

Stephen Cowin Which of these feedback loops allow Brånemark's patients to lift that egg; it is light, and you hardly know when you are around it. Yet the person stops at the right time without squashing the egg. Would it be possible to feel the weight?

Richard Skalak Do you have any idea where the sensors are in the patient who has the exteriorized tendons? Are they in the muscle?

DC I think it mainly comes from the muscle, but obviously some of it comes through the skin. The reason I think it comes from the muscle and the soft tissues is that when you replace a hip joint or a knee joint or something, you do not lose your proprioception, and you still have pretty good proprioception. I suppose in physiology, when I went through physiology, they said: Well, you have these sensors in your hip and in your knee to tell you where it is, but you can take those all out and you can still tell where your joint is, so you get this information, I think, through the muscles and soft tissue. Now, Beasley suggests that all the nerve roots and nerves to the skin are important, so that when you pull on this here you feel a certain place there. But I think it comes from muscles.

Elof Eriksson That was a great talk, very enlightening. A question and a comment: the question concerns the title "amputation prostheses." I think a lot of what we do depends on what we call things. Are we ready to call this "prosthetic reconstruction" instead of "amputation prostheses"? That is my question. My comment is: What comes to mind when watching this is the principles of surgery in tetraplegics. This is really a close cousin of surgery in tetraplegics, when you make an inventory of what is there and then start thinking about what you can reconstruct to create the basic functions of keypunch and grips.

DC In response to the question, please note I do have another title, so your point is made well. Perhaps we do need to think in terms of changing the terminologies that we use. I am not sure about the issue of tetraplegia because of the fact that you have so much loss of sensory input. In polio cases, we have sensory possibilities, and you see very good results with technical equipment. It is not quite so good when you go to a person who becomes tetraplegic from an accident or something of that nature.

GL How much excursion can you get on one of these externalized tendons?

DC The excursion depends on what the surgeon does. I think that you need to have the soft tissues free. You want to make sure that the muscle can remain mobile. Today we have electric prostheses and electric control systems, so whatever the surgeon gives me, I can match the patient's needs. But I think we can assume to make it work because we can just change the electronics. But I think it works better if you have bigger excursions and bigger forces.

GL Can you reproduce the excursion of, say, full extension to full flexion of the fingers?

DC Yes, you can in the prostheses. Of course, the excursion here may be slow. It is very slow generally.

GL In the United States and Sweden, electronics may be suitable, but in Mexico, or maybe in Africa, these electronic devices are going to be more of a problem to maintain.

DC Of course, access to competent maintenance is essential.

GL It would be nice if there is a possibility: Have you ever seen anybody who has enough excursions to actually open the hand completely?

DC In a nonelectric prosthesis, no. In fact, the cineplasties in the forearm all failed in the United States, or we dropped them because physicians and users did not think they had enough excursion or enough force.

GL Do you know which muscle is used?

DC I don't know in general. In the cases that are shown here, you need superficial muscles.

Tommy Hansson I was very impressed by your video where the person had the racket that was attached to his arm and he played squash right away. I think I could use something like that. The reason is I roll my wrist too much. But I don't want my arm amputated. *(laughter)*

DC You could go to arthritis surgeons. They can make it rigid for you.

TH You demonstrated how you used your hand and shoulder and finger and then the prosthetic device to write your name. You know that there was a fairly famous calligrapher who could write simultanously with his teeth biting on a pen and then with two hands and two legs at the same time. I always was very impressed, but after hearing you, maybe it is not that difficult. But the main question is that I think there is an integration of the proprioceptive stimuli, and I think this goes back to the question about visual feedback. The question is, in these patients, do they increase their visual feedback effectiveness, like a blind person who compensates by auditory mechanisms? Do they improve their visual feedback and also the muscle proprioception? Is there any compensation?

DC We don't know. We have only had one patient so far and so we don't know what happens, but I think that this will happen. Our objective is to control these prostheses without watching them, because if you have to sit and watch your arm and watch your food and every motion you make, you have to use your brain too much and you are not going to enjoy eating with somebody. When I sit and have dinner with you, I sit there and talk and use my hands. I don't use my cerebral cortex, at least not knowingly, to manipulate my arm. That is what we want to do for the amputees. Now, to your first point, the newspapers always say: Oh, he can even write his name with his prosthesis. Well, it is nothing because he is just using his distal motions. In the same way with painting, you can paint with your mouth, then say: Well, this is painted by mouth. But it is really just painted by the brain. Even Monet, who had arthritis, used his mouth. He held the paintbrush in his mouth sometimes for painting.

EE Question regarding self-image: I think that self-image, particularly body image, partly is more important to an amputee than to us in general. What do you think? What is your experience when we start modifying, for instance, the upper limb amputation stump? It is going to look quite odd. Is that well accepted by the amputee?

DC No, and you know almost the whole personnel of the establishment is against my work on this direct muscle attachment because it has a messy appearance. I think that to some extent it does. We need to be able to show it increases effectiveness and need to create more and better-appearing types of connections, and this is where I think surgery comes in. For example, you could release the muscle and you could take the soft tissues out from under the skin and put the skin to cover the muscle sheath. We have all seen this in amputations just happening naturally. There is also a possibility of hooking the muscle to the skin without making a loop. There are lots of problems here. It is not directly straightforward, especially if you look at the amputees after World War II. The Americans went to Germany and learned about the prosthetics treatment and applied it to amputees after World War II. After World War II, there was a navy commander, Canty, in San Francisco Bay. When an amputee came in, he'd say, "Son, you've lost your limb. Do you want a prosthesis and do you want a Cadillac or a Ford?" The amputee would say, "I want a Cadillac," and the commander would say, "Give that man a biceps cineplasty." That is what he got; that's all it meant. Now, we don't like the appearance of those shapes today, but I think we might be able to improve them.

References

1. Shurr DG, Cook TM. Prosthetics & Orthotics. New York: Appleton & Lange; 1990.
2. Bowker JH, Michael JW (eds). Atlas of Limb Prosthetics, ed 2. St Louis: Mosby–Year Book; 1992.
3. Hall CW. Permanently attached artificial limbs. Bull Prosthet Res 10-20 1973 Fall:337–345.
4. Hall CW. Developing a permanently attached artificial limb. Bull Prosthet Res 10-22 1974 Fall:144–157.
5. Hall CW. Permanently attached artificial limbs. Bull Prosthet Res 10-23 1975 Spring:321–327.
6. Hall CW. Permanently attached artificial limbs. Bull Prosthet Res 10-24 1975 Fall:243–249.
7. Hall CW. Skeletal extension development: Criteria for future designs. Bull Prosthet Res 10-25 1976 Spring:69–96.
8. Hall CW. Permanently attached artificial limbs. Bull Prosthet Res 10-27 1977 Spring:165–168.
9. Hall CW. Permanently attached artificial limbs. Bull Prosthet Res 10-28 1977 Fall:188–191.
10. Hall CW. Permanently attached artificial limbs. Bull Prosthet Res 10-29 1978 Spring:155–158.
11. Hall CW. Permanently attached artificial limbs. Bull Prosthet Res 10-30 1978 Fall:245–248.
12. Hall CW. Permanently attached artificial limbs. Bull Prosthet Res 10-31 1979 Spring:136–145.
13. Hall CW, Rostoker W. Permanently attached artificial limbs. Bull Prosthet Res 1980 Fall;17(2):98–100.
14. Hall CW, Cox PA, McFarland SR. Some factors that influence prolonged interfacial continuity. J Biomed Mater Res 1984;18:383–393.
15. Hall CW. A future prosthetic limb device. J Rehabil Res Dev 1985;22(3):99–102.
16. Murphy EF. History and philosophy of attachment of prostheses to the musculo-skeletal system and of passage through the skin with inert materials. J Biomed Mater Res Symp 1973;4:275–295.
17. Mooney V, Predecki PK, Renning J, Gray J. Skeletal Extension of Limb Prosthetic Attachments—Problems in Tissue Reaction. J Mater Res Symp 1971;2(I):143–159.
18. Sauerbruch F. Die Willkürlich Bewegbare Kunstliche Hand: Eine Amleitung für Chirurgen und Techniker, ed 1. Berlin: Julius Springer; 1916.
19. Kessler HH. Cineplasty. Springfield, IL: Charles C Thomas; 1947.
20. Klopsteg PE, Wilson PD (eds). Human Limbs and Their Substitutes. New York: McGraw-Hill; 1954.
21. Beasley RW. The tendon exteriorization cineplasty: A preliminary report. ICIB Inter-Clinic Inf Bull 1966;V(8):6–8.
22. Childress D, Grahn E, Weir R, Heckathorne C, Uellendahl J. Modification of a Bock electric hand for E.P.P. Control by exteriorized tendons. J Proceed Am Acad Orthotists Prosthetists 1993;13.
23. Childress D. Medical/technical collaboration in prosthetics research and development. J Rehabil Res Dev 1993;30(2):vii.

Chapter 13

Osseointegrated Metacarpophalangeal Joint Prostheses in the Hand

Göran Lundborg

Power grip as well as precision movements of the hand require mobile, stable, and pain-free joints. Patients suffering from rheumatoid arthritis, osteoarthrosis, or posttraumatic/postinfectious arthrosis often present with considerable impairment in hand function. In rheumatoid patients, ulnar drift of the fingers, combined with volar subluxation and an inability to open the hand grip, may result in severe impairment of hand function.

Many principles have been suggested for the treatment of destroyed joints, such as perichondrial transplantation, soft-tissue interposition, and various arthroplasties using cemented or noncemented artificial joints.[1] The most commonly used technique today for replacement of metacarpophalangeal joints is the use of silicone rubber implants, originally introduced by Swanson.[2] This type of arthroplasty usually results in pain relief, correction of deformities, and some degree of preserved range of motion. The Swanson implant is regarded as a dynamic spacer that is surrounded by a fibrous capsule that is functionally adapted and oriented by guided postoperative motion. Thus, the purpose is not to fixate the silicone spacer to the bone—instead the implant is allowed to piston in the bone marrow cavities to spread the stresses of the implant over a broader area. Although initial results usually are satisfactory, the friction of silicone moving against bone may, in the long run, result in progressive bone destruction with subsidence and fracture of the implant as well as progressive stiffness of the hand.[3] In addition, local reactions induced by the silicone may occur.[4-5]

Based on the osseointegration principle, originally proposed by Brånemark in 1969, we developed an arthroplasty for the metacarpophalangeal joints to permanently fixate a joint implant to the phalangeal and metacarpal bones without the use of cement. The principle of osseointegration of titanium implants has been used for more than 25 years for anchorage of dental prostheses in edentulous patients.[6,7] If one uses a carefully controlled surgical protocol, titanium screws can integrate into bone tissue with no interposing connective tissue layer in the interface zone between bone and titanium.

For the past several years, the osseointegration concept has been used for fixation of joint endoprostheses at the metacarpophalangeal joint level in the hand of a large number of patients referred to the Department of Hand Surgery, University Hospital, Malmö. The purpose of this paper is to review current concepts regarding indications for the use of this technique as well as the surgical and postoperative protocol. Some typical cases are described and the results in terms of improved hand function are discussed.

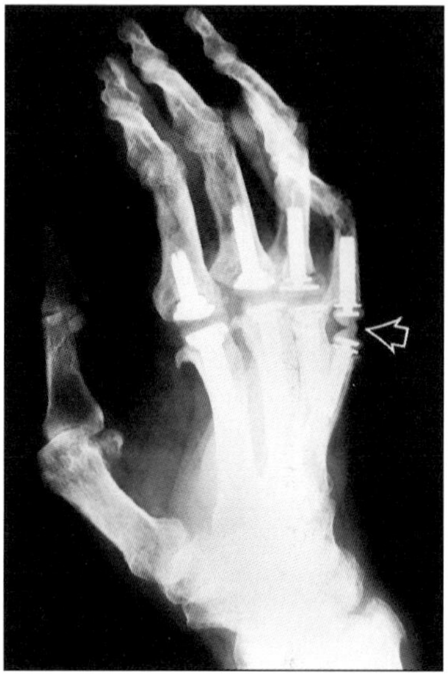

Fig 13-1 Arthroplasty of second to fifth metacarpophalangeal joints in a 55-year-old male rheumatoid patient. Note the contour of the silicone spacer at the level of the fifth metacarpophalangeal joint (arrow).

Patients

Between 1988 and 1994, 150 metacarpophalangeal joints in 45 hands (38 patients) have been operated on. Of these, seven patients underwent surgery on both hands and, in most of the cases, in the second to fifth metacarpophalangeal joints. The youngest patient was 22 years old, and the oldest was 74 years old. The majority of the patients suffered from rheumatoid arthritis, but the technique also was applied to several patients suffering from primary osteoarthrosis, posttraumatic osteoarthrosis, and postinfectious osteoarthrosis. The longest follow-up is now 6 years. Preoperatively, all cases were subjected to a careful evaluation of hand function with respect to grip strength, pain, leisure activities, and ability to perform work tasks. The same evaluation was performed at regular intervals following surgery. Radiography was carried out after 3 and 6 months as well as at yearly intervals.

The Endoprosthesis

The artificial joint consists of two components, the joint mechanism and the titanium fixtures. The screw-shaped titanium fixtures (Institute for Applied Biotechnology, Gothenburg, Sweden) were essentially of the same design as the anchorage elements used for edentulous patients. The artificial joint mechanism (ATOS Medical, Hörby, Sweden) consisted of a constrained flexible silicone spacer, each end being mounted on a titanium plate with a short stem, designed to fit into a central longitudinal cylindrical channel in each fixture (Fig 13-1).

Surgical Procedure

Usually, surgery was carried out while the patient was under general anesthesia. The surgical procedure initially followed the protocols described by Swanson.[2] A dorso-ulnar longitudinal incision was used over each joint. The extensor hood was incised to the ulnar side of the extensor tendon and appropriate amounts of the metacarpal heads were resected. Longitudinal cylindrical channels were carefully drilled by hand into the medullary cavities of the metacarpal and phalangeal bones. Special care was taken not to induce any heat; motor-driven drills were never used.

Rheumatoid patients usually present with fatty, loosely textured marrow, and thin, very dense, and often brittle cortical bone in the diaphysis. In these cases, the marrow cavities were always packed with grafts of cancellous bone and marrow from the iliac crest. The fixture sites were then prepared as described previously. Titanium screws of appropriate width and length were then introduced into the drillholes of the metacarpal and phalangeal bones so that absolute stability of the screws was achieved. A silicone spacer was connected to the titanium screws by introducing a short titanium stem from the spacer into the central channel in the screw (see Fig

13-1). An important part of the concept is that the spacer, if necessary, could be replaced at a later time, and that a new spacer at that time could be fixed to the already osseointegrated screws. Following closure of the wounds, a fluffy dressing and a splint were applied for 5 to 7 days.

Rehabilitation Program

It was thought that the artificial joint could not be left unloaded for an extended time since such a period might lead to stiffness of the hand. According to our protocol, the mobilization program, therefore, was started on the fifth to seventh postoperative day. Under supervision of an experienced hand therapist, controlled motion was initiated with the help of dynamic splints. The program initially included a limited number of passive and active exercises twice a day, increasing in intensity with time. Full range of motion and light functional activities without splints were encouraged by the sixth week, but moderate-to-heavy activities were not permitted until 3 months postoperatively.

Results

We have presented a detailed follow-up on our first 68 metacarpophalangeal joint endoprostheses.[1] At an average follow-up of 2.5 years (range, 6 to 54 months), the average range of motion was 57 degrees in the rheumatoid cases, and 50 degrees in all cases. Radiologic and clinical osseointegration occurred in every case, and there was no clinical sign of loosening. Fracture of the spacer occurred in 6%. Patient satisfaction was high with pain relief, increased range of motion, improved hand function, and good cosmetic appearance.

Over the following years, the results have been consistent with respect to osseointegration (Fig 13-2) of the fixtures as well as clinical results. Fractures of the joint mechanisms

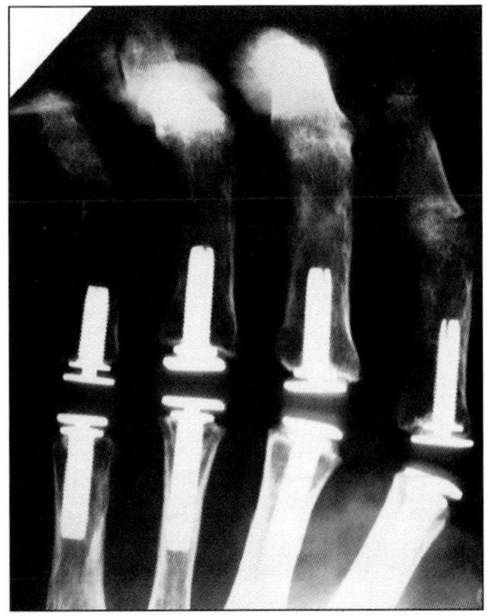

Fig 13-2 Close-up view of osseointegrated joint prostheses at metacarpophalangeal joint level 2.5 years postoperatively in a 41-year-old male suffering from severe rheumatoid arthritis. The bone marrow of metacarpal and phalangeal bones was very fatty and loosely textured, and all bone marrow cavities were packed with cancellous bone grafts from the iliac crest before introduction of screws. Note the excellent osseointegration of the titanium implant with absence of any local reaction around the screws.

seem to occur at about the same rate as reported before, usually in cases in which the proximal and interphalangeal joints had been fused, and the whole range of motion occurred at the metacarpophalangeal joint level.

Case Reports

Case 1

A 25-year-old woman presented with juvenile rheumatoid arthritis, severe pain, and advanced subluxation of the third, fourth, and fifth metacarpophalangeal joints with ulnar drift of the corresponding fingers (Fig 13-3a). The bone marrows of the metacarpal and phalangeal bones were packed with cancellous bone grafts from the iliac crest. Arthroplasty resulted in complete pain relief,

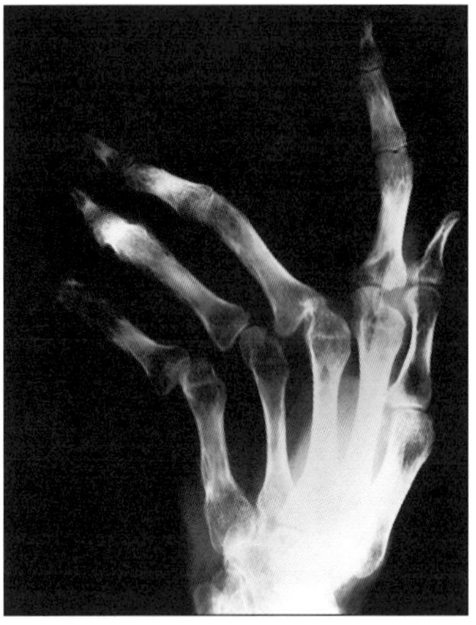

Fig 13-3a A 25-year-old female with juvenile rheumatoid arthritis and severe subluxation of the third to fifth metacarpophalangeal joints with ulnar drift of the fingers.

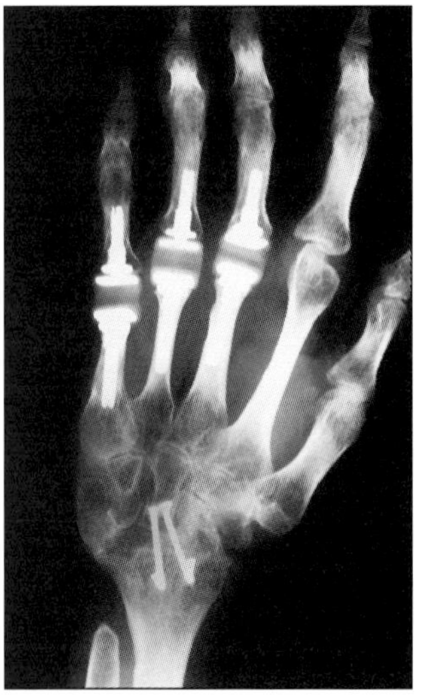

Fig 13-3b Radiograph 2 years following arthroplasty.

Fig 13-4 A 37-year-old woman suffering from advanced juvenile rheumatoid arthritis.

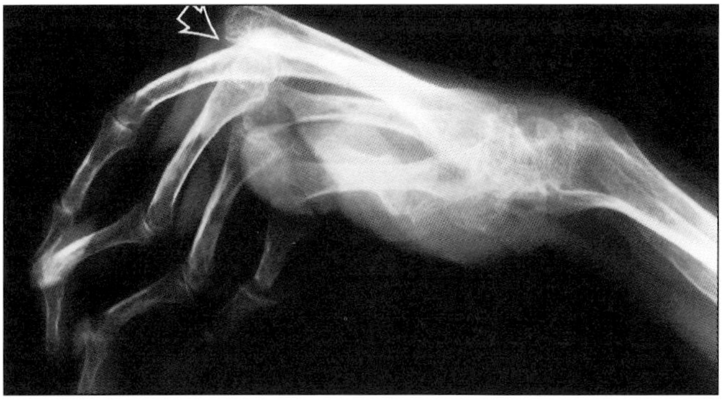

Fig 13-4a Preoperative radiograph showing fixed swan neck deformities in fingers, and advanced subluxation of all, especially the second metacarpophalangeal joint (arrow).

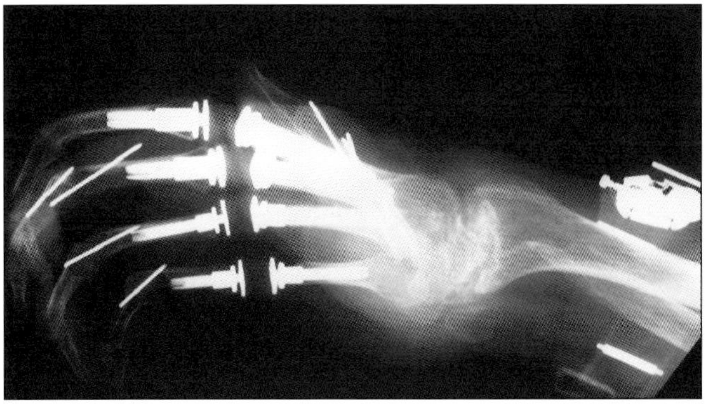

Fig 13-4b Radiograph at 1.5 years following grip reconstruction, including arthroplasties of the second to fifth metacarpophalangeal joints and fusion of the third to fifth proximal interphalangeal joints in physiologic position.

Fig 13-5 An 18-year-old man with a severe hand laceration, including substance loss of the fourth metacarpal bone treated with bone graft from iliac crest (*).

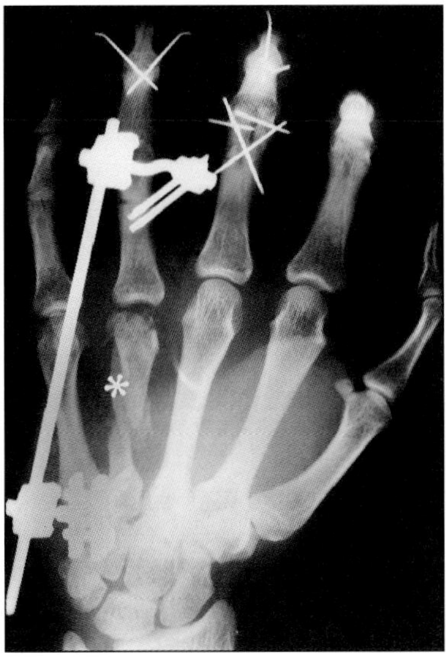

Fig 13-5a Bone block in place, stabilization with an external fixation device.

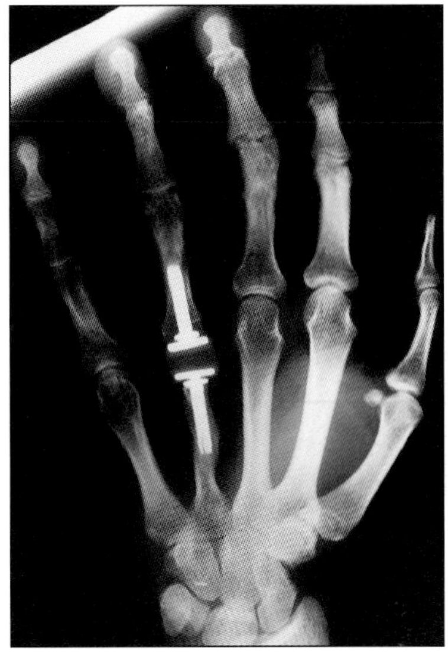

Fig 13-5b Radiograph 1 year postoperatively, with the proximal screw being introduced into the grafted bone block. Excellent osseointegration was achieved.

correction of deformities, and range of motion at 50 degrees with 15-degree extension defect. A radiograph obtained 2 years postoperatively shows perfect osseointegration (Fig 13-3b).

Case 2

A 37-year-old woman presented with juvenile rheumatoid arthritis, advanced deformities in the right hand with fixed hyperextension in the proximal interphalangeal joint ("swan neck deformity"), and advanced subluxation of the second to fifth metacarpophalangeal joints (Fig 13-4a). The patient had no useful hand grip. She underwent a major grip reconstruction, including arthroplasty of the second to fifth metacarpophalangeal joints, and fusion of the second to fifth proximal interphalangeal joints in physiologic position (Fig 13-4b). Complete pain relief was achieved, and the patient enjoyed a major improvement in grip function of the hand.

Case 3

An 18-year-old man had a laceration of the left hand that resulted in severe soft-tissue damage and laceration of the fourth metacarpophalangeal joint. He was initially treated with a posterior interosseus artery island flap to cover soft-tissue defects. A bone block from the iliac crest was grafted into the fourth metacarpal bone because of substance loss (Fig 13-5a). External fixation provided stability in the system. After incorporation of the bone block, an arthroplasty was performed, the proximal screw being fixed to the transplanted bone. Osseointegration was achieved (Fig 13-5b) with a range of motion measuring 55 degrees.

The application of the osseointegration principle to arthroplasty procedures aims at permanent fixation of the implant to the bone without use of cement. In our series, such osseointegration has been regularly achieved, even though the implant has been subjected to load already after the fifth postoperative day. In our material, osseointegration occurred in practically all cases, even when patients were receiving permanent steroid or cytotoxin medication.

Although there is a potential risk for damage of the flexible spacer in a longer time perspective, the procedure allows replacement of this part, if necessary, and fixation of a new spacer to the osseointegrated screws.

No progressive bone resorption with time has been observed so far in our material. This fact is reflected in the very good range of motion (57 degrees in rheumatoid patients), which is equal to or better than that reported in the literature from other types of arthroplasties (25 to 66 degrees).[1]

Application of the osseointegration principle for joint replacements in the hand is a natural step based on experiences from the intraoral region[8-10] and temporal bone.[11] Although we think that the problem of fixation of an implant to the bones of the hand now is solved, efforts to further improve the joint mechanisms should be encouraged. Such development must not necessarily include attempts to imitate nature, but rather to find simple and reliable solutions allowing stability as well as good range of motion. The constrained hinge is a theoretical solution for this purpose, but we expect development of improved materials in the future, offering an optimal combination of mechanical strength, mobility, and biocompatibility.

DISCUSSION

Richard Coutts Just a technical question. How many different sizes of screws do you have available?

Göran Lundborg Well, at present, too many sizes are needed, I am afraid. You never know until you are there. We have diameters varying from 3 to 6 mm and lengths from 15 to 30 mm. We are trying to establish, of course, the most commonly used size and stick to that. But to gain absolute stability, you can never be sure in advance. You can never read from the preoperative radiograph exactly what size you will have to use. So, a great number of sizes are in fact required.

RC Our experience from the hip and the knee is that when the plastic material fails, it also causes failure of the stem fixation or the metal fixation. Are you getting any kind of a reaction in the finger joints that is affecting the fixation of your osseointegrated screws?

GL No, we have not seen that.

RC In the more or less catastrophic failure were they fractures of the bone or Silastic?

GL They were fractures of the Silastic spacer. They were always at the same place in the Silastic. But we could not see anything in the radiograph indicating reaction or loosening of the screws.

William Maloney I don't do hand surgery anymore, but with the Swanson's implant, soft-tissue balancing was very critical in terms of having any success over the medium to long term. With this type of implant, is soft-tissue balancing these deformed rheumatoid hands as critical, or does the implant become more stable?

GL Yes, balance is always critical. You have to restore ligaments and centralize extensor tendons. But that is the same type of precaution as in other arthroplasty procedures.

Tommy Hansson You know that single plane radiography is not a good technique for depicting loosening or predicting longevity or anything like that. Even so, it seems very promising. The issue is, of course, that you have this type of loosening in the hips and in the knees, for example, which may occur within the next 3, 4, 5, or 10 years or something like that. You haven't planned to add biplanar radiographs?

GL We haven't done that yet. It is a very good idea, though. The technique is available in Malmö as well as in Gothenburg. We will think about that.

Åke Carlsson I know that your experience with the wrist is limited, but I see that you have implanted a fairly extensive device. If it happens to fail, do you have a second line of defense?

GL Yes, of course, you can always fuse the wrist. That requires extensive bone grafting, but that is possible to do. We have done that in several cases of failed Swanson prostheses. The problem might be, of course, that the fixation screws are so fixed to the bone that it is not easy or almost impossible to get them out. But there are ways of doing this if it is required.

TH When I was practicing hand surgery, many years ago, it was very important that you started proximally to stabilize whatever you should do. This technique promised good function, good stability, and a well-functioning metacarpophalangeal joint replacement.

GL Here there is no difference as compared to other techniques. It is a general rule that you should, if possible, fix the wrist first to get the right balance, and then treat the distal problems. In the long run, of course, we do not know what happens. We have follow-up times only since 1988. Rheumatoid arthritis is a systemic disease that produces continuous forces trying to again deviate and subluxate the fingers. I think you need 10 to 15 years of follow-up to assess the final outcome.

WM It seems to me that if you had an osseointegrated peg in the radius and in the metacarpal and you had to fuse the wrist, that you could simply couple those two pegs. Have you thought about that or tried to do this? You know we use intermedullary rods in the knee, for example, to fuse the knee. It is sometimes more difficult in the small bones, but if you had an osseointegrated screw in the metacarpal and in the radius, you could just make a connection between them to get them together.

GL Yes, certainly. We have done that in several cases in the proximal interphalangeal joint when we had problems. So, of course, that is a possibility.

References

1. Lundborg G, Brånemark P-I, Carlsson I. Metacarpophalangeal joint arthroplasty based on the osseointegration concept. J Hand Surg 1993; 18B:693–703.
2. Swanson AB. Flexible implant arthroplasty for arthritic finger joints: Rationale, technique and results of treatment. J Bone Joint Surg 1972;54A: 435–455.
3. Hagert CG, Eiken O, Ohlsson NM, Aschan W, Movin A. Metacarpophalangeal joint implants. I. Roentgenographic study on the silastic finger joint implant, Swanson design. Scand J Plast Reconstr Surg 1975;9:147–157.
4. Ferlic DC, Clayton ML, Holloway M. Complications of silicone implant surgery in the metacarpophalangeal joint. J Bone Joint Surg 1975;57A: 91–94.
5. Khoo CTK. Silicone synovitis: The current role of silicone elastomer implants in joint reconstruction. J Hand Surg 1993;18B:679–686.
6. Brånemark P-I. Introduction to osseointegration. In Brånemark P-I, Zarb GA, Albrektsson T (eds): Tissue-Integrated Prostheses: Osseointegration in Clinical Dentistry. Chicago: Quintessence; 1985: 11–76.
7. Worthington P, Brånemark P-I. Advanced Osseointegration Surgery: Applications in the Maxillofacial Region. Chicago: Quintessence; 1992.
8. Brånemark P-I, Breine U, Lindström J, Adell R, Hansson BO, Ohlsson Å. Intra-osseus anchorage of dental prosthesis. I. Experimental studies. Scand J Plast Reconstr Surg 1969;3:81–100.
9. Adell R, Lekholm U, Rockler B, Brånemark P-I. A 15 year study of osseointegrated implants in the treatment of the edentulous jaw. Int J Oral Surg 1981;10:387–416.
10. Adell R, Eriksson B, Lekholm U, Brånemark P-I, Jemt T. A long-term follow-up study of osseointegrated implants in the treatment of the totally edentulous jaw. Int J Oral Maxillofac Implants 1990;5:347–359.
11. Tjellström A, Rosenhall U, Lindström J, Hallén O, Albrektsson T, Brånemark P-I. Five-year experiences with skin-penetrating bone-anchored implants in the temporal bone. Acta Otolaryngol 1983;95: 568–575.

Chapter 14

Amputation Prostheses and Osseoperception in the Lower and Upper Extremity

Björn L. Rydevik

Amputations have been performed for several hundred years and originally, as today, these surgical procedures have often been related to war injuries. Causes for amputation on the lower extremity are mainly related to peripheral vascular disease, such as atherosclerosis, and diabetic microangiopathy, which together represent 85% to 90% of amputation cases in the lower extremity. Trauma is related to 5% to 8% in various clinical materials. However, traumatic amputation is more prominent in countries at war. Tumors constitute 4% to 6% of cases, and infections and congenital deformities are caused by 1% to 2% of cases.[1]

Prostheses have been available for many years, but with varying designs. New materials have been introduced to optimize prosthetic design and function. However, there are still considerable problems in certain patients who are rehabilitated with conventional socket prostheses. The load from the ground is transferred to skin areas that are thin and have hair, which may result in skin ulcerations and sometimes infections. Subcutaneous tissues and muscles gradually undergo atrophy. This may lead to repeated exchanges of the prostheses because of stump alterations. Some conditions with stump pain may be related to neuroma compression by the socket pressing on the soft tissues. Furthermore, there is often inexact motor control due to poor prosthetic fitting and poor sensory and tactile feedback. Thus, it is not only in the upper extremity, but also in the lower extremity that the feedback system is of significance.

Based on identification of some of the problems as outlined, a clinical project has been initiated in Gothenburg, Sweden, applying the osseointegration principle in prosthetic orthopedics to anchor external amputation prostheses in selected cases[2] (Fig 14-1).

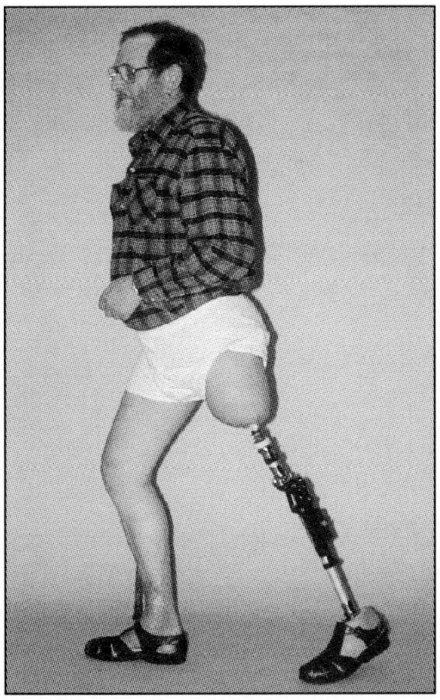

Fig 14-1 This patient has been rehabilitated with an osseointegrated amputation prosthesis following transfemoral amputation. The prosthesis provides a stable fixation to the skeleton and allows exact control of the prosthetic limb without a socket, which could interfere with nearby joint motion. The cosmetic component of the amputation prosthesis has been removed in order to demonstrate the direct anchorage to the stump.

Clinical Observations

Reconstructive surgical procedures involving installation of titanium fixtures in the medullary cavity of the remaining skeleton have been performed mainly on short amputation stumps. The surgical procedures have been performed in two stages. At stage one, the fixture was installed in the medullary cavity, and at stage two, the skin penetration was performed to allow connection of the external prostheses via an abutment connected to the osseointegrated fixture. The time between the first and the second operation was about 4 to 6 months to allow osseointegration to take place. Special attention was paid to exact surgical techniques at the area around the skin penetration.[3,4]

The advantages with transcutaneous prosthetic anchorage via osseointegrated fixtures are:

1. Direct transfer of load to the skeleton
2. Restitution of some sensory and tactile function (osseoperception)
3. No risk for mechanical skin irritation or nerve compression by the socket prosthesis
4. No need for prosthetic exchange due to stump alterations
5. Optimized control of prosthetic movements due to rigid prosthesis fixation

Osseoperception

Patients who have undergone surgery with amputation prostheses fixed to the skeleton via osseointegrated fixtures have generally reported a remarkable ability to discriminate various types of mechanical stimuli to the prosthetic foot. Anecdotal observations include reports of the ability to identify, in considerable detail, the kind of surface that the prosthetic foot is coming in contact with. To evaluate the tactile thresholds, a research program has been initiated in which vibrometric analyses are performed using a vibrometer, modified after that of Lundborg and coworkers.[5]

The overall aim of the present research is to determine the perception around osseointegrated implants in the human bone. Research on the tactile and nociceptive function around implants is of primary importance to investigate the relationship between peripheral feedback and the central pattern generator in jaw and limb motion, and to understand avoidance mechanisms for overloading of implants. Motor control of both limb and jaw muscles is determined by a central pattern generator and peripheral feedback (muscle spindles, tendon organs, joint receptors, and for the jaw also periodontal receptors).[6] By loading a limb or a tooth, exteroceptive inputs of cutaneous or periodontal receptors, respectively, inform the nervous system of the stimulus characteristics. This peripheral feedback can thus modulate the motoneuron pool to avoid overloading. Around percutaneous or permucosal osseointegrated implants, exteroceptive inputs of the skin or the periodontal ligament are lacking, and thus a potential risk exists for overloading the implants.

Nevertheless, these patients seem to function quite well, and perhaps the loss of cutaneous information is compensated for by other pathways. This previously unknown perception-phenomenon, or "osseoperception," was introduced by Per-Ingvar Brånemark to describe a phenomenon related to the perception of certain stimuli around a titanium implant osseointegrated in the human bone. Mechanisms able to compensate for the sensory loss of amputees could be of proprioceptive or exteroceptive origin, depending on the stimulus characteristics.

Proprioceptors of the muscles are known to be very sensitive to vibratory stimuli.[7,8] The principal function of proprioceptors around the joints is detection of extreme joint movement. They may also respond to vibration by transmission through the bone.

Periosteal mechanoreceptors of the Golgi-Mazzoni type are very sensitive to vibration at around 100 to 300 Hz.[9] Cutaneous mechanoreceptors also respond to different stimuli applied to the skin.[10] Slowly adapting units are activated by low-frequency stimulation (below 5 Hz). The detection of frequencies between 5 and 40 Hz is mediated by activity in the fast-adapting units of the Meissner type. Pacinian units are most easily excited by mechanical oscillations of the skin of frequencies ranging between 100 and 300 Hz.

The objectives of the present research are to assess the vibratory threshold level around osseointegrated implants in human bone by means of psychophysical threshold determination, and to compare the vibratory perception threshold of implant-supported and ordinary prostheses.

Vibratory perception around implants in the femoral, tibial, ulnar, radial, and (meta)carpal bones is assessed by means of the psychophysical threshold determination of passive stimuli applied to the implants, whereby the subject has to answer whether he detects the stimulus or not. According to Lundborg,[5,11] the idea of the presently applied noninvasive method corresponds to the previously described semiautomatic system for vibrotactile assessment of hand function. A number of modifcations were carried out, however, allowing a better quantification of the perception through receptor activation more distant of the implants. In a pilot study, experimental setup and stimulus characteristics were established.

Experiments were carried out on two groups of patients who had suffered limb amputation. Group 1 consisted of amputees who had been rehabilitated by means of prostheses supported by osseointegrated implants. Group 2 consisted of amputatees wearing conventional socket prostheses. For each subject in group 1, a subject of the same gender with a comparable age and amputation level was selected to participate in group 2.

The vibrator was driven by a 486 IBM-compatible computer through an A-to-D converter card. The applied force level was continuously monitored by means of an interconnected force transducer. The amplitude of the vibrations could be influenced by the signals of a remote control operated by the patient. The remote control consisted of a press button, which was controlled by the hand or foot, and connected to the computer by means of the same A-to-D card. As a psychophysical method, the up-down or staircase method was applied, which allowed the stimulus amplitude to be kept near the detection threshold (Stenfelt et al, unpublished data, 1996). The latter implied that the subject could modify the amplitude of the present vibrations by pushing on the remote control button when he perceived the vibrations and release it after this sensation had disappeared. At that moment, the amplitude started to increase again.

Output was recorded on an X-Y graph, with the stimulus frequency on the X axis, and the perception threshold on the Y axis. After 16 consecutive runs, the threshold could be calculated as the midrun estimate (mean value of the oscillations).

Vibratory threshold determination was carried out on the control limb and the prosthetic limb of all amputees. For above- or below-knee amputees, the great toe (I) and the little toe (V), or the metatarsal, were tested. For above- or below-elbow amputees, the thumb, the index finger (II), and the little finger (V) were tested (Fig 14-2). This method was also applied for threshold determination of the stump of amputees in group 2 to compare these values to implant stimulation threshold, as explained below. Subjects were seated comfortably in a chair in a quiet room, and provided with headphones to reduce any extraneous noise. When testing the stump of above-knee amputees, subjects were lying on a bed to establish contact between the stump skin and the vibration contactor. All subjects were instructed to keep their eyes closed during the evaluation. They were instructed to place their hand or foot on the testing table, with the midpoint of the palmar aspect of the distal phalanx of the finger

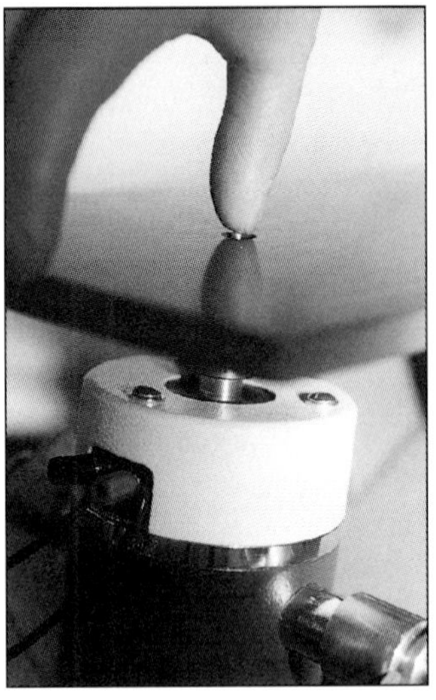

Fig 14-2 Prosthetic hand (thumb) of a patient with a below-elbow prosthesis. The prosthetic thumb is in close contact with a small pin, connected to a tone generator. This experimental setup allows evaluation of vibration perception of various frequencies.

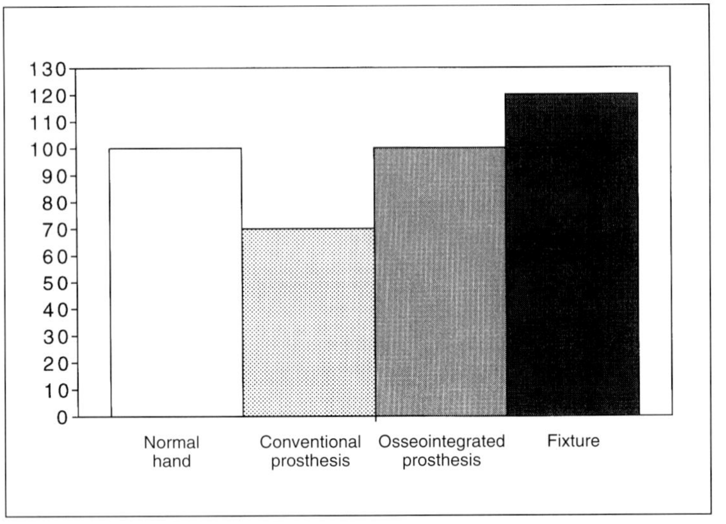

Fig 14-3 Diagram indicating the perception of vibration in a patient with a below-elbow amputation prosthesis. The vibration perception of the normal, contralateral hand is set to 100 on an arbitrary scale. Note that the perception of vibration in the hand of an osseointegrated prosthesis corresponds to that of the normal hand. The conventional prosthesis has about 70% vibration perception as compared to the osseointegrated amputation prosthesis. Note also that vibrations applied directly to the fixture (abutment) are felt even more strongly than vibrations applied to normal hand or osseointegrated prosthesis.

or toe touching the tip of the vibration exciter in the center of a small hole in the testing table. The pulp of the finger or toe was thus subjected to sinusoidal vibrations at seven frequencies, ranging from 8 to 500 Hz.

All frequencies were randomly administered for an equivalent time during all trials (8, 16, 32.5, 65, 125, and 500 Hz). The vibrations were delivered by a vibration exciter (vibrator Model V4, Gearing & Watson Electronics, Hailsham, East Sussex, England) that was driven by a power amplifier and digitally controlled. A force transducer (LIVM, dynamic force sensor, Dynamic Transducers and Systems, Chatsworth, CA) was always present between the vibrator and the vibratory pin, allowing a reliable read-out of the amplitude of the vibratory stimuli (dB-scale relative to mN). The measured perception of vibration with an osseointegrated amputation prosthesis in place was generally comparable to that of the normal contralateral hand (Fig 14-3). This was different from the corresponding measurements obtained with a conventional amputation prosthesis.

The setup for testing an isolated implant (group 1) is very similar to the aforementioned setup, although a few modifications have to be carried out. Subjects lie comfortably on a couch in a quiet room. The prosthesis is removed and a connector, similar to the abutment-design of the orthopedic implants, joins the vibrator on the implant. The design of the miniature vibratory stimulator is very similar to that applied for bone-anchored hearing aids.[12] It was rebuilt allowing production of frequencies as low as 8 Hz. The actual force level is registered by the aforementioned force transducer, which was placed in the abutment by means of a screw-type extension and fixed to the vibrator with a clip system.

Direct stable and permanent anchorage of amputation prostheses to the skeleton via osseointegrated fixtures and skin-penetrating abutments shows promising initial clinical results. The phenomenon of osseoperception, ie, the ability of patients to discriminate in detail various mechanical stimuli applied to the prosthetic hand or foot in such situations, is an important characteristic of osseointegrated amputation prostheses.

Acnowledgments

The collaboration of Dr Kjell Olmarker, University of Gothenburg, and Dr Reinhilde Jacobs, University of Leuven, in researching the evaluation of osseoperception is acknowledged.

DISCUSSION

Richard Coutts When you have a screw in the bone, do you know the load to which it is being subjected? Do you have any analysis of what the torsional and bending moments are on such a fixture in the medullary canal?

Björn Rydevik There was a considerable amount of work done on that. The question needs to be directed to Dr Skalak. This was briefly alluded to before in this workshop. Dr Skalak, would you like to comment on this, please?

Richard Skalak Yes. First about the screw itself, which is made of titanium: Its bending and torsional strength is well above that of the femur, so I think that if there is a problem, it is of the exterior connection. It has been designed to fail in a safe way. I have forgotten the numbers exactly, but if you catch your foot in a crack and fall sideways, the prosthesis is designed so that it will break outside the body and hopefully not inside. This is due to the design of the prosthesis itself.

For the bone, and the stress analysis of it, we have not made extensive compilations. There are some by Rik Huiskes, which are for the hip implants. We have made approximate computations, but there are no very detailed studies. These computations can be made, and as far as we have gone, it seems quite safe. But the question of the bone resorbing and changing the bone structure that we are dealing with is much more serious. We can compute for the original bone and the original installation, but it is clear that, clinically, some of these cases have already resorbed to a large extent, so that the initial computation is no longer the critical consideration. I think again, Rik Huiskes' computation of resorption is probably the most complete kind of computation that is available, and needs to be developed into a reliable design method.

BR Maybe one should also emphasize that these initial cases are, of course, very carefully followed up at regular visits to identify any problems. There have been no fractures of the femur and no mechanical failures of the integration, although they have been weight-bearing with the full body weight.

Dudley Childress Yes, it is the interface between the screw and the bone that should be of concern. You would lose your integrated relationship if you exceeded certain limits of stress that the bone might have for load and load transmission across that interface.

BR Dr Rickard Brånemark, would you like to comment on this in view of your evaluation of the mechanical stability and the anchoring capacity of bone?

Rickard Brånemark Well, I could say that if you correlate the findings from the animal experiments to estimate torsional capacity in the femur, you will end up with a figure of about 50 Nm. A normal femur has a torsional capacity of roughly 70 Nm, so it would be in the order of the normal femur in torsion. If you perform a gait analysis on one of the amputees, you find that it would produce torsions of 15 Nm when walking with a conventional sort of prosthesis.

DC So you are saying that the normal torsional force generated in walking is about 15 Nm?

RB Yes, if you have an above-knee prosthetic replacement.

DC But it depends on how you are walking and how fast you are walking.

RB Yes, of course, but it is an indication of the order of magnitude involved.

DC I don't think the question now is what kind of loads the femur will take. The question is, what kind of loads will the interface take? I mean, it is the relationship between the repetitive loading of that interface and what its tolerance for load is. If it is 15 Nm of load with every step over time repetitively, can the interface between the bone and the prostheses withstand that?

RB From an experimental viewpoint, we don't know anything yet about such dynamic situations, but we know that we have had patients walking on a prosthetic bone-anchored limb for several years and they still have the fixture in place.

RS Could I comment on the screw shape and the interface? Even if you have osseointegration, sufficiently large loads will produce failure. But we have not seen cases either in the dental experience or other experiences where the integration fails. The resorption that does occur is usually on the external surfaces of the bone rather than at the interface, so the interface seems very secure. In the tests Rickard Brånemark has done on rabbits, for example, the loads of pullout or pushout failure are very high. It is clear that the screw is a very secure form, and that the interface, biologically, once it is integrated, rarely seems to get into trouble. That is observational experience.

I also wanted to comment that I think the prosthetic amputee cases are all a little bit different from knee implants and hip implants, where you have a physiologic system that tends to protect the bony structures and to keep loads within proper ranges. With these joint prostheses, you have built in those kinds of protections that nature provides. I think that in the amputee prosthesis cases, you can't just transfer the socket prosthesis data to prosthesis cases with bone-anchored fixtures. I think there could be quite a difference.

Lars Magnus Bjursten I want to return to the topic of osseoperception again. I think it is important to consider how the brain interprets different signals. I am specifically thinking about the visual system where you have the primary cortex, a coupling release center, and then you compute the images deeper in the cortex. If you had vision and you lost it, you would still think in pictures when you sensed things, so you could visualize for yourself what you are sensing. If you haven't had vision, you wouldn't do that—you have other modalities to vision with. So I think the same occurs on the subconcious level with osseoperception; you use whatever information is available. Try to picture for yourself what's experienced from the environment. So the question is, what kind of information is available to you through the prosthetic device?

BR This is a very good comment and just emphasizes that we need to explore osseoperception. Dr Lundborg and I had an interesting discussion yesterday. You said that when you were in Tokyo, you noticed that they have indications on the streets for people who are blind so they could, with their stick, feel where they should walk. Maybe that could be evaluated with somebody who has an osseointegrated amputation prosthesis.

Dick Heinegård Not knowing much about how the prosthesis is designed makes it difficult to relate to the question posed before. In the normal leg, of course, you have a lot of cushions in the form of cartilage and muscle, which takes up load and acceleration profiles so that the impact is slightly different. I don't know whether that is incorporated in these prostheses, but that could interfere with osseoperception. Now, if you don't have those kinds of things, I think you may be developing a very nice model for osteoarthritis in the hip. Because if you have this impact loading, which will be fairly high, there may be microcracks that develop under a very high-impact loading. This may lead to the arthritis.

BR There are dumping mechanisms in the systems. I don't know if the engineers designing them are aware of the danger of developing osteoarthritis. But to allow for a more normal, natural situation with the heel and the other cushions is important.

Elof Eriksson Just a comment: you ought to obtain your data in darkness, in absolute darkness, and in that fashion, figure out how much vision contributes to your data when there is normal lighting.

Steven Garfin Relevant to the bone mechanisms, some comments may be made based on the spine experience where the focus is on the screw-bone interface. The inital designs of the screws had great big threads and smaller inner diameters, and they would always fail where the screws went from thread to smooth. I was impressed by the large size of the threaded component used in the femur. Then it went to a smaller diameter, at least as I saw it coming out of the skin. In pedicle screws in the spine, that junction keeps getting bigger and bigger. In the region of the hand, there was never a problem because they were controlled patients, and everybody was being watched. Even in a loose system, like Chuck Edwards had, interdigitation with the screw was just through a loose hook. The screws continue to break, and so that junction keeps getting beefed up as years go by. A small area is a hazard, and I am sure that has been taken into consideration, but I just raise the issue.

BR It is not only a matter of the actual design, but also the material properties of the metal involved, which may explain some of the differences of the design performance.

William Maloney How have you chosen the time between the first and second steps of the procedures? It seems that if you can extract, at least roughly from the noncemented hip experience, you could shorten that. It seems like the 6 months between the time you go from stage one to stage two is rather long. I would expect, at least based on the hip data and the animal data, that you have osseointegration probably at 6 to 8 weeks.

BR Well, if you remember the first discussion and presentation about the dynamics of the gradual integration and the load-bearing capacity of the bone tissue, I think this time is based on experience from surgeries in many different other areas of the human skeleton. Considering the loads that are being or will be applied, I think that has been found to be a time one would not like to reduce, at least not too much at this time. I don't know if Dr Brånemark would like to give any further comments on the significance of the time factor here.

Per-Ingvar Brånemark I agree we should minimize the time when the patient is without prostheses. In the mouth, we have experience from 1,200,000 fixtures. We found out in the early days that if the healing period was shorter than 6 weeks, we had problems in some patients because of the diversity. We believe that this is so important. We give them ample time, and think that 3 months is the minimum, really. Particularly, quite often these sites have to be loaded for many decades, so this bone is not normal in most of the cases. But when we plan a case, all the fixtures are planned so that the bone can take them and then, of course, it is a different situation.

RS We need to follow up on those comments because initially, you are going to have the trauma of the insertion, which will stimulate the bone to heal and to osseointegrate. But then, the longer you wait without loading, then that interface is not going to see the mechanical environment in which you are asking it to function. If it has not seen any load, theoretically you could get resorption simply because it doesn't have any functional load. It doesn't see any functional need and functional demand. We know this is true in using allografts or any bone grafts. You need to protect it until you begin to get some union, but then you need to apply some mechanical load to continue the process of bone formation.

PIB The information from the dental experience is that up until 18 months, there is no resorption of bone, even without loading. With the graft, fixture insertion should be done quietly so as not to disrupt the graft. The anchorage prevents resorption; because of surgery, 10% is lost, that's all. But we have shortened the time after grafting until we apply the load from 8 months to 6 months. After a few more patients, we might end up with even shorter periods.

WM I guess we have to be a little careful about extrapolating from my experience with the hip and your experience with the maxillofacial reconstructions. I think at least in the femoral diaphysis, when we are getting good contact, even good three-point contact, there is permanent incorporation of bone. We all have experienced cases when our patient can walk out of the hospital, put their crutches down, and never pick them up again. They are putting on loads anywhere between three and four times their body weight. We have shown that those implants have all of the characteristics that have been described as osseointegrated. I think it would be much different if we were trying to place an implant in bone of the distal femur where you are not going to get the same type of inital stabilities. Certainly, we have to pack the graft to restore bone stock, but this again is a much more difficult situation.

PIB Because we are looking for lifetime function for 15 or 20 years with most of the hips, I still believe it would be better for the patient if we stay with the more conservative biologic protocols. But let's wait for the 10 years and see.

TH We have heard that bone subjected to disuse for a very long time is completely different compared to the hip. They didn't find more demineralization around the implant here, but apparently it certainly almost occurs at the proximal level when it comes to demineralization, just because of the stress.

References

1. Murdoch G, Donovan R (eds): Amputation Surgery and Lower Limb Prosthetics. Edinburgh: Blackwell Scientific; 1988.
2. Brånemark P-I. Osseointegration in orthopedic surgery. In Dolci G, Favero GA (eds): Proceedings of First World Congress of Osseointegration, Venice, 1994:17–18.
3. Tjellström A, Jansson K, Brånemark P-I. Craniofacial defects. In Worthington P, Brånemark P-I (eds): Advances in Osseointegration Surgery: Applications in the Maxillofacial Region. Chicago: Quintessence; 1992:293–312.
4. Holgers K-M. Soft tissue reactions around clinical skin-penetrating titanium implants [thesis]. Gothenburg, Sweden: Gothenburg University, 1994.
5. Lundborg G, Lie-Stenström AK, Sollerman C, Strömberg T, Pyykkö I. Digital vibrogram: A new diagnostic tool for sensory testing in compression neuropathy. J Hand Surg 1986;11A:693–699.
6. Jacobs R. Sensory onput and jaw motor control [thesis]. Leuven, Belgium; University of Leuven, 1993.
7. Brown MC, Engberg I, Matthews PBC. The relative sensitivity to vibration of muscle receptors in the cat. J Physiol 1967;192:773–800.
8. Burke D, Hagbarth K-E, Löfstedt L, Wallin BG. The responses of human muscle spindle endings to vibration during isometric contraction. J Physiol 1976;261:695–711.
9. Sakada S, Aida H. Electrophysiological studies of Golgi-Mazzoni corpuscles in the periosteum of the cat facial bones. Bull Tokyo Dent Coll 1971;12: 255–272.
10. Vallbo ÅB, Johansson RS. Properties of cutaneous mechanoreceptors in the human hand related to touch sensation. Hum Neurobiol 1984;3:3–14.
11. Lundborg G, Dahlin LB, Lundström R, Necking LE, Strömberg T. Vibrotactile function of the hand in compression and vibration-induced neuropathy. Scand J Plast Reconstr Hand Surg 1992;26:275–279.
12. Håkansson B, Brandt A, Carlsson P, Tjellström A. Resonance frequencies of the human skull in vivo. J Acoust Soc Am 1993;95:1474–1481.

Soft-Tissue Management in Extremity Surgery and Amputation Prostheses

Elof Eriksson, Dennis P. Orgill, and Julian J. Pribaz

Dramatic improvements in the management of soft-tissue defects of the extremity have led to a decreased need for amputation and increased preservation of function. Current principles of soft-tissue management in the trauma patient include eliminating life-threatening injuries and using fluid resuscitation, adequate debridement, rigid fixation of bony fragments, and prompt soft-tissue closure. Advances in soft-tissue defects with loss of bone include replantation surgery, coverage of exposed bone with vascularized tissue, and bone lengthening procedures.

Hans von Gersdorff (circa 1480–1540) was reported to have performed more than 200 amputations for erysipelas or gangrene. In his book (*Feldbuch Der Wundartzney*, 1517), a diagram shows surgeons sawing a leg over a large bucket. The amputation wound was left open to heal by contraction over several weeks, often resulting in chronic draining sinus tracts. Since that time, technical advances including sterile technique, anesthesia, anastomoses of blood vessels, rigid fixation techniques for bone, myocutaneous flaps, and microsurgery have enabled surgeons to reduce the morbidity and mortality from amputation procedures.

Current surgical therapy dictates amputation for a variety of conditions, including certain cases of extremity trauma, tumor, infection, and peripheral vascular disease.[1] Critical to the decision to amputate is the blood supply to the extremity. Both the arterial inflow and venous outflow must be assessed. With the aging population, arteriosclerosis has become a common cause of decreased blood flow to the extremities, particularly in the diabetic population. Imaging of the vascular tree through angiography, Doppler, or magnetic resonance imaging has allowed surgeons to understand the anatomy of blockages in the major arteries of the extremity, and to plan to augment distal arterial flow through angioplasty or bypass grafts. For the purpose of this chapter, we will divide our discussion into the treatment of soft-tissue defects with and without bony loss.

Treatment of Soft-Tissue Defects Without Loss of Bone

In treating a variety of soft-tissue defects of the lower extremity, it is important to fully assess the patient's medical history and cardiovascular fitness. For example, in the trauma patient, a severely deformed fracture of the arm can direct attention away from a life-threatening airway problem, a pneumothorax, or a ruptured spleen. The multiple-trauma patient must be fully evaluated by a trauma surgeon. For mutilating injuries, we prefer to have a team approach with early involvement of surgeons skilled in vascular surgery, orthopedic surgery, and plastic surgery.

In contrast, the diabetic patient who presents with a nonhealing ulcer on the bottom of the foot must be carefully evaluated as to the status of the pedal pulses, often with noninvasive vascular diagnostic tests.[2] The cardiac history of these patients and an electrocardiogram need to be carefully reviewed since this patient population has a very high risk of intraoperative and postoperative myocardial infarctions. In these patients, a plain radiograph of the foot is often quite helpful in determining the presence of osteomyelitis.

The first surgical priority is to define the soft-tissue defects. For defects caused by trauma or infection, a wide debridement with meticulous hemostasis is usually indicated. When possible, a margin around the entire wound is excised so that the surgeon is confident that the wound is decontaminated. When a vital structure, such as a nerve or blood vessel is in the field, copious irrigation is necessary to reduce the bacterial load. The presence of exposed bone in the wound requires special evaluation since osteomyelitis can cause prolonged morbidity. Conventional therapy requires that the infected bone be removed, followed by adequate soft-tissue coverage and antibiotics; however, selected cases of osteomyelitis of the foot have been successfully treated with long courses of high-dose intravenous antibiotics.

Investigational studies have shown that reducing the bacterial contamination level to 10^5 or less for most types of bacteria will provide a high probability of successful closure. If the surgeon does not believe that this level of decontamination has been achieved in the operation, the wound should be left open and on dressing changes until the contamination is reduced before closure is attempted. Clostridial myonecrosis, streptococcal myonecrosis, or Meleney's progressive synergistic gangrene must to be treated promptly and aggressively with radical debridement, often requiring an open amputation above the level of the infection.

Mammalian wounds, with adequate blood supply and nutrition, will heal by a process of wound contraction and scarring in the absence of infection. However, an open wound is a liability with risks of blood loss, protein loss, and infection in the wound. For small wounds, normal saline wet to dry dressings are often adequate to close wounds in a period of days. For larger wounds with a vascularized muscle or soft-tissue base, skin grafting can be an effective method of closing wounds. For wounds with exposed vital structures, such as major blood vessels, nerves, bone, or tendons, flap coverage is often preferable. A flap differs from a graft in that it requires an independent blood supply. A local pedicle flap is supplied from axial vessels and is rotated or advanced into position. A distant pedicle flap is initially supplied from a distant site such as the superficial circumflex iliac vessels in a groin flap attached to a mutilated hand. By 2 to 3 weeks, the flap will be perfused by a dual blood supply from the surgically attached hand, and the flap can be divided safely from the groin. The advent of microsurgery revolutionized care for difficult wounds by allowing this two-stage process to take place in a single operative setting. A vascular pedicle to a given segment of tissue is dissected prior to performing arterial and venous anastomosis at the recipient site.

Soft-Tissue Defects with a Loss of Bone Length

Replantation

Several techniques have been developed to treat soft-tissue defects of the extremity with associated bony loss.[3] Replantation was first popularized by Malt in 1962, when an upper extremity was successfully re-attached. Many trauma centers are now equipped with a team of surgeons skilled in these techniques.

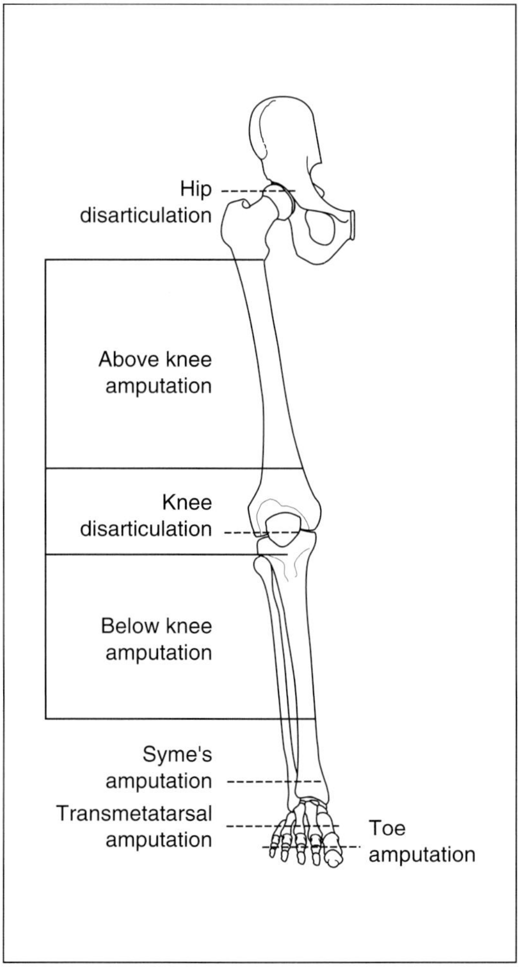

Fig 15-1 Amputation levels of the lower extremity.

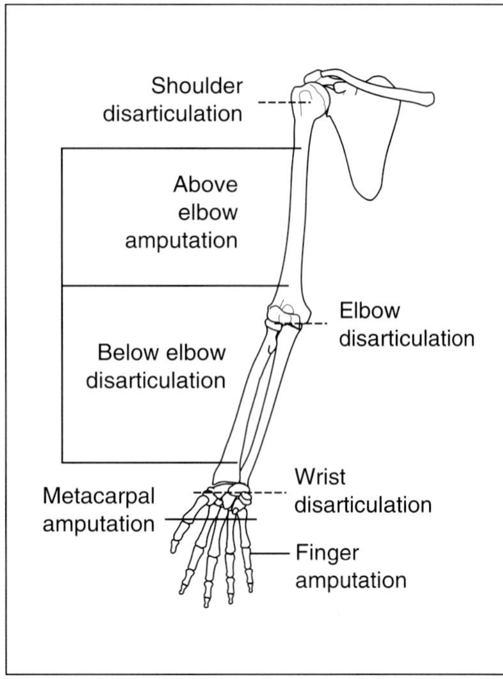

Fig 15-2 Amputation levels of the upper extremity.

Critical to the success of replantation is the time of ischemia present in the limb. Ischemia times can be prolonged with the use of cooling. Upper and lower extremity traumatic amputations containing significant muscle have significant complications when the ischemia time is greater than 3 to 4 hours. In contrast, there are successful reports of replantation of digits (lacking muscle and stored at 4°C) for up to 48 hours. The technique of replantation utilizes the principles of debridement; rigid fixation of bone; and repair of a dominant artery, nerves, and draining veins.

Conventional Amputations

Many factors contribute to the decision as to the level of amputation. In general, more distal amputations provide improved function for the patient. In patients with lower extremity amputations, there is an increased cardiovascular demand because of lack of function of the leg. A long-term study of World War II veterans showed that above-knee (AK) amputations had a mortality rate due to cardiovascular disease 1.4 times the mortality of a below-knee (BK) amputation. It has been estimated that the increased work of walking for a patient with a BK amputation is 25% over normal and 40% over normal for an AK amputation caused by BK amputations. The diagrams for amputations of the lower and upper extremity are shown in Figures 15-1 and 15-2. Below-knee amputations remain one of the most commonly performed surgical operations today. Motivated patients can perform extremely well with a properly fitted prosthesis. The transmetatarsal amputation is a common amputation (Fig 15-3), particularly for the diabetic population. It retains good function of the foot for walking. More proximal foot amputations, such as the Chopart amputation (at the tarsometatarsal junction), can be effective in selected patients. The Syme amputation is at the level of the malleolus and is suited for mutilating amputations of the foot. This amputation has the advantage that the patient does not require a prosthesis to walk short distances. The prosthesis tends to be more bulky than that of conventional below-knee amputations.

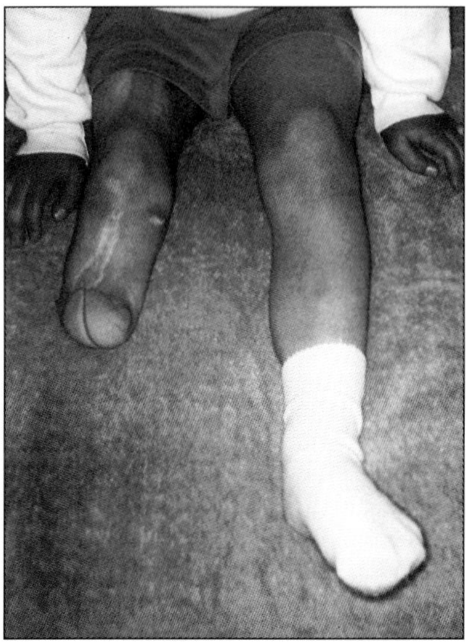

Fig 15-3 Conventional amputation.

Modified Amputations

Many clinical situations, particularly traumatic, do not exactly correspond to the typical indications for amputations. Modifications based on basic principles of amputation can be applied to each patient. In many cases, there is inadequate soft tissue to cover the bone. This can be accomplished by transferring graft or flap tissue to the area to cover the bone and preserve bony length (Fig 15-4).

Spare Parts Surgery

When a portion of a limb is amputated and not amenable to replantation, the amputated part should be kept at 4°C in the event that parts from the amputated limb can be successfully used[4] (Fig 15-5). These parts include skin grafts, tendons, blood vessels, nerve

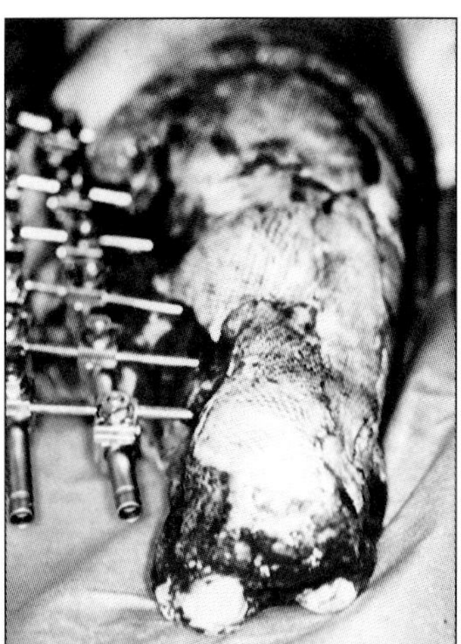

Fig 15-4a Patient with degloving injury and tibial fracture.

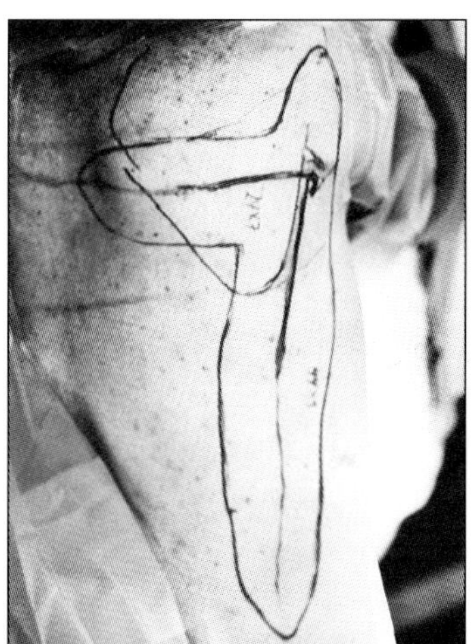

Fig 15-4b Injury was salvaged with a scapular/parascapular flap.

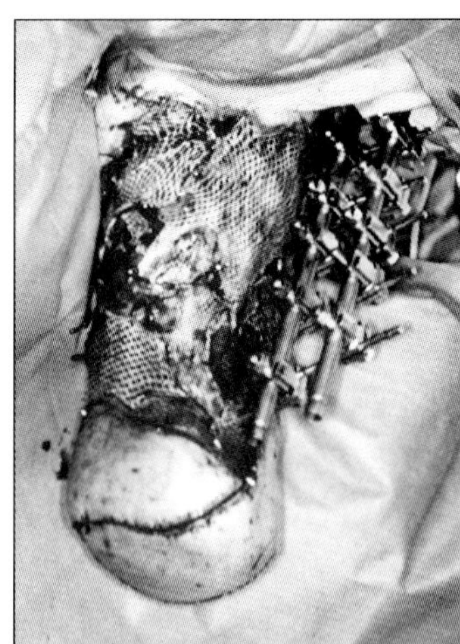

Fig 15-4c Flap was transferred by microsurgery to provide stable soft-tissue coverage of the stump.

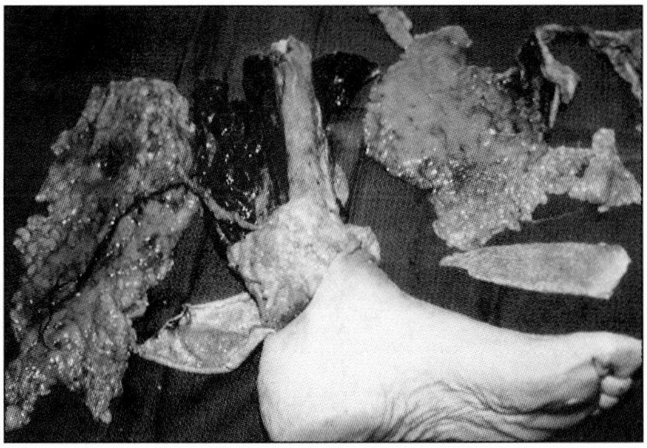

Fig 15-5a This 40-year-old woman suffered a traumatic amputation of the right foot.

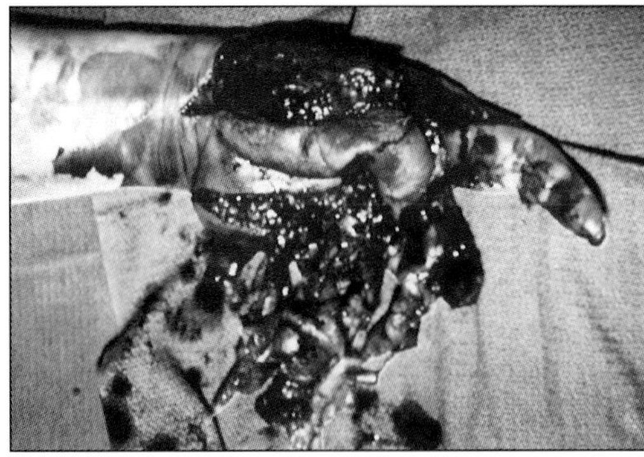

Fig 15-5b The digits of her right hand were also amputated.

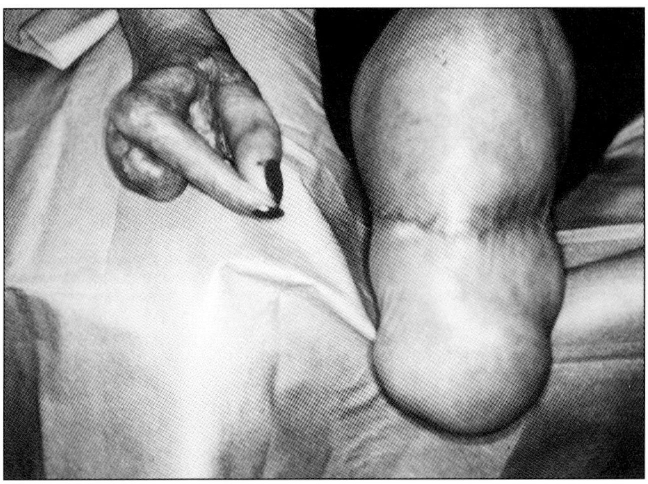

Fig 15-5c The amputated foot was used as spare parts to cover the leg amputation stump and for a toe-to-thumb transfer.

grafts, and flaps that can be dissected out and transferred by a vascular pedicle to the recipient site. By use of this technique, it is often possible to preserve a BK amputation over a knee disarticulation or AK amputation.

Ilizarov Technique

In 1951, Ilizarov from Russia described a technique for bone lengthening. Using this technique, a complex external fixator device is placed on the bone in proper alignment. A subperiosteal cut is made, and the external fixator device is distracted at about 1 mm/day.[5] Over a period of weeks, a significant increase of bony length along with the intervening soft tissues occurs.

Osseointegration

The use of an osseointegrated implant connected to an amputation stump carries some promising possibilities in prosthesis design.[6] Not only can the prosthesis be securely attached to the stump, but the patient can walk on the implant and thus avoid repeated

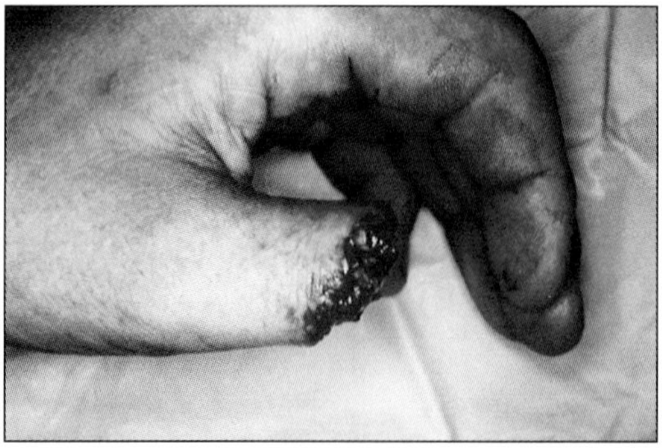

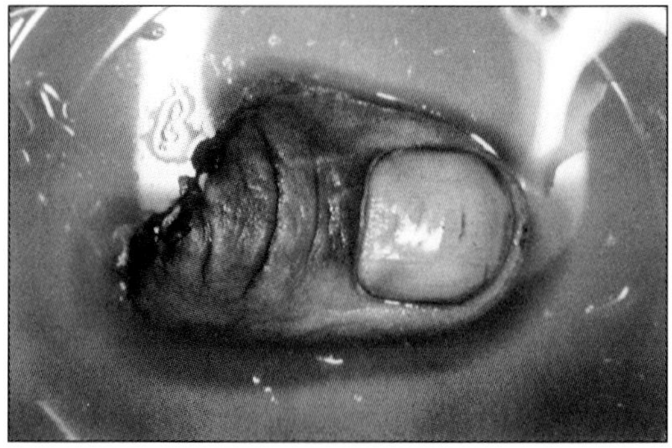

Fig 15-6a and b This 21-year-old man had a traumatic amputation of the left thumb.

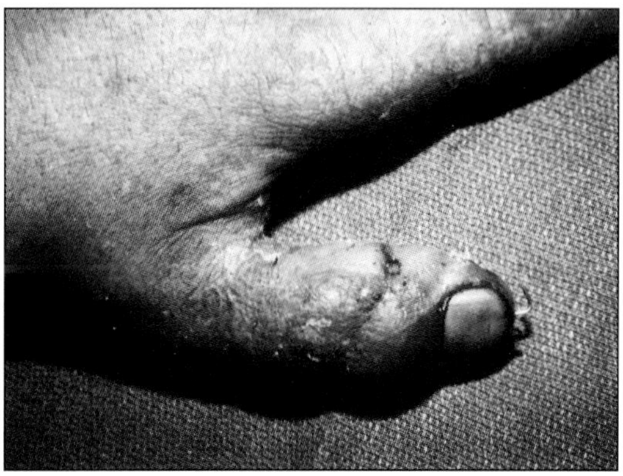

Fig 15-6c The injury was treated with replantation using microsurgery techniques.

trauma to the skin, which causes significant problems in many cases. As discussed by Dr Rydevik (see Chapter 14), this arrangement provides osseoperception, which appears to be a kind of proprioception. The soft-tissue coverage over the bone next to the osseointegrated implant needs to be immobilized in relation to the bone and the implant. Particularly in the lower extremity, this can be a problem because of the relative thickness of the soft tissues in combination with sometimes inelastic skin. In the upper extremity (Fig 15-6), where there is usually less soft tissue and more elastic skin, rotational forces appear to have less of an impact. Work is currently underway to attempt to define the optimal situation for soft-tissue coverage at the time of insertion of the osseointegrated implant or the abutment.

DISCUSSION

Per-Ingvar Brånemark Could I ask a question concerning grafting of skin in the oral situation when there is a need for soft-tissue coverage? When skin was grafted, there were concerns about the conditions around penetrating fixtures as compared to the situation of the normal periosteum. There have been some problems when the skin grafts are down to almost a split-skin graft. But then the question is how to make that skin survive in the maxillofacial region. But if we look now upon the really tricky case of the high penetration, how likely is it that the skin graft would survive? We have only done a few, so we don't have enough experience. Would you, with your background, have any comments on the prognosis of a split-skin graft placed directly on the fascia, or the muscle, or on the bone integument?

Elof Eriksson It is a very good question. I think that once the skin graft has vascularized, it is likely to survive. So then I think the key problem is to provide the vascularized bed to which you can apply a skin graft. I have a suspicion that you also have to think in terms of modifying the shape of the bone primarily as a first step, then as a second step perhaps graft bone with existing muscle if it is thin enough, or use the fascial transplant or something of that sort left to heal in perhaps the buried position and then put the skin graft there. I think if you do that and the skin graft is viable, it is likely to survive long term. We think that reduced trauma and tension on a skin graft promotes longevity.

PIB Would you suggest in these cases, perhaps a three-stage procedure: first install the fixture in bone, and after some proper healing and bone grafting and so forth, apply the split-skin graft and wait until that stabilizes, and then put on the prosthesis?

EE I think that the negatives of multiple-stage procedures is getting greater all the time because even if everything else is the same, they tend to increase cost and time of the treatment. So I think that one should try to avoid multiple stages. I would think it would be possible to shape the bone, probably just around edges, and we could place, for instance, a fascial graft at the time of placing the initial implant. We could then expect the fascia to be properly revascularized at the time we went back to put in the abutment. I see nothing wrong with placing the skin graft at the same time as placing the abutment. So I think it could still be a two-stage procedure, and I think we should try very hard to get it done in two stages because, if not, it is just going to decrease the utility of the procedure.

PIB That is exactly what we have been doing with the craniomaxillofacial region, where one has put off thinning the skin until the abutment time. But also there is no question of several years, experience of putting in the fixture abutment and split-skin graft at the same stage, as a one-stage procedure.

Björn Rydevik Do you have any comments about the possible role of controlling the local environment in wound healing in the extremities, and maybe even in the skin penetration, by growth factors or other modalities?

EE I like your question. You know I have interest in these particular issues. I think the simpler you can make this, the better. Of course, we want to design a solution that essentially every surgeon can technically execute. I think there are possibilities, such as those I have been developing using a healing liquid environment. This can be applied to the situation where, in addition to providing the wound with the growth medium, you can deliver, for instance, antibiotics. I have a strong suspicion that we will find it useful in cases of an infected wound. You can avoid infections and treat them in that situation. For instance, with this system, we could decontaminate a wound bacterially within about 48 hours. So I think it may very well have some utility. Together, with Dr P-I Brånemark and his co-workers, we are exploring these possibilities. Once we have that chamber system and the medium and the antibiotics under control, we can start transplanting cells as suspensions, etc. But we have to keep it simple.

Dudley Childress I find the area of the interface between the prosthesis and the skin to be most fascinating. You have to understand, for the nonorthopedics in this group, there is skepticism because skin penetration has always been a major source of problems for us in terms of external fixations and traction. We have problems with the skin. I was wondering what work has been done in looking at the nature of the post that comes through the skin itself. Should that be highly polished and smooth, or should it be rough? I think what Dr Brånemark said earlier is that you don't want any movement transmitted between the post and the skin, which would imply that you'd want a very highly polished and smooth post to try to avoid any attachment of the skin. Does that fit in with your concepts of how this should work?

EE My only significant experience lies with the soft tissues. I have treated only a limited number of cases where we have used implants. My practical experience from the head and neck area is not very extensive. I think it is conceptually given that there will be a microwound at the junction between the bone, skin, and implant. I think the smaller we make that microwound, the lesser in general is the impact. As to the surface of the implant, I would like to pass that question on to Dr Brånemark and the others to hear what they have to say about it.

PIB We believe that the periosteal level is important, so that 1 or 1½ mm above the bone is critical. At least that has served for the intraoral and extraoral applications, and even for the finger prostheses. And it seems to be the same for the lower extremities. Furthermore, it is technically possible in the high-femur amputees to reduce the soft tissue as much as desired. Now we feel quite sure about the anchorage, even over long periods, and have tried to redesign the anchorage element to fit these stress-shielding ideas. We believe that we could go into more aggressive resections of soft tissue. We should talk to Dr Eriksson, of course. I think that we will end up in a situation where we try to have a split-skin graft facing bare bone for about 1 cm around the fixture. Then we allow, according to Dr Skalak, the skin and soft tissue to move outside of this so that we could use it to adapt to the rotation requirements. That seems be the situation in those cases where it seems to work very well.

Richard Coutts Do you think it makes any difference as to whether the surface of the implant of the post that goes through the skin is rough or smooth?

PIB Yes, I think it is very important. I think that the surface of the abutment penetrating the periosteum should be similar to ones we have in bone because we know how those behave. It is interesting that periosteum and bone are close together. I think we shouldn't forget the marrow of either because we put marrow on top of the list of progenitors. So I think that we are coming close to a solution of this situation of split-skin grafts—toward a bone that seems to work. The real trick is to handle this rotational movement; this may be a matter of selecting the patients first—for instance, a girl with a very high amputation. We did it years ago for a very high femur amputation, and Dr Ericksson, who is experienced in soft tissues, suggested we do a radical resection of the soft tissue. I am still a little bit cautious. But I think that's the real trick, and their excision would be very important to consider.

RC Could you give us an idea of how many lower extremity amputees have this system in place?

PIB We have completed work on 17 femurs.

RC What was the selection process? Over what period were the 17 treated?

PIB The longest follow-up for this procedure is 5 years. So it is still in its infancy.

RC How many have been done in the upper limb?

PIB We have seven, I think, completed in the upper limb.

RC Over what time?

PIB That is now about 4 years' follow-up, maximum.

RC Have there been any deep infections or osteomyelitis in any of those patients?

PIB No, not related to the presence of the fixture. We had some problems with skin penetration, but those were taken care of. The only problem we had was with a fellow who had one femur with some kind of unexplained infection that was taken care of, but it was not in the interface. That seems to be safe. We had one man who could not control his movement. He got stuck three times when leaving the elevator; he rotated his limb the first two times and nothing happened. The third time, he rotated the prostheses 30 degrees to the femur, and then we had to replace it.

RC Is that the only one that had to be revised?

PIB Yes. But there may be more, I guess, because of those who had undergone amputation many years ago. The bone is so bad at the time of the installation that even if we provide autologous bone and spongeous bone grafts, I think we must optimize the healing time. I think it is very important to have adequate healing time in those cases. We are planning to try Dr Hansson's ideas about vibrations to see if we can stimulate bone modeling and remodeling. I think we are still in the very early phase for the lower limb. For the upper limb, I think we are much ahead.

Göran Lundborg We just heard somebody mention the Ilizarov technique of lengthening bone. There is, of course, a principle to lengthen skin also. I am thinking of the skin expansion principles that can provide areas of skin with good sensibility. This, in some instances, can be an ideal way of covering soft-tissue defects, and perhaps also to cover amputation stumps. Do you have much experience with tissue expansion in principle to deal with these problems?

EE Yes, I think it is one of the alternatives that we have to consider, and I think it is a good method. In my experience, tissue expansion of skin and skinlike structures is the method of choice for the areas of high priority; it is a good choice for the face and for the trunk. Then, the more distal you get on the extremities, the less attractive it becomes because the utility goes down. Here, you gain less through the expansion techniques as compared to the other areas, and the complication rate goes up. We should remember that the basic rate of complications when you deal with tissue expansion is about 25%. It is a very complicated procedure, on the average.

RC What kind of complications do you refer to?

EE Exposure of the expander and infections.

RC Is there skin death?

EE No, that usually doesn't happen, but it can happen. But things like technical failure of the expander do happen, which requires that you start over. So it is a very complicated procedure to begin with, and you would encounter the highest complications in the distal parts of the extremities, particularly the lower extremities. That doesn't mean it can't be done. Maybe we can standardize it to a point where it is useful. But let's go back to this idea of the skin moving, and to the idea of moving the point of relative motion back from the area of the implant to more a proximal location. Having extra skin gives us an opportunity to do that because less of the relative motion will be transmitted through the skin.

Another alternative in the skin graft would be to wrap a thin flap around the distal portion of the bone, thus moving the juncture between immobile skin and mobile skin in relation to the bone. Actually, the motion is the opposite: it is the bone that moves under the skin. But let's say we move that junction back more proximally. That is probably the best overall strategy, and we could probably achieve it in perhaps five or six different ways, tissue expansion being one of them. But it is rigorous and very highly complicated, so I hope we don't need to use that often.

BR I have a question I would like to direct to Dr Lieber regarding the muscle function, particularly in the lower limb amputee. If we have a patient with a standard amputation who is fitted with a socket prosthesis, we may compare that situation to a situation that we discussed here with a rigid anchorage with load transfer directly to skeleton. We think this may produce a situation that will create new demands of the musculature. I think we have some clinical observations of at least some initial muscle soreness in these patients when they start to use their musculature with new mechanical demands on them. I would like to ask you, first of all, do you think that there are reasons to believe that there may be, from a physiologic point of view, new demands on the musculature? And a second question: How about the possibility, some time after surgery, of considering muscle transfers to optimize the length of the muscle, and to provide the best possible function of the remaining musculature?

Richard Lieber I think it is reasonable to expect new demands. For example, one of the most simple changes would be to alter the relative length of muscle fiber to the amount of connective tissue. This is a good predictor of how much muscle length will change when it is forcibly lengthened. If, in a situation with an amputation, you have lost any serious connective tissue, even if that muscle was doing exactly the same thing as it had before, you would expect to see more muscle injury. That would be a very simple situation compared to an alteration in functional demands that might be placed on the muscle by the nervous system. I think that after altered neural activity, it is reasonable to expect they would adapt well.

The second issue was regarding the optimal transfers; unfortunately, we have little experience in the lower extremity. We have so little information on the normal relationship between the muscle length changes that occurred during normal movement and joint angle rotations in simulated kind of motions that I wouldn't have anything specific to say about transfers in the lower extremity.

BR How about your experience with the upper extremity? Do you have some modality or methods of evaluation of that kind of soft-tissue surgery?

RL Yes. Suppose in a situation you know for sure that a given muscle is going to experience higher excursions, or experience higher loads than before. That would be a pretty important discovery. There are relatively specific and probably appropriate transfers that you could make to accommodate the altered function. But in my mind, the understanding of what the muscle is going to be required to do is necessary.

For example, if you were looking at the function of the rectus muscle with normal sarcomere ranges, and were transferring without the patella, you could do some measurements of minimum length. Then, the rectus could be placed with some kind of device to enhance its lever arm, so that as the joint rotated, the sarcomere would be moving the right amount. You could also alter the insertion site of the muscle, so that when the joint is rotated, the sarcomere is at a certain length, or changes length over a specified range. All that is certainly possible, assuming that you knew or you had a pretty good idea of what you were going to require of the muscle.

DC Dr Gunterberg introduced a very interesting term yesterday. I think he called it the "glass leg," or the "glass limb." I wonder, of these 17 femurs that you have done, do any of them have a glass limb?

PIB I don't know offhand, but I presume not, as far as we know.

DC In other terms, how vigorously do they walk? Do they walk as if they have a glass limb?

PIB No, no, they don't walk that way. They really are walking quite normally. We have very ambitious therapists, so they are walking as much as they can. There doesn't seem to be any reluctance from the patients' point of view. But I think that what you point out is very important, and the next time we should report on the gait analysis of these patients. I am impressed by the way some of them walk anyway. Some of them, I know, are jogging 10 km and so forth, and then they have some muscle pain. There's a fellow who belongs to the Tax Department in Sweden, so he is very ambitious.

RL Is there any long-term goal to achieve active knee extensions? Is there a goal of taking the remaining quadriceps muscles and crossing the knee joint to distribute moments across the knee and hip?

DC Obviously, one of the nice things to do would be to take the quadriceps and use it to control the stage of the knee, whether it was locked or unlocked. The only person that had struggled with this is Dr Boundlager in Münster, Germany. He was not successful, but that doesn't mean that it could not be done. I wanted to raise an interesting point that a few people have hypothesized about muscle action. If you have an external bending moment on the bone, that muscle action applies the opposite bending moment so that the bending moment of the external force is compensated for by muscular action immediately. And this is one thing that perhaps you don't have in this amputation. I don't know whether that is true or not, but that has been hypothesized as one important mechanism.

Lars Magnus Bjursten I wonder what the experience is using these amputational prostheses in patients with diabetes and vascular disease as a reason for the amputation? Are any of those rehabilitated? Are any of these patients who have a lower-extremity prosthesis diabetics?

PIB We have one case. She was 55 and she had the prosthesis placed by Dr Åke Carlsson. She had the integration and no skin complication for 4 years, I think. It didn't take much load in the early phase, but then she was really able to walk with the prosthesis. She had serious problems, but we didn't see any effects in this isolated case in the bone or soft tissue. She died of another disease, so I obtained her femur and adjacent tissue, which we analyzed. So, next time we meet, I will give a report on that. Clinically, the diabetes did not seem to matter.

Åke Carlsson I can confirm that the healing of the skin penetration was completely uneventful, and thus, to my astonishment, there were no signs of infection or local irritation whatsoever.

PIB I have one more case I could mention in this context. He was a 60-year-old man who had a vascular problem, and they tried to do some vascular surgery. They gave him a graft of the artery, like a bypass. Then, unfortunately, they had to amputate successively, so he had a very short stump and very bad soft tissue. We put in a fixture, and it healed in very well. The skin was very bad, so we used a second granulation procedure for him, and he has had that now, I think, for 3 years. He is walking, and I am surprised that there are no skin problems. We see him at 6-month intervals. The only thing that disturbs him is that this graft skin is sensitive to cold, and he is from northern Sweden. But that's all.

LMB He should come to San Diego.

PIB That is a good proposition. He has a very short stump.

Björn Gunterberg Coming back to the soft-tissue problems on the stump: I would like to mention that nature provides us with some examples of soft tissues that would be ideal for this kind of procedure. Occasionally, we have amputees with an infection of the stump, and they sometimes end up with skin that is healed to the bone. There you have a situation where the skin is absolutely immobile in relation to the bone.

PIB The man that I reported about was exactly in this condition.

BG So this is something to learn about. There is another example of children who have amputations, and they have an overgrowth of the bone in relation to the soft tissues, so they have the bone pointing out almost through the skin. There you have another example of an ideal situation.

LMB But those are bad candidates for treatment of burns on the skin. I mean the skin that's burned through.

BG Yes, they are not supposed to have stress on the skin but on the bone.

Richard Skalak In regard to some of the comments that have been made, I don't think we are advocating a slip ring. That's a mechanical solution for machinery. I think in this case, you do wish to attach the skin firmly to something, preferably the prosthesis, or if not, to the bone. I think Dr Gunterberg's suggestion and experience are good for guidance. That means you probably want some roughness on the prosthesis surface so you stay in touch. I think what we need right now is a study of skin integration as we have had on osseointegration. Now we need to know more about the soft tissues. In any case, you must seek the development of firm connections. I think of the one on the teeth as a good example. There is a connection of the gingiva to teeth, which the dentist tries very hard to preserve and monitor against infection. I think the notion of slip is definitely not to be considered, but that you need a firm connection.

RC So you would like to see the skin not only attached to the bone where the implant exits, but you would like the skin also to attach to the post that is coming through the skin.

RS Yes, I think you need that as a biologic seal. If you could reliably produce that, you would have a very good situation.

Dick Heinegård I think nature has some things to say about that. There is a disease with a genetic deficiency that actually provides the anchor of the skin through underlying tissues. One could make use of those types of genetic agencies. I think one could get anchoring of skin to whatever part one decides.

References

1. Schwartz SI, Hoagland FT. Amputations. In Schwartz SI, Shines GT, Spencer FT (eds): Principles in Surgery, ed 5. New York: McGraw-Hill; 1989 2024–2027.
2. Liu PY, Eriksson E, Mustoe TA. Wound healing: Practical aspects. In Foundations of Plastic Surgery Practics: Wound Healing Concepts and Applied, Practical Principles. PSEF Instructional Courses, vol IV; 1991.
3. Bunke HJ. Microsurgery: Transplantation-Replantation. Philadelphia: Lea and Febiger; 1991.
4. Pribaz JJ, Morris DJ, Barrell D, Eriksson E. Double filet of foot free flap for emergency leg and hand coverage with ultimate great toe to thumb transfer. Plast Reconstr Surg 1993;91(6):1151–1153.
5. Jupiter JB, Kour AK, Palubo MD, Yaremehuk MJ. Limb reconstruction with free tissue transfer combined with Ilizarov method. Plast Reconstr Surg 1991;88(6):943–951.
6. Eriksson E, Brånemark PI. Osseointegration from the perspective of the plastic surgeon. Plast Reconstr Surg 1994;93(3):626–637.

Managing the Failed Total Hip Arthroplasty

Richard D. Coutts

The principal differences between a revision vs a primary total hip replacement are that, in a revision, one must remove the existing implants and any cement that may be present, and compensate for a variable amount of bone loss.

These two issues are sometimes related, in that in the process of implant removal the surgeon may contribute to the bone loss if the extraction is difficult. Another difference is that, in the majority of primary replacements, the procedure is fairly routine, and it lends itself to a reasonably regimented operative protocol. To the contrary, hip revision is highly variable and requires considerable preoperative planning to ensure that an appropriate array of implants and instruments is available to manage the variables that may be encountered. The revision procedure takes longer to perform, has a higher complication rate, and generally results in a lowered outcome prospect than a primary procedure.

Surgical Approach

The surgical approach is one of the first decisions that has to be made. The previous surgery often results in scarring, loss of tissue planes, and stiffness. This can make exposure of the acetabulum and the femur difficult, especially for the purpose of removing cement where full visualization is desirable. A transtrochanteric approach gives the best exposure of both the acetabulum and the proximal femur. The downside of this approach is the potential for nonunion, especially if the proximal bone has been thinned by osteolysis. On the other hand, trochanteric osteotomy affords the opportunity to advance the trochanter and increase the abductor muscle tension, thereby aiding in stabilization of the reconstructed joint. The surgeon must weigh the needs for exposure against the potential complications, and decide which approach will be best for any individual patient.

Component Removal

Cup Removal

The removal of a cemented, all-polyethylene cup used to be the common procedure. However, metal-backed acetabular components have existed since the early 1970s, when Harris produced such a design. The removal of both is similar. A metal backing presents the additional difficulty of presenting a very hard barrier to accessing the cement, which does not exist with an all-polyethylene component. Arched or curved gouges are introduced between the cement and the cup, with the purpose of disengaging the two. If the cup is all polyethylene, extraction can be facilitated by drilling a hole through the cup, but not through the cement. Insertion of a screw into the hole and advancing the screw when it strikes the cement will force the cup out of the cement bed.

Once the cup has been removed, the cement is cracked with an osteotome and removed from the bone bed in pieces. In the case of a metal-backed cemented cup, it is necessary to actually free the cup from the surrounding cement and remove it by a combination of pulling and levering before the cement itself can be removed. The screw extraction method just described cannot be utilized.

A cementless ingrowth metal cup must likewise be separated from bone using curved gouges, after first removing any fixation screws. The risk is that the acetabular rim will crack or be destroyed by the gouge. However, once removed, there is no cement to be removed, although there frequently is a thick fibrous lining that can be scraped off the bone surface. A pencil-thin high-speed bur can be used to create the initial separation of cup rim from the bone, which may help minimize the risk of major bone loss.

Stem Removal

Femoral stem design has evolved from the fixed head cemented stem to the modular cemented stem to the modular cementless stem. Extraction of these devices usually is not difficult if the loosening process has freed the stem from its incarcerating cement or bony bed. However, even a fibrous fixed cementless stem can prove extremely difficult to remove, as can a cemented stem that has a methyl-methacrylate precoat on its surface.

The stem extraction process for a cemented stem begins with the careful removal of cement from around the upper portion of the stem. Wherever it can be visualized, the cement is cracked with an osteotome and removed. If necessary, a shielding metal collar can be cut off with a diamond-cutting disk to remove proximal medial cement. If the head is fixed to the stem, a collaring device can be placed around the neck and a slap hammer used to gently pull the stem out. If the head is modular, special grasping devices are placed around the Morse taper, and the stem is slap-hammered out.

Once the stem is out, the remaining cement can be removed by cracking it and lifting it off the bone with osteotomes. Special long-shanked instruments have been made for this purpose, with a variety of cutting ends to facilitate working down the medullary cavity. The use of a headlight is essential for visualizing down the canal as the cement removal process progresses.

A number of adjunctive methods have been developed to make the cement removal process more rapid and easier. The SegSes system is based on the principle that new methyl-methacrylate will bond to previously cured cement. New cement is placed in the track remaining after stem removal. A threaded rod is placed in the middle of the new cement and removed after the cement is set. A series of short-threaded rods are placed into the threaded track and slap-hammered out. This method removes segments of cement 2 cm in length. This same system has an intramedullary drill guide that allows cement distal to the old stem to be drilled and extracted in the same manner as the more proximal cement. This also works well for removal of cement plugs.

An ultrasound system has been developed for cement removal. This energy form is applied to the cement through metal cutting tips. The cement virtually "melts" in front of the tool and can be scooped out. Irrigation must be used during the cutting process, which somewhat obscures visualization. This system also has a "Christmas tree" attachment that can be embedded in the distal cement plug. Once power is turned off, the cement resets around the tool, which allows the device to be slap-hammered out with the surrounding cement plug.

High-speed burs have also been recommended for removing cement. Although very effective in this application, they must be used very carefully since the burs can penetrate bone just as easily as they can remove cement. The burs must be irrigated during use, and image intensifier control is recommended to avoid bone penetration.

If the new stem is going to be cemented, it is not essential to remove all the old cement, especially if it has adhered well to the underlying bone.

The removal of a porous ingrowth stem has its own problems. Even if loose, the cementless stem may require tremendous force to remove because of fibrous adhesions to the ingrowth surface. The greater the amount of porous surface, the more difficult extraction can be. The ultimate difficulty is the removal of a fully porous-coated AML device.

Engh has devised a technique wherein a window is made in the femur, and a diamond bur is used to cut the stem in half. The proximal, wedge-shaped half of the stem is removed after releasing its surface from the bone using burs and osteotomes. A coring sleeve is then powered over the distal cylindrical half of the stem, which can then be extracted.

The ultimate exposure for retrieval of a porous device involves bivalving the femur, and elevating the incised segments off of the stem. The bone segments are then re-approximated and cabled together to restore a tubular shaft for the new stem implantation.

Once the components have been removed, the remaining bone stock is assessed, and final decisions are made on the reconstructive procedure, even though many of these decisions may have been made preoperatively during the planning process. Based on the bone deficits present, the decision of whether to utilize structural or morselized bone graft, and whether autograft or allograft will be used, is made. Whether or not to utilize cemented or porous ingrowth components, and their size and configuration, will have to be decided. The following description will deal with the major considerations regarding these decisions.

Table 16-1 Classification of Acetabular Deficiencies*

Type I.	Segmental deficiencies
	Peripheral
	Superior
	Anterior
	Posterior
	Central (medial wall absent)
Type II.	Cavitary deficiencies
	Peripheral
	Superior
	Anterior
	Posterior
	Central (medial wall intact)
Type III.	Combined deficiencies
Type IV.	Pelvic discontinuity
Type V.	Arthrodesis

*From D'Antonio JA, Capello WN, Borden LS, et al. Classification and management of acetabular abnormalities in total hip arthroscopy. Clin Orthop 1989;243:126–137. Used with permission.

Acetabular Reconstruction

It is now recognized that acetabular loosening is the result of a progressive erosion of bone secondary to the biologic response to particulate wear debris. When an acetabular component loses its fixation to bone, additional bone resorption may be caused by the mechanical motion of the cup eroding the bone.

The classification most commonly used to describe acetabular bone loss was devised by the AAOS and is shown in Table 16-1.[1]

Cemented cups have been supplanted by the use of porous ingrowth cups. The results of these two cups are comparable at the time of follow-up. The cementless cup has an added advantage in that its position can be changed after implantation, should the surgeon find the hip unstable. It is a more costly prosthesis, and its continued use may be curtailed by considerations of price.

There is as yet no evidence that cementless

cups will experience a lower incidence of late loosening compared to cemented cups. There is the requirement of having adequate amounts of host bone to contact a porous ingrowth cup when it is used in a revision situation, and if bone grafting has been performed.

A single deficit, regardless of whether it is segmental or cavitary, usually has limited impact on the reconstruction of the acetabulum. Basic research on cup fixation has shown that the best stability is achieved by capture of the rim of the prosthesis by the acetabular perimeter. A single peripheral segmental defect will not compromise the perimeter containment of the prosthetic cup. As a rule, if greater than 50% of the acetabular perimeter remains intact, it will capture a press-fit cup. Likewise, a cavitary lesion involving a single segment of the acetabulum will rarely require a modification of standard cup insertion methods. Of course, assuming that a patient may once again be a candidate for revision, any deficit that can be grafted easily should be, particularly if autogenous bone graft is available from the local site. Under most circumstances, a cavitary lesion will be bone grafted with either autogenous or allogeneic bone products to restore bone stock and give greater support for the cup over time.

With combined defects, a more challenging reconstruction must be performed. With a segmental deficit encompassing two quadrants of the periphery, it is highly unlikely that more than 50% of the perimeter will remain, and capture of the acetabular component will not occur. Likewise, a combination of a peripheral segmental lesion and a cavitary lesion will compromise stable fixation, with cup migration being the most probable outcome.

Combined lesions will almost always require grafting, and usually with structural bone allografts, especially for perimeter defects. A structural allograft can usually be fixed to the ilium with lag screws and washers. The graft, whether a proximal or distal femur part, can be directed inferiorly to replace a deficient anterior or posterior column. Alternatively, reconstructive plates have been advocated for fixing a graft to the anterior or posterior columns. It can be quite challenging to expose and fix a plate to either the ischium or the pubis, and both bones are usually soft and hold screws poorly.

It is my preferred practice to have the allograft fixed to the ilium, and the extensions secured if possible to any surrounding and remaining bone. In this circumstance, a cementless cup with multiple screw holes can function much like a bone-fracture plate. Screws securing the cup to the host bone, and additional screws through the cup into the allograft bone, help to hold the graft in place. However, a cementless cup is not recommended unless 60% of its surface makes contact with host bone, since bone ingrowth will not occur from the allograft contact.

In the circumstance of insufficient host bone, the cup should be cemented. A cemented cup with screw holes can be cemented in place, and the screw holes used for additional fixation of the graft by a screw through a hole in the cement. Alternatively, an augmentation ring can be utilized, wherein screws affix the ring to the host bone, in turn screwing into the allograft, affixing it to the pelvis. A standard all-polyethylene cup can then be cemented into this augmented, reconstructed acetabulum.

Structural allografts have been very effective in the short term for restoring deficits and re-creating circumstances in which acetabular components can be secured to the bone, particularly in the cementless, but also in the cemented situation. The long-term results of allografting suggest a failure rate of up to 40% at 7 to 8 years.

Various strategies have been devised to provide alternative reconstructive options. Cups have been designed to fill the defects, such as a double cup, wherein a second hemisphere is attached to the major cup to fill a large combined peripheral and cavitary lesion. Also, there is an augmented cup that will fill the space created by a protrusion into the acetabulum. An alternative strategy is to shift the hip center proximally and secure the cup in new bone proximally in the ilium. Hip function is not compromised biomechanically as long as a long-necked femoral component compensates for the shortening of the limb that would otherwise occur by shifting the acetabular center proximally.

Another prosthesis has emerged recently that is a modification of one of the earliest acetabular cup designs. Ring designed a stemmed acetabular component in 1964. The stem was screwed into the iliopubic bone of the ilium. McMinn has modified the original ring design. The McMinn prosthesis has an 85-mm tapered stem that is press-fit into a track drilled into the same iliopubic bone, aiming towards the sacroiliac joint. The cup has a polyethylene liner, as opposed to the metal-on-metal articulation of the ring, and it has a hydroxyapatite coating to assist bone ongrowth, compared to the smooth surface of the original ring design. The stem achieves fixation in the iliac bone, which sometimes is all that remains after extensive bone destruction has taken place. By providing initial stability to the cup through the stem, it is possible to bone graft around the cup using *morselized* bone. This type of bone revascularizes more easily and more rapidly, and has a better chance of being replaced by host bone.

Femoral Reconstruction

Femoral bone loss has been classified by the AAOS, following much the same pattern used for the acetabulum (Table 16-2).

Patterns of bone loss around the femur are not as amenable to stratification into treatment responses as are the acetabular defects. If the goal is to insert a cementless prosthesis, the considerations for prosthetic containment are similar to those described for the acetabulum. In the circumstance of a femoral component, the stem must not only be contained, but must achieve contact with most of the host bone where the porous surface is applied.

Table 16-2 American Academy of Orthopaedic Surgeons Classification of Femoral Defects

Type I.	Segmental deficiencies
	A. Proximal
	1. Partial
	2. Complete
	B. Intercalary
	C. Greater trochanteric
Type II.	Cavitary deficiencies
	A. Cancellous
	B. Cortical
	C. Ectasia
Type III.	Combined deficiencies
Type IV.	Malalignment
	A. Rotational
	B. Angular
Type V.	Stenosis
Type VI.	Discontinuity

There are modular stems that allow one to customize a prosthesis at surgery, with a large porous module that can fill a vacuous proximal femur mated to a smaller stem that will fit the uninvolved distal femur. Because the amount of proximal femoral bone loss will be highly variable, the use of modular components would seem to be the best method for using a porous ingrowth prosthesis.

Alternatively, one can opt to use a fully porous coated stem that engages unaffected bone distally, with the expectation that prosthesis fixation will occur at the distal end. The proximal deficient segment can be packed with morselized bone graft (either autograft or allograft or a mixture) in an attempt to restore the lost bone. The disadvantage in this circumstance occurs when there is a large amount of proximal bone loss and a narrow distal canal. A considerable amount of bone must be sacrificed to fit a standard dimensioned prosthesis into the femur in an effort to fill the proximal segment, while at the same time fitting the distal portion. It also requires the use of a thicker and stiffer prosthesis, which runs the risk of causing stress shielding.

Cementless fixation is a very attractive option when the surgeon encounters a medullary canal that is lined by dense cortical bone. This can occur as a result of osteolysis, or as a result of loosening and mechanical motion. A membrane forms between the prosthesis or cement and the bone. This membrane has the capacity to lyse the bone at the interface, particularly the cancellous bone that may have existed at a cement-bone interface.

Oftentimes, motion of the stem in the canal leads to a "polishing" of the medullary canal, leaving only a dense smooth cortical shell. This is not a good surface for cementing, because cement interdigitation into irregular voids is necessary for adequate fixation. In this circumstance, ie, a smooth cortical walled medullary canal, a cementless stem has a better chance of fixation and ultimate survival.

Another option exists, which is to cement the stem into the femur. The area of osteolytic bone can be debrided thoroughly, presenting a new surface for cement capture. This is a viable option, particularly if the bone loss has not been severe. Stems longer than the primary stem can be used to extend fixation further down the shaft, engaging previously unaffected bone and increasing the area of fixation. This approach becomes problematic if the revision is of a previous revision. One can envision the unfortunate patient who may end up with several revisions, each failure resulting in extension of the bone loss and involvement of more and more of the femur.

When the proximal femur is reduced to a thin shell because of osteolysis, and especially if large defects are present in the cortex, it is advisable to use a structural allograft. The allograft can be shaped to fit a variety of deficits and can be fixed to the femur in a number of ways. An onlay graft to replace a cortical segment can be cabled to the host shaft. Likewise, a graft replacing half or more of a cortical deficit can have an extension that overlaps the host shaft, and can be attached with cables. If the defect being grafted is circumferential, graft fixation can be accomplished using the stem of the femoral component. The graft should be cemented onto the stem. The stem in turn can be cemented or press-fit into the host, depending on the design chosen.

Although cortical allografts characteristically are vascularized and replaced by host bone very slowly or not at all, it is common for them to heal to the host bone at the junction of the graft to host. This has the advantage of protecting the prosthetic stem at the junction of the allograft to host bone, and preventing fatigue fracture of the stem.

A newly introduced technique and supporting instrumentation for femoral revision allows the surgeon to densely pack the medullary canal with *morselized* allograft while creating a space for a replacement stem. With this method, bone stock is restored to the femur, and it is unnecessary to use a long stem that reaches distally into the femur. If there is a segmental defect, it can be sealed with wire mesh or with allograft strips, which I prefer.

This chapter presented a review of the principles related to the performance of total hip revision. The concepts of implant and cement removal, and the subsequent reconstruction, have been covered. The major issues revolve around the type of implant(s) chosen and the amount of bone loss. A variety of techniques are available, and the surgeon must weigh the relative merits of prosthetic substitution vs graft replacement of the different bone defects encountered, while at the same time creating a reconstruction that will work in a stable and functional fashion for a long time.

DISCUSSION

Björn Rydevik What would you say would be the approximate figure or an average revision vs primary replacement of the hip joint?

Richard Coutts It ranges between 10% and 25%. As a general rule, you could figure that the revision rate is 1% per year. We are now getting the results of the 20-year clinical experience, and that's about what it is.

Tommy Hansson Is that a stable figure, or is it increasing? Have you reached a plateau?

RC I think the series that Callahan reported on of the Charnley implants showed that, even though the implants survived, most of those patients died. Of those who survived, there were certainly radiologic appearances of loosening so that you could anticipate that if the patient continued to live, he or she would continue to require revisions.

Steven Garfin You talked about a lot of work to replace hip joints, and Dr Maloney said in his talk that he would observe his patients yearly because there is often an asymptomatic loosening. For those asymptomatic cases, are you going to put them through what you just described? Why do you want to know if you are not going to do anything?

RC I think you want to know particularly from the standpoint of the osteolysis because the process could make the revision more difficult. So you would like to be able to make the decision as to whether or not you need to go ahead in a timely fashion, or wait for the catastrophic event that brings the patient into an emergency room.

SG Do all patients have pain when the radiograph shows loosening? If they don't and you rule out infection, are you going to revise?

RC It would depend on the bone loss situation. Do you agree, Dr Maloney?

William Maloney I would agree that for extensive lesions we recommend revision surgery; if you allow the process to progress, even in the absence of symptoms, you are going to end up with a situation that is nonreconstructable. So I think, depending on the location and size of lesions, that revision should be considered. We don't know exactly when to advise that. We have seen some real catastrophes with lesions that have gone on too long.

RC Cells don't have an enzyme system that can deal with plastic. What we need are cells that can process plastic.

WM I would like to make a comment and maybe get some feedback from Dr Brånemark and others. In the hip and knee situation, as well as in revision cases with poor bone quality in the metaphysis, we often graft the metaphysis. But in the acetabulum, the bone graft is a problem. Normally, we do not get osseointegration in areas that have been bone-grafted. I think that is quite different than your experience in the mandible, and that may be one of the issues of not loading the mandible initially. In the acetabulum, the graft will heal and become revascularized, but it doesn't grow into the implant; it doesn't osseointegrate. That may be related to the fact that we are loading these implants right away. We are not giving time to heal as in the mandible. So that is a problem we have in the femur, for instance; we can graft and reconstitute bone stock, but we still have to get fixations somewhere else. So we keep going farther and farther down the diaphysis.

TH Would you comment on that, Dr Brånemark?

Per-Ingvar Brånemark I believe that you are right there. On the other hand, with the thumb prosthesis and in all the finger joint replacements, we provide bone graft and there is osseointegration. No doubt about it.

WM I was fascinated to hear that.

PIB Maybe it is due to the technique used. Some exchange of ideas on that for the long bones may be useful.

WM I think it makes a difference as to what the thickness of the graft is. In a knee replacement, a slurry of autologous bone around the prosthesis at the time of the implantation assists in the osseointegration.

Åke Carlsson I recall that some kind of two-step procedure has also been tried. As a first step, you fill the defects with autologous bone and let them heal, and then you proceed with the second step. This might improve the chances for bone ingrowth.

WM That might be a better situation, but our patients and our economic system will not tolerate it.

PIB I'd put the shoe on the other foot. How would it appear to patients if somebody told them that they had to have two operations on their hip to have it fixed? How would they feel about that?

ÅC They go and see another surgeon.

TH But I know, for example, Peter Herberts in the group at Sahlgren Hospital in our department has tried a two-step procedure. Maybe we don't have to think so much about the economy.

PIB You said that the bone capacity would be lost. How about adding bone growth factors? What is Dr Eriksson using in his method of reducing infection and promoting growth? Does it have a potential for growth and repair?

WM I think that it is an interesting possibility.

PIB Thank you.

Reference

1. D'Antonio JA, Capello WN, Borden LS, et al. Classification and management of acetabular abnormalities in total hip arthroplasty. Clin Orthop 1989;243:126–137.

The Implant-Bone Interface and the Effects of Particulate Debris

William J. Maloney and Charles A. Engh

Assuming an implant material is not overtly toxic, two variables appear to be important in determining the type of tissue found at the interface between implant and bone. Initially, implant stability is critical. An implant that is rigidly fixed to bone will have a distinctly different interface from one in which motion is occurring between implant and bone under load. This emphasizes the importance of implant design and surgical technique. The second important factor relates to access of particulate wear debris to the interface. The importance of this factor becomes more apparent in the intermediate and long-term follow-up studies done on patients with total hip arthroplasties.

To understand the effect of wear particles on the implant-bone interface, it is first helpful to determine the conditions that exist in rigidly fixed implants. Only then can one understand the consequences of particulate wear debris and subsequent biologic reactions at the implant-bone interface. We have had a long-standing interest in the analysis of autopsy specimens from patients who have previously undergone cemented and cementless total hip arthroplasties.[1-9] This provides the unique opportunity to examine specimens that have been cycled in vivo. This experience as well as ongoing clinical studies form the basis for this report.

Implant Retrieval

At autopsy, whole femora were harvested from patients who had previously undergone cemented and cementless femoral replacement. The duration of implantation ranged from 1 to 16 years. To assess radiographic stability, radiographs of specimens were obtained in multiple planes. Implant stability was assessed based on both the radiographic analysis and mechanical testing.

Specially designed jigs were manufactured, and the specimens were loaded in a servo-hydraulic testing machine in both simulated single-limb stance and stair-climbing positions. Implant stability was measured by using extensometers (electrical displacement transducers) that recorded relative motion between implant and bone under load. Following mechanical testing, the specimens were embedded in methyl-methacrylate and sectioned using a high-speed, water-cooled saw. Radiographs of each section were obtained. Representative sections were either ground thin for histologic studies or examined using scanning electron microscopy.

Autopsy Studies: The Implant-Bone Interface

Mechanical testing of both the cemented and cementless femoral components demonstrated remarkable stability. In those implants graded radiographically stable, none demonstrated axial displacement of greater than 30 µm when loaded in the single-limb stance position. In the transverse plane, displacement exceeded 40 µm in only two cases. All displacements were elastic. This is comparable to in vitro studies that have evaluated the stability of femoral components cemented in the laboratory.

In the stable cemented specimens, there was a paucity of fibrous tissue seen at the cement-bone interface. Despite the frequent occurrence of radiolucencies noted on clinical and specimen radiographs, fibrous tissue at this interface was rarely seen.

When fibrous tissue was present, it most commonly occurred in the proximal 1 to 2 cm and was an extension of particulate-induced membrane formed at the hip joint. On direct examination of the transverse sections, there was no evidence that the cement-bone interface had failed. Over the majority of the surface area of the implant, the bone was in intimate contact with the cement, with no intervening fibrous tissue. Osseointegration with the bone cement was a common finding. Only one specimen, retrieved at 101 months following surgery, had any significant fibrous tissue present distal to the proximal 2 cm of the femur. Even in this specimen, the fibrous tissue was thin and sporadic. Occasional areas of fibrous tissue could be found in the other sections, but they were uncommon. Normal marrow elements were commonly seen at the cement-bone interface. Foreign body giant cells were not seen at this interface, and there were no areas of macrophage aggregates commonly seen in the membrane around loose cemented components.

As previously noted, numerous radiolucencies were seen on clinical and specimen radiographs. These radiolucencies did not represent fibrous tissue formation, but were representative of the remodeling process that occurred at the cement-bone interface. In the osteoporotic specimens, an inner cortex was noted. A second medullary canal formed between the inner and outer cortex. Radiographs obtained tangential to the second medullary canal appear as a radiolucency. This phenomenon accounted for almost all cement-bone radiolucencies seen in this study.

The histologic makeup of the implant-bone interface with the cementless specimens (AML, Depuy, Warsaw, Indiana) was similar (Fig 17-1). In a group of eight circumferentially porous-coated, chrome-cobalt stems, the mean area density of bone ingrowth was approximately 25%. The metal-bone interface in these specimens was osseointegrated in numerous areas (35%). In the areas that were osseointegrated, 65% of the pore space was occupied by bone. The most consistent area of bone ingrowth occurred at the end of the porous coating. Here, the porous surface was clearly integrated with the cortex.

Gaps between implant and bone were most commonly occupied by normal marrow elements. In isolated areas, there was fibrous tissue ingrowth into the porous surface. The fibrous tissue in these regions was a paucicellular, collagenous stroma. The fibrous tissue typically extended to the substrate.

High-powered light microscopic evaluation failed to demonstrate any evidence for particulate debris in these regions. A second pattern of fibrous tissue was seen around the smooth portion of the stem distally. The fibrous tissue around the nonporous area of the implant was well-organized, collagenous tissue, the fibers of which ran parallel to the implant surface. In some of the long-term specimens, the fibrous tissue was partially mineralized.

These autopsy studies have demonstrated that osseointegration is not only possible, but common in stable cemented and noncemented chrome-cobalt femoral components. The implants demonstrated no progressive displacement comparing clinical radiographs with specimen radiographs. Under physiologic loads, no slip or permanent displacement was seen. Histomorphologic analysis demonstrated viable bone in direct contact with cement and chrome-cobalt alloy without intervening fibrous tissue or gaps. Normal marrow elements were also seen at the interface. Similar findings have been noted in stable cementless acetabular components.

The Effect of Particulate Debris on the Implant-Bone Interface

Particles can gain access to the implant-bone interface by local implant failure (corrosion/abrasion/fragmentation), or by dispersion in the joint fluid via the "effective joint space." The concept of the effective joint space has been helpful in explaining and predicting the effect of wear particles on the implant-bone interface. As described by Schmalzried et al,[10] the effective joint space is defined as the entire implant-bone interface and surrounding tissues that is accessible to joint fluid and thus particulate debris. Our autopsy studies have demonstrated that both cement and circumferential porous coatings are highly effective in limiting the effective joint space in the femur. In these specimens, there was no histologic evidence of particulate debris in the diaphyseal region.

The cement-bone interface of cemented acetabular components appears to be particularly prone to particle/fibrous tissue invasion, and is a good example of the effective joint space.[11]

Intimate apposition of bone with cement in the acetabulum was noted, although it was less common than in the femur. Coronal sections and contact radiographs revealed soft-tissue interposition between cement and bone to varying degrees in every specimen. The soft tissue was generally thickest at the intra-articular margin.

Histologically, the pseudocapsule and the peripheral interface membrane were essentially identical, characterized by macrophages with abundant cytoplasmic reticulum and containing multiple particles of polyethylene. The joint pseudocapsule was in continuity with the soft tissue between the bone and the cement at the intra-articular margin.

Toward the dome of the implant at a variable distance from the intra-articular margin, transition zones from areas with membrane interposition between the bone and the cement, to areas of intimate apposition of bone and cement without any intervening soft tissue, were identified. These regions were characterized by a cutting wedge of bone resorption that appears to progressively disrupt the cement-bone interface. This cutting wedge consists of collections of plump macrophages laden with submicron polyethylene wear debris in association with evidence of active bone resorption. The three-dimensional extent of the soft-tissue interposition was directly related to implant stability.

The effect of particulate debris on the implant-bone interface in the femoral autopsy specimens was noted in two areas. In the metaphysis, the proximal 1 to 2 cm of the implant-bone interface was occupied by a soft-tissue membrane. Despite the fact that the femoral components were mechanically stable, there was a soft-tissue membrane between implant and bone in some of the specimens. This membrane appeared to be an extension of the pseudocapsule. Histologic analysis revealed the characteristic membrane seen in loose implants with a combination of macrophages and fibroblasts. Polarized light microscopy showed evidence of polyethylene debris.

The second example of the effect of particulate debris on the interface represents areas of focal osteolysis (Figs 17-2 and 17-3). Two of the cemented specimens in this series

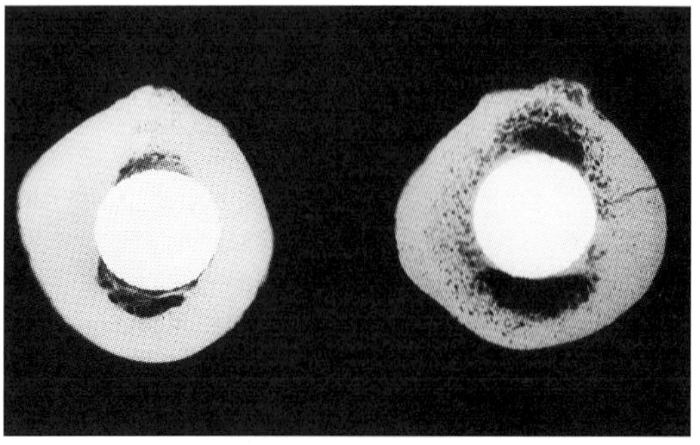

Fig 17-1 Diaphyseal cross sections through paired femora harvested at autopsy from a patient who had a total hip arthroplasty for unilateral hip disease. *Left,* The implant was placed in the laboratory, demonstrating the situation that exists immediately after surgery. *Right,* The implant had been in place for 6 years. Note the cortical remodeling and interface between implant and bone.

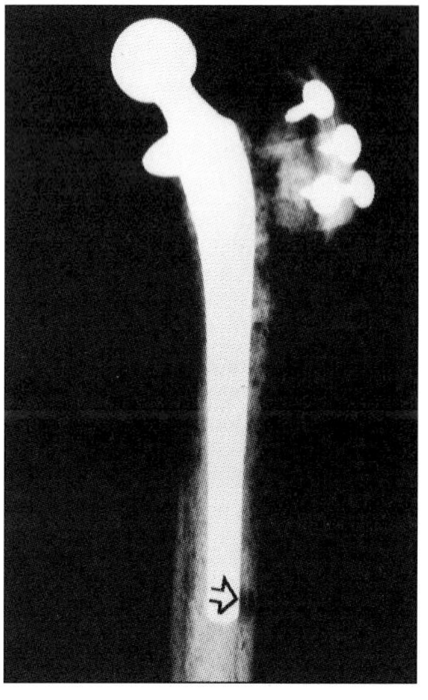

Fig 17-2a Focal cortical osteolysis in an otherwise stable cemented component.

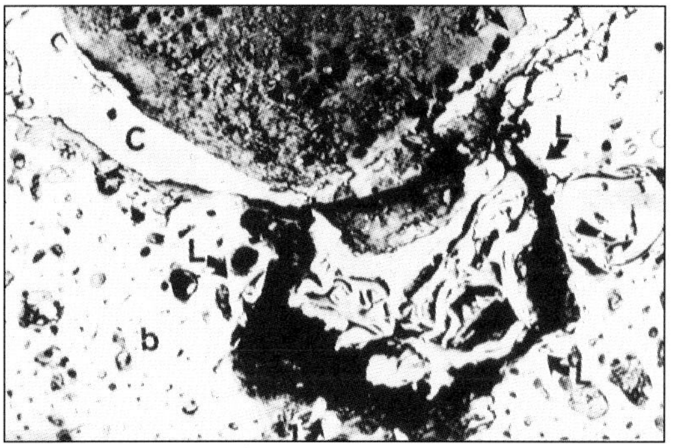

Fig 17-2b Scanning electron micrograph in the same femur in cross section, demonstrating fragmented cement adjacent to an area of focal cortical osteolysis.

demonstrated focal cortical osteolysis despite being radiographically and mechanically stable.[12] In one of these specimens, focal osteolysis at the tip of the implant was noted. Localized debonding of the tip of the prosthesis from the cement mantle had occurred. This, combined with a thin cement mantle in this region, led to numerous fractures through the voids in the cement and generation of particulate debris. In both specimens, the area of focal osteolysis was restricted to areas of localized acrylic fragmentation.

In the cementless specimens, focal areas of osteolysis were noted in the area of the greater trochanter. Histologically, these lesions had the characteristic features of a foreign body granuloma. Macrophages predominated. Polarized light microscopy demonstrated polyethylene debris.

Clinical Studies

The potential for wear debris to have a deleterious effect on the implant-bone interface is not a new idea. Early work by Willert and Semlitsch[13] noted the foreign body reaction that occurs in response to wear particles. Since then, numerous studies have examined interface tissue, and have shown that this tissue has the capacity to produce substances such as interleukin-1 and prostaglandin E2. These substances are known to play a role in bone resorption. More recent analyses using immunohistochemistry and in situ hybridization have been useful in clearly identi-

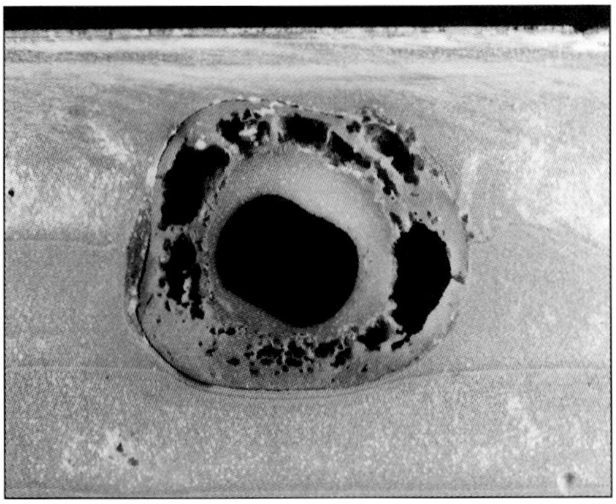

Fig 17-3a Focal cortical osteolysis.

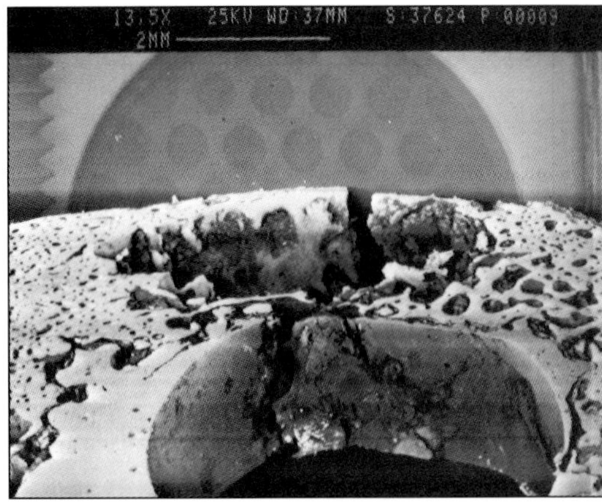

Fig 17-3b Scanning electron micrograph after the implant has been removed demonstrated cement fragmentation adjacent to the area of cortical lysis.

fying the specific cell types and cell products (eg, cytokines) involved in these processes.

Our clinical studies illustrate the importance of implant design on limiting the effective joint space. This is clearly demonstrated through our review of two different cementless stems: the Harris-Galante (HG) and AML components.[14-17] The HG stem is a patch porous-coated stem, whereas the AML is circumferentially coated. These design features should theoretically influence the effective joint space and thus the location of osteolytic reactions.

In the patients with HG stems, the incidence of femoral osteolysis increased from 22% at 3.5 years to 52% at 6 years. Two thirds of the osteolytic lesions occurred around the tip of the implant (Fig 17-4). In contrast, there was no osteolysis in the diaphyseal region of the femur in patients who had AML stems. All osteolytic reactions occurred in the metaphysis at or above the level of the lesser trochanter.

This information is important for future implant designs. In the femur, both cement and porous-coating, provided it is circumferential, are relatively effective barriers for migration of particulate debris along the implant-bone interface. Patch porous coating should be avoided.

Analysis of post-mortem specimens from patients who had clinically successful cemented and cementless total hip arthroplasties has demonstrated that implants become osseointegrated. They are radiographically and mechanically stable. There is no evidence of progressive displacement or slip at the implant-bone interface. Histomorphologic analysis reveals that bone is in direct contact with both cement and cobalt-chrome alloy, proving that osseointegration is not unique to titanium. This evidence strongly suggests that osseointegration may occur regardless of implant material, as long as the initial implant stability that exists at the time of surgery is sufficient to withstand site-specific loading, minimizing interface motion sufficiently to allow healing.

In the hip, production of wear debris is the number one long-term problem.[18] The biologic response to wear particles may compromise the implant-bone interface, resulting in aseptic loosening and osteolysis.[19,20] Research efforts are currently aimed at minimizing wear through new materials and examining the cellular mechanisms involved in osteolysis.

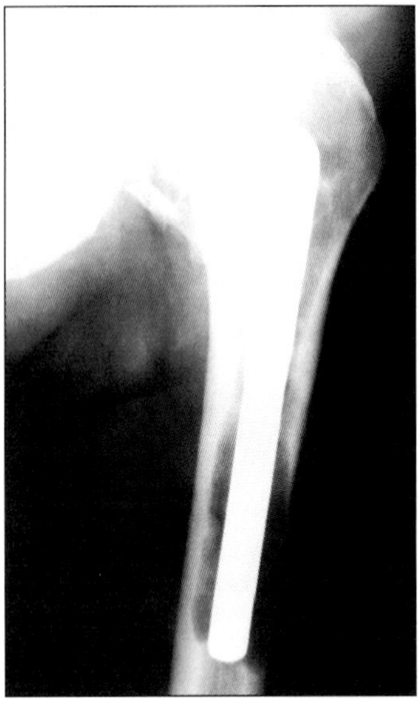

Fig 17-4a Diaphyseal osteolysis in a patch porous-coated implant.

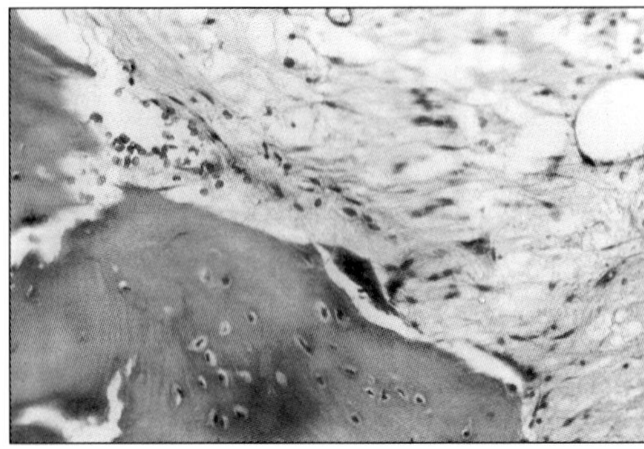

Fig 17-4b Femoral biopsy specimen demonstrating osteoclastic bone resorption at the soft-tissue–bone interface.

DISCUSSION

Robert Pedowitz Can one design a prosthesis that eventually develops a collar that seals off the debris particles from the prosthesis?

William Maloney Particles and joint fluid tend to follow the path of least resistance. The body can handle a tremendous amount of particles, but when the system is overloaded, it becomes a problem. We have seen cases where they go out into the soft tissue. Patients presented to the general surgeons with large masses in their groins. There have been resections of what would appear to be sarcomas, and in fact it's just a granulation from the hip joint. So the particles are going somewhere, but we prefer them not to go along interfaces. We really need to create implants that do better in terms of particle production.

Richard Coutts This is really a question for Dr Brånemark. Was it a titanium alloy prosthesis? If the concept of osseointegration is correct, and that bone likes titanium, one would anticipate that it might like this titanium alloy surface. If so, this leads to the question of how critical it is that one have pure titanium vs the titanium alloy as an integration surface. Clearly in this prosthesis it did not integrate.

Per-Ingvar Brånemark For osseointegration in the oral cavity, it was very important to have this particular material—because of the bacterial access to the surface. But as I mentioned in my introduction, Lars Lindhe showed that if you take revised hip joint and use cement, they feel that there is osseointegration. There is close contact and no relative motion, as far as I understand it. So I don't believe that osseointegration is exclusive to one particular material or one particular surface. It's just that we don't know whether other surfaces work safely or not until after 10 years of a large number of follow-up studies. And that leads, I suppose, to the general discussion of what happens at the interface over very long periods. That is a question for Dr Heinegård and Dr Bjursten, most certainly.

So that is my concern. What will happen over many decades if you have materials that don't have that particular kind of corrosion resistance that pure titanium has?

RC So you are saying there is a difference between pure titanium and the titanium alloy.

PIB Yes, I think there is. But that is important for the older hip-joint patients, those 65 to 70 years of age. I don't know, but I think it is something very important to consider.

Björn Rydevik I wonder about the clinical results and the follow-up results. Would you comment briefly on that? Particularly, you seem to equate cemented prostheses to noncemented prostheses in the hip joint. Do you see any differences in terms of the indications for those different kind of systems? Do you have any comments on the overall clinical results and the follow-up results available to judge by?

WM First of all, I don't think you can compare cement with cementless generically. All cemented implants are not the same. I think design issues are important. If the cementless implant that I talked about had a good cemented femoral component at 10 years, the results are equivalent in terms of clinical function. I think the issue of what is best in any given individual is more confused now than it has ever been.

I just returned from the NIH conference in Washington, DC. Peter Herberts was there and presented a lot of the data from Sweden. We have seen some late failures now with cement that we didn't anticipate in terms of loosening, and we are seeing that some of the cementless designs are doing much better than we anticipated. For instance, Dr Engh does the same implant for all patients, old or young, and it doesn't matter what bone quality it is. If he gets what he calls a scratch fit in the diaphysis, the implants become osseointegrated. They become radiographically osseointegrated 99% of the time. It is a rare patient that fails. I have never seen a single case that becomes unosseointegrated. Now, that is not true with cement. Cement becomes osseointegrated, but over time, the cement itself fails; we have seen that in our autopsy specimens. Even though the cement-bone interface osseointegrates, we have failure at the metal-cement interface.

I now cement most of my implants, but my thinking on this is changing, and I am concerned that if the implant is failing mechanically, it will eventually fail clinically. That is in direct contrast to the cementless experience that I have been looking at. There is no evidence of failure or toxicity at the implant-bone interface. I think when we are looking at a patient who is 40 years old, and has a life expectancy of 45 years, it is critical to have a situation where the implant is osseointegrated and can take those kinds of loads over that period. I think we can get implants osseointegrated with good surgical techniques and designs 99% of the time, but what remains to be seen is whether we can design an implant that won't break, that the substrate won't fail, and that the beads won't come off in 40 to 45 years.

BR Would that include the series that you are talking about that shows long-term prognosis indicating stable permanent anchorage 99% of the time, but eventually fails?

WM Yes, more than 90% of the cases are that. That has been published in the Journal of Bone and Joint Surgery.

BR Do you mean noncemented or cemented?

WM Cemented. The results with the cemented implants at 10 to 15 years in terms of implant survival; on the femoral side, is about 90%. Some institutions, like the Mayo Clinic, are at 93%, but they are all in the same range, 85% to 93%. The acetabular side in this country with cemented is not doing as well. At the same time, cemented acetabular components are unsatisfactory radiographically in about 50% of the cases, and about a third of those patients have had the cup revised.

Åke Carlsson I would like to point out a possible alternative mechanism. Even if there is an integration between cement and bone in the femur, there might be a space up to 100 µm in width between the cement and the prosthesis. This is another pathway down to the distal femur. There might be some kind of pumping mechanism allowing fluid, and later on debris, to be pumped to a distant site, provided there is a defect or a fracture in the cement. The effect might be the same.

WM Yes, I agree with you 100%. We have seen some evidence of that in the autopsy specimens of revisions. You often see a thin fibrous layer or thin membrane around the femoral component between the metal and the cement. If you look at that, you can see debris.

In the British Journal of Bone and Joint Surgery in 1991, Anthony et al actually reported on a group of specimens that had local femoral osteolysis with otherwise stable implants. Their interpretation was that there was indeed a mechanism for those lesions, and that a pumping action drove fluid and debris between the metal and cement out through cement fractures. I think I have seen that. It is a hard thing to prove in retrospect, but I agree with your point. I think it is a mechanism for osteolysis.

ÅC There is also another thing. Most hydroxyapatite-coated cementless prostheses are only partly coated, and very often you see a sclerotic line around the smooth distal part, which is an indication of motion. As far as I can recall, I have seen no report that this phenomenon leads to mechanical stimulation of mediators attracting osteoclasts. Osteolytic lesions may appear with time, but up to 7 years we have not seen them.

WM I think that the radiographic picture you describe parallels observations around the smooth portion of partially coated cementless implants.

Dennis Osmond I was intrigued by the formation of the rather extensive marrow cavity around the diaphyseal implant in your studies. You described this as a secondary medullary canal. Is this in the sense that originally the prosthesis would be in apposition with the cortical bone, so this cavity is located in what was originally cortical bone?

WM That is correct. We have tried to model that on the computer, but we have never been able to make the bone do that. That is quite a uniform finding in the more osteoporotic specimens. The inner cortex remains, and you get the second medullary canal that Charnley described 20 years ago with radial struts of bone that go from the inner to the outer cortex.

I think it must be stress-related. I can't hypothesize any other mechanism. I wish that would happen in the cementless implants. Again, it gets to the point that all osseointegrated interfaces aren't the same in terms of their ability to transfer load. You can have osseointegration up against a smooth surface that is not very strong, as opposed to osseointegration against the porous surface or a screw that has indentations on it, that interface will resist a lot of shear. I think it is a way the load can be transferred to the endosteal bone. The cemented specimen appears to be at least favorable. I can explain it by postulating that there is a small amount of load transferred over a large surface area, whereas in the cementless case, you have a large percentage of load transferred over a relatively small surface area. You never see that second medullary canal forming in this case.

DO One wonders if stress is necessary. As we discussed previously, ectopic locations bearing no stress at all tend to develop a marrow cavity.

WM Yes, that brings up another issue. What drives the remodeling processes? We don't know. Using the intact side as a control, you will realize that the patient has lost some bone compared to the normal side. The only thing we can correlate with how much bone is lost after a cemented or cementless total hip replacement is how much there was to start with. The patients that have the most marked osteoporosis after total hip replacement are the patients who start with poor bone.

DO There was an extensive apposition between the cemented prostheses and marrow elements. It seems to me that your autopsy series provides a very nice opportunity to look at that interface in some detail. I wonder if you had any observations as to the cellular or molecular nature of the interface.

WM That is getting out of my area of expertise. I can tell you that the marrow elements were viable. They looked like healthy marrow elements. We had histologists looking at them saying, "You know, that's normal marrow to me." Beyond that, we haven't the characterized cellular nature of those elements.

PIB It seems to me that we are coming to a situation where we could grade osseointegration according to the quality of mechanical capacity and the quality of the biologic component, whether there is normal or abnormal bone, normal or abnormal marrow, multinuclear giant cells, and so forth. I sincerely appreciate your presentation because you bring up the situation where we have not looked at one particular type of hardware. It is a reconstruction system. I believe in the marrow with the cementless prostheses, which works very well. If you don't remove all the marrow, you don't kill. You allow the marrow to be there. I think that could be one reason why it works.

There is another possibility, as you indicated. You use a very gentle surgical technique. When you introduce the cementless prosthesis, you might not use exactly that meticulous procedure. So I think it is extremely important now that we move into a discussion on different subgroups of osseointegration and that we emphasize the very long-term follow-up studies. Thank you very much for your contribution.

WM I just wanted to make one comment. I think you have site-specific differences. The femoral diaphysis is really a friendly surface for any material. We can get away with rougher surgical technique. I related the experience of having residents put the reamer on the drill and smooth it, and it still works. But I don't think you are going to do that in the metaphysis, and I don't think you are going to do that in the jaw. I just think that the femoral diaphysis happens to be the most friendly area in the body for osseointegration.

PIB Could I just make a comment there? I believe that the more you kill the bone, the longer it takes for it to react and to reject. Charnley said that many years ago when I visited him.

Lars Magnus Bjursten I was intrigued by your findings of completely different responses to the smooth as opposed to beaded metal surfaces. Do you have any specific feeling on which is the optimal surface?

WM This is purely speculation, but I am concerned about any smooth implant that is put in a cementless mode for several reasons. One is that I think you are going to get more areas of fibrous tissue, as I showed you. I think it is more difficult to put it in with stability from the beginning, and in the hip we can't unload it. It is going to be loaded from the beginning. I think the debris that has been generated is going to be forced down the interface much more easily than if you have a porous surface.

More than that, I can't tell you. I think that these beaded surfaces do become osseointegrated in almost 100% of cases. For at least up to 11 years, we haven't seen any evidence on the autopsies that the substrates are failing, or that the implant itself is failing. There is absolutely no evidence of toxicity at that interface up to 11 years after surgery.

PIB What's the kind of surface on the porous material of the beads? Is it a very smooth surface, or is it slightly irregular?

WM It is smooth, but they are not polished. Some people do make polished beads, but after these beads have been grit-blasted, they are as smooth as the implants.

PIB Are there irregularities?

WM There are small irregularities. I don't know the roughness of the beads.

ÅC I would like to say a few words about the mechanisms behind osteolysis around the acetabular component. The most common opinion is that debris produced in the bearing causes bone resorption at the edges of the socket, which later becomes loose. There might, however, be other mechanisms starting the events. The initial stability of an implant is critical. We know from at least three radiostereophotogrammetric studies that it is possible to predict if an implant is going to fail or not. There are a few reports of this in the interface between pelvic bone and implant, and it is possible that in some cases this is the primary event leading to socket loosening. Polyethylene may partly be, in such cases, a secondary phenomenon.

WM I agree that there is probably more than one mechanism for failure. I can tell you that on the autopsy studies of cemented acetabular cups, there is a linear correlation between volumetric wear of polyethylene and the stability of the cemented acetabular socket.

ÅC What is your opinion on the the holes in the metal shell?

WM There is no evidence to suggest that the holes in the acetabular component are a significant problem. I have looked at some 900 cups with a minimum 5-year follow-up. They all have holes in them. As you know, the occurrence of osteolysis is 3% in that group of patients. We have a lot of cups with no holes that have osteolysis, so although it is a potential pathway for fluid and debris, it certainly doesn't tell us the whole story. I think what we need is to limit the effective joint space and to get osseointegration around the periphery of that implant. Maybe with the press-fit techniques that people are using to get better bone contact at the mouth of the acetabulum, we will get better osseointegration there and have a better long-term success rate.

You mentioned the wear on the backside of the cup. Backside wear is a little different than wear generated by the articulation. The size of the particles on the backside seems larger, based on limited analyses that we've done, when compared to the size of the particles generated from the articulation itself. Those are about one half micron on average, with the ones on the backside being quite a bit larger than that. That probably influences the biologic reactions. I think backside wear is something to be concerned about, but not the major issue.

References

1. Schmalzried TP, Maloney WJ, Kwong LM, Jasty M, Harris WH. Autopsy studies of the cement-bone interface of well-fixed cemented total hip replacements. J Arthroplasty 1993;8:179–188.
2. Engh CA, O'Connor DO, Jasty M, McGovern TF, Bobyn JD, Harris WH. Quantification of implant micromotion, strain shielding, and bone resorption with porous-coated anatomic medullary locking femoral prostheses. Clin Orthop 1992;285:13–29.
3. Engh CA, Hooten JP, Zettl-Schaffer KF, Ghaffarpour M, McGovern TF, Bobyn JD. Evaluation of bone ingrowth with proximally and extensively porous coated AML prostheses retrieved at autopsy. J Bone Joint Surg (in press).
4. Engh CA, Zettl-Schaffer KF, Kukita Y, Sweet D, Jasty M, Bragdon C. Histological and radiographic assessment of well-functioning porous coated acetabular components: A human post-mortem study. J Bone Joint Surg 1993;75A:814–824.
5. Jasty M, Maloney WJ, Bragdon CA, Haire T, Harris WH. Histomorphological studies of the long term skeletal response to well-fixed cemented femoral components. J Bone Joint Surg 1990;72A:1220.
6. Jasty M, Maloney WJ, Bragdon CR, O'Conner DO, Haire T, Harris WH. The mechanisms in the failure of fixation of cemented femoral components. J Bone Joint Surg 1991;73B:551–558.
7. Kwong LM, Jasty M, Mulroy RD, Maloney WJ, Bragdon CA, Harris WH. The histology of the radiolucent line. J Bone Joint Surg 1992;74B:67.
8. Maloney WJ, Jasty M, Burke DW, et al. Biochemical and histological evaluation of cemented total hip replacements: A study of autopsy retrieved femurs after in vivo cycling. Clin Orthop 1989;249:129.
9. Maloney WJ, Schmalzried TP, Jasty M, Kwong LM, Harris WH. The cement interface: Retrieval studies. In Morrey BF (ed): Biological, Material, and Mechanical Considerations of Joint Replacement: Current Concepts and Future Direction. New York: Raven Press; 1993.
10. Schmalzried TP, Jasty M, Harris WH. Periprosthetic bone loss in total hip arthroplasty: The role of polyethylene wear debris and the concept of the effective joint space. J Bone Joint Surg 1992;74A:849–863.
11. Schmalzried TP, Kwong LM, Jasty M, et al. The mechanism of loosening of cemented acetabular components in total hip arthroplasty: Analysis of specimens retrieved at autopsy. Clin Orthop 1992;274:60.
12. Maloney WJ, Jasty M, Harris WH. Bone lysis in association with well-fixed cemented femoral components. J Bone Joint Surg 1990;72B:966–970.
13. Willert H-G, Semlitsch M. Reactions of the articular capsule to wear products of artificial joint prostheses. J Biomet Mater Res 1977;11:157–164.

14. Maloney WJ, Jasty M, Harris WH, Galante JO, Callaghan JJ. Endosteal erosion in association with stable cementless femoral components. J Bone Joint Surg 1990;72A:1220.

15. Tanzer M, Maloney WJ, Jasty M, Harris WH. The progression of femoral cortical osteolysis in association with a cementless total hip arthroplasty. J Bone Joint Surg 1992;74A:404.

16. Woolson ST, Maloney WJ. Cementless total hip arthroplasty using a porous-coated prosthesis for bone ingrowth: A 3-1/2 year follow-up. J Arthroplasty 1992;7:381–387.

17. Zicat B, Engh CA, Gocken E. The radiographic assessment of osteolysis around total hip arthroplasty using a cementless femoral component evaluation of cases after 7 years. J Bone Joint Surg 1995; 77A:432–439.

18. Maloney WJ, Smith RL, Schmalzried TP, Chiba J, Huene D, Rubash, H. Isolation and characterization of wear particles generated in failed uncemented hip arthroplasty. J Bone Joint Surg 1995; 77A:1301–1310.

19. Jasty M, Bragdon C, Jiranek W, Chandler H, Maloney WJ, Harris WH. Etiology of osteolysis around porous coated cementless total hip arthroplasty. Clin Orthop 1994;308:111–126.

20. Maloney WJ, Peters P, Engh CA, Head WC, Chandler H. Severe pelvic osteolysis in association with uncemented acetabular replacement. J Bone Joint Surg 1993;75A:1627–1635.

Chapter 18

General Discussion

Göran Lundborg and Richard Skalak, Co-chairs

Göran Lundborg In these past hours, we had a chance to bring up topics that have not been sufficiently covered before and to delineate some wider perspectives in the field of osseointegration. We defined three main topics.

First, basic mechanics and mathematical models. What is the relevance of this?

Second, we have been examining the osseointegration process. How should it be defined? How does it work, with reference to material, techniques, age, and bone marrow? How does it work in the long term? What are the risks and negative factors (eg, aging, radiation) that may make the osseointegration process difficult? What are the positive factors? How could we influence the process? We have heard about the matrix molecules and the growth factors. Are there ways to stimulate the osseointegration process by, for instance, vibration? Are there perhaps other ways? What is the role of bone grafting in this process? Are there possible complications of our implants in terms of malignancy and allergic reactions in the long run?

Third, we all like to hear more about the clinical applications. Two main fields are tumor resection and spine surgery. We have more to discuss here, I feel, than what was done before. What is the role of osseointegration in this kind of major reconstructive surgery?

Regarding limb prostheses: when are osseointegrated prostheses used? What are the indications—when are they superior to the ordinary types?

We could spend hours discussing many interesting points here, and I suggest that we start with the concept of osseointegration. There are various ways to define osseointegration from the viewpoint of the patient; macroscopic and microscopic biology and medicine; macroscopic biomechanics; and microscopic biophysics. So, I think we should start with the discussion on this.

Richard Skalak I would say that these definitions are developed by historical evolution and attempt to answer different groups who ask for more or less exact definitions. These are really suggestions to be considered, and we would be interested in any comments or further suggestions from the group.

Per-Ingvar Brånemark I think we have to at least recognize that there might be some common denominator, and then different people could use the term related to their particular interest and understanding. As I mentioned before today, it is a systems approach. It is a system of hardware and software, and I believe the only meaningful way to verify if it does or doesn't work is long-term clinical studies. That is the aim of the game anyway. We know much too little about it.

Let me give you an example. There has been an attempt to simplify the matter by getting some kind of numerical appreciation of integration in sections of bone and titanium to measure how much of the bone is facing the titanium. I don't think that matters very much at all. I think the qualitative aspect is much more important. So that's why I think we brought it up. It is important to remember that this is complex, and no single parameter can encompass the idea.

I would like to say, in addition, that the ideas on this should be used in clinical reality. Patients are not alike. For instance, the work we have done with Dr Lundborg on finger joints is certainly not on normal bone tissue. That is why it is so difficult to transfer clinical reality from controlled biologic experiments.

GL The fixture is osseointegrated if it provides stable and apparently mobile support over the lifetime of the patient. But are we able to define something as osseointegrated while the patient is still alive, or must he be dead to provide a biopsy specimen? Is there a time limit when you can be sure that things are osseointegrated: 10 years, 15 years?

PIB We know that we waited for 10 and 20 years in the maxillofacial region, and we didn't see any difference between 10 and 20 years. Now we don't see the difference between 20 and 30 years. So it may be that after 10 years, we have identified the level of success, and that is something that is very important to bring up here. Is there a potential for long-term problems because of changes in the marrow cellularity, or because of allergic reactions or tumors and so forth? I think that is the key issue here. When do we know for sure that it will not do any harm? That is why a lifetime limit is suggested.

Dick Heinegård To me, integration means that the prosthesis becomes part of the body. It doesn't swim around as a foreign body. Of course, the way to assay that is functionally. An assay is just as good as the sensitivity, so the true outcome is permanent function.

Now, the problem is, of course, that there are bone changes. We have osteoporosis in the elderly, and we don't understand the mechanics all that clearly. But of course, that is a different situation. So, if we place an implant in an individual of a particular age, that implant may be well suited to interact and integrate in the bone at that age. But what happens 50 years later when the bone is different? Will it still be osseointegrated? I think there are a number of issues that one needs to think about in this case.

GL Yes, we have the age aspect . . . How about the osseointegration of prostheses when you perform an implantation in a young individual vs an aging individual? What are the differences, P-I?

PIB I think we have a friendship between the implant and bone in the craniofacial regions. Age is important. But then, in this part of the body, we don't see any negative effect of that long interval, even in those who are edentulous for 50 years. They heal as well as the young ones. That might be peculiar for the mandible. This is something we should relate to osseointegration prognosis. The different cases of long-term amputees, rheumatoid patients, and others must each be considered separately.

William Maloney In at least the porous-coated systems, where you are getting fixation to the bone, the age of the patients doesn't seem to be a factor either. As I said, Dr Engh takes all comers. Radiographically, they have signs of being osseointegrated, despite having some cases with very advanced osteoporosis. I think, at least in the hip situation, that if we get an implant that is stable, it will osseointegrate.

I think one of the major unanswered questions for us is, what surface area do we need to osseointegrate to be able to take specific loads over the lifetime of the implant and not get failure within the bone itself? I think that is somewhat dependent on the interface mechanics, whether we use screws, porous coating, or cement. That is important for implant design, especially cementless implants, because the longer the surface of porous ingrowth, the more distal you get fixation and load transfer, and the more bone resorption you have proximally as a result of that.

Tommy Hansson Dr Maloney almost said it: osteoporosis hasn't shown to have any effect on remodeling in itself, so even if something is breaking loose because of osteoporosis, that doesn't necessarily mean the principle for osseointegration is changed. We talk of osteoporosis as a normal condition, but rheumatoid arthritis is a pathologic condition. So, it is hard to say that age or osteoporosis necessarily affects the remodeling process when we talk about osseointegration.

WM It appears that despite osteoporosis, if the bone is being loaded by the implant, it will favorably remodel to actually increase stability over time.

Lars Magnus Bjursten The question is, are we allowed to use the term *osseointegration* for an animal experimental system, or is the term confined to a clinical setting?

DH I don't know the answer to that question, but I think it is fundamental if we consider it as a clinical procedure. That tells us that we have to put it in the perspective of the functionality of the patient, not to a specific histologic feature.

PIB I believe it is very important to consider that this is intended to provide stability for prosthetic replacement, whether it is a tooth or a foot. We have come into a conflict here; namely, we would like to achieve as good a primary stability as possible, but on the other hand, we don't—we want to avoid stress shielding.

GL If you achieve primary stability, can various materials, including cement, then become osseointegrated?

PIB I don't know if it has to wait for a 10- to 15-year follow-up, but it might. I understand from Dr Heinegård, Dr Bjursten, and others that, yes, it might well be so. We are not married to one particular material. We are just trying to recognize the systems approach.

Anon I guess I am little confused because prior to this meeting, I had in mind that *osseointegration* was sort of a chemical-biologic term. Now I sense that you are adding the biomechanical aspect, which means, as Dr Maloney has been using it and Dr Coutts used it, that it could be cement. We put down cement, and the bone is not moving. It is critical using the term *osseointegration*, because there is no osteolysis. My sense is that you create a different entity by placing titanium, and this may have a delayed reaction.

PIB I think it is what we are getting—instead of how we do it—that's important. Because this is a clinical problem, the aim of it is a clinical rehabilitation procedure. I don't think we can limit it to one particular situation.

WM Methyl-methacrylate in bone becomes osseointegrated too, by your definition.

PIB I don't know yet; let's wait 10 to 15 years.

TH We have the time aspect to consider also.

WM The oldest cemented specimen we tested is 17 years old, and I don't think you have a titanium specimen that is 17 years old.

PIB Longer, 30 years.

WM The cemented prosthesis lasted for the life of that patient; she died 17 years later; it meets your criteria for length of service.

PIB But that is why we add the other criteria here.

WM The cemented prosthesis meets all four of them.

LMB I want to make a functional definition of biocompatability. It should perform the intended function for the required time, which is usually the life of the patient. I think the best way to judge is in functional terms.

Björn Rydevik If you look at the available clinical data in various systems, like the osseointegrated implants according to the Brånemark principle and many others from the orthopedic field, there are not many reports. Some information available, as Dr Maloney pointed out, may suggest you achieve osseointegration, but I think there may be differences in terms of the predictability.

We have to look carefully at clinical evaluation series to see what percentage of cases of a given procedure you could expect to achieve stable long-term fixation osseointegration. There may be differences as judged from clinical data. There is very strong evidence about the high predictability sink using commercial pure titanium, as it has been applied mainly in the maxillofacial region and the cranium region. I think it is not only a matter of a possible time factor, but also the level of predictability for the outcome that should be considered.

WM In noncemented hip implants, titanium has been a poor performer for a variety of reasons. Coated-porous titanium has poor properties in terms of its ability to withstand load if it is not osseointegrated. The "dirtiest" of all the joints we have seen in revision surgery are from titanium implants. Large amounts of fine titanium debris are found at distal sites including the lymph nodes, the spleen, the liver, and so forth. In the short run, with our current implant designs, titanium has no advantage; in fact, there are several disadvantages.

PIB We have to identify the hardware components, but particularly what are we using titanium for. I still believe that the particular kind of titanium is really good for the anchorage to the bone when it comes to wear debris and so forth. That is a very different issue, of course.

WM You can't separate them clinically.

PIB No, not for these cemented prostheses.

Robert Pedowitz The mechanism must not wear out if lifelong success is expected. Is it realistic to expect the hardware to necessarily last if it is osseointegrated over the life of the patient? With the teeth, you would expect some to break eventually. So it seems like it is an overly rigid expectation for lifetime function.

PIB You know, osseointegration was defined with that time schedule for the craniomaxillofacial region, and there we have reason to believe it will last a lifetime. When it comes to other parts of the body, we should consider using the same definition, or have different definitions for skeletal and extremity bone, and different load requirements.

TH For expected purposes, that is a very good definition. I mean that there is no definite time limit to the expected use of it.

WM I could say one thing about the cemented components, which is that on the femoral side, even if it is osseointegrated, it can still get into trouble over time. We see the debonding at the metal-cement interface and fractures propagating from the metal-cement interface through the bone cement.

Fracture is associated with pores in the bone cement. This has been published in the British *Journal of Bone and Joint Surgery*. It is not the cement-bone interface that fails initially; it is a long-term sequela of failure of the material itself. I think there are some material deficiencies with cement that may be inevitable, and we may have reached our limitations in terms of the use of bone cement.

GL I suggest we move to discuss factors that could influence the course of osseointegration. Say negative factors here, such as poor bone quality. What are the limits? What clinical experience do we have as to when osseointegration can still occur? I am talking about conditions, such as irradiation, patients receiving a lot of drugs, rheumatoid patients, etc. I know, Dr Brånemark, that you have a lot of experience with very bad bone quality.

PIB Yes, the intriguing thing is that osseointegration works in almost all cases. Even in severely resorbed bone it works, and I have been an observer in dentistry for many years. I would like to add the risk factor of smoking. Smoking, at least in this part of the body, really has a very strong negative impact on establishing and maintaining osseointegration. That is something one should really take note of.

WM In the hip, the only contraindication we have seen is radiation. I have reviewed 900 acetabular cups, and the only ones that were not radiographically stable and failed were two cases that had radiation therapy in the pelvis.

PIB Dr Maloney, did you check on those who failed as to whether they were heavy smokers?

WM No.

PIB It might be wise to do that.

GL How about the issue of the loading? We have discussed this many times. On one hand, we hear that 6 weeks of unloading of the skeleton makes bone density decrease enormously and it never comes back. On the other hand, we hear that the implant should not be loaded during the first phase. So how can we reconcile these things?

PIB I don't think that is correct. What we say is there should be no relative motions, that it must be absolutely stable so the interface can be created in a proper manner. How that is done is a matter related to where in the body the fixture is located and the circumstances.

GL Dr Worthington was very keen on having an unloaded period.

PIB Yes, in the mouth. But there is some load there. If you put on a denture there is a load. If you look to the maxilla for bone grafting, and if they put in the dentures prematurely, we lose the fixture. So I believe it is very important to understand there is a load, less controlled than that in finger replacements, but there is still a little.

GL So you can tolerate some load but, by controlling it, it becomes a positive thing.

PIB Yes, by controlling the circumstances so there is no relative motion.

WM Would you agree that site-specific, local conditions would dictate your initial requirements for stability?

GL Yes, there is the requirement of no relative movement, which may require site-specific measures.

RS I think I should call in my colleague, Dr Cowin, to comment on this because he has worked on resorption and accretion theories.

Stephen Cowin We always assume we are dealing with healthy bone in models. We always assume that we have normal tissue, whatever that is, and we never considered the problem of having some kind of fragile tissue there. That is something from this meeting to think about. In fact, I was wondering how you assess the tissue before you operate. How do you know the degree of fragility of the tissue?

TH Clinically, we have no idea, ie, we have no quantitative scale. We just estimate in each case whether it is strong enough or not, and usually it is. I mean, in the virgin case of a hip or spine, or whatever we are talking about, rather than any revisions.

GL If we think it is not strong enough, we add healthy bone by doing an autologous bone graft.

SC Let me go on and try to get to Dr Skalak's point. There is a range of strain that the bone accepts as normal. Above that range it will deposit, and below that range it will resorb. If the strain gets very high, the bone again begins to resorb. This is some sort of overstrain necrosis. So, if you have the actual structure of the bone, you can then calculate which of these ranges you have, and that might be a way to assess beforehand if you know what the fragility of the bone is to begin with. Then, with increasing load, you may be able to compute whether or not you would get a good response.

I think this loading, when you are looking at the interface, boils down to what kind of stability exists at interface. If you can provide a load that will not disrupt the contact between the interface cells and the material surface, then you are well off. So that is the limiting factor from getting and maintaining a good contact between the material and the bone tissue.

GL Speaking of loading and mathematical models, what we have not discussed sufficiently is the optimal design of the implant. We heard Dr Skalak describe the various kinds of stresses in longitudinal and transverse directions. How should these results be applied to the design of the implant? Is it an advantage to have a long implant or a short implant? You could think of forces applied sideways or longitudinally. We thought, in the beginning at least, that the very long implants were providing the best stability and the best osseointegration in the long run, but I am not sure if that is true. Dr Skalak, do you have any comments on that?

RS Yes. Of course when one sees the very large resorption of bone in the distal portion of the stem of a hip prosthesis in the femur, the logical reaction is that the stem could have been shorter in the first place. Then the resorption might not be so extensive. I think that the designs could be improved in this direction, although I don't have a detailed design on hand. But that strikes me as one of the things that should be given a lot of attention to right now—the biology of the bone resorption to get more accurate loss data and to try to shorten the prosthesis. I think it would be adequate, less traumatic, and it would give you a lot more flexibility. So, my opinion is that there are possible improvements in this direction.

PIB Could I ask the orthopedic surgeons here if, when you do a revision, you use shorter stems? I noticed that Dr Coutts showed one revision where there was a fairly short stem as compared to the medium one. Is that a rule or is that just one case?

Åke Carlsson In early years, there was a tendency to use longer and longer stems for every new revision. Today, you want to use as short a stem as possible. This can of course be compromised by bone loss. If you have bone resorption, this must be taken into account. As a general rule, in my opinion, there is no reason to go for longer and longer stems.

WM I agree 100% to that general principle. You have to achieve initial implant stability. You use the shortest stem that you can and still achieve that goal.

TH I just wondered who decided the length of hip prostheses. Was it Charnley or Ferar? Then we have this stress shielding by the tip of the stem. Actually, you would think a shorter prosthesis might be better.

WM With cementless designs, it again gets back to your point about reliability. You know if we can do it in one case it is not as important if we can do it in 95% or 99% or 100% of the cases. With cementless designs, the highest success rate is when you get fixation in the diaphysis. You must get down at least to the metaphyseal-diaphyseal junction. When you use an implant that is designed to be fixed above that junction, then your failure rate goes up. The average surgeon has to be able to do it. If it is not simple, you will have a lot of disasters.

DH I believe the material properties, and the difference in material properties between the bone and the implant, will be of importance. I guess that you may end up having osseointegration, and then there is a firm interface in at least one location on the implant. When you start out, you may have a good fit over a long region that will distribute a load over the initial phase, and perhaps one of the functions of that would be to unload the interface, much like you do in your dental implants. For that, you really need a fairly long stem. It is something that needs to be thought about.

GL So do we know what the interface looks like? Is that defined, Dr Bjursten?

LMB I think what it looks like when you have a biologically inert material is fairly well defined. Then it usually looks about the same. You have an interface to the bone that always has an oxide on the metal, and often that oxide layer is growing over time. It also incorporates calcium and phosphorus.

I would say that you have an amorphous layer of thickness that tends to vary a little bit, but is usually about 100 Angstroms (A). The nature of this layer is not known at present. I think that this layer may very well function as a kind of shock absorber or transition between the different material properties between the bone and the implant material. I wouldn't be surprised if the thickness of this layer depended a little bit on the loads applied to the implant as well as the dimensions of the implant.

I would say that you usually have a well-classified bone structure without any fibrous tissue present immediately adjacent to the 100-A amorphous layer. That is the general outline of the interface to an inert material.

PIB What about the possible special function of titanium in forming a peroxide gel? Is this a real event, or is it just an academic observation?

LMB If you look at different surfaces, I think that you have to distinguish between ones that are inert and ones with biologic functionality. Many materials have a negative biologic functionality, for instance, degrading materials. One example is when you have particle formation. The particles themselves have a biologic functionality often associated with bone resorption. If you are looking at titanium, it has one feature that distinguishes it from most other materials, and that is its handling of free radicals because it functions as a trap for free radicals. Maybe that is of importance in the healing period of an implant. But of course, in the future, we could look at other more sophisticated surfaces like the ones that Dr Heinegård presented here. We may have a positive kind of biofunctionality.

PIB But if you have a systemic disease or a local disease so that you have a lot of free radicals, might this then be more important than in the normal level found in healing tissue?

LMB Yes, I think so. It may be significant if you have an infection because the handling of oxygen by titanium interferes with the formation of radical species that are in fact bactericidal products. So it may be an issue when you have infections associated with implants.

GL In addition to considering the material, loading, and surgical technique, what else could we do to facilitate the osseointegration process? Dr Heinegård, you mentioned matrix molecules and adhesion factors. How about growth factors here? Could you comment on that?

DH Yes, I think there is new technology developing. The BMPs or a number of the BMPs have been cloned and some of them are in clinical trials, trying to recruit cells that make bone.

There is another approach, and that is using mesenchymal stem cells. With age, we have fewer of them. There are techniques to identify these mesenchymal stem cells, take some of them out, and expand them in culture. Then we could potentially put them back because they can go on to form bone, or they can go on to produce cartilage, depending on how we trigger them. So I think there are many possibilities to modify events occurring at this interface.

TH That is actually what happened with the cartilage transplantation in Gothenburg. They took out cartilage cells, not stem cells. They grew them in culture and triggered them, so they continued to be cartilage cells.

DH I think that is different. I think that the key feature in their experiment was the periosteum that they used to cover it. There you have the mesenchymal stem cells. This is an old technique by Coburn or someone else.

TH But just using the cells doesn't help.

DH They do form cartilage.

TH Yes, but people have been trying that for many years. The surgical technique is not the key because that has been worked out. It has been tried for many years and it hasn't worked.

DH It could just be a matter of the cells surviving in the system.

TH Yes, it could be.

PIB Would it differ between old and young patients when you refer to the periosteum? Because in the older patients, periosteum is not supposed to regenerate.

DH The older patient has less mass of stem cells?

PIB Yes.

DH So the idea of this new approach is to really take the mass of mesenchymal stem cells. You can take them from peripheral blood and expand them in culture, which has only become possible very recently.

PIB Would it be even better to take blood from the marrow?

DH Perhaps.

GL I think we should turn to the discussion of the bone marrow now.

Dennis Osmond This discussion raises the point that the same principle could well apply to stem cells from bone marrow. There are techniques for culturing and expanding bone marrow stem cells capable of supporting normal hemopoiesis.

We are also just beginning to understand the nature of cell surface interactions within the marrow microenvironment, and a number of cell adhesion molecules are being identified. One could expand stem cells and insert them with the implant. On the other hand, one could theoretically treat the implant surface so that it might attract the marrow stem cells. It is interesting that the bone marrow has been taken so much for granted that it has been assumed it will develop normally at the implant site. By and large, that seems to be what is happening. But maybe we could actively encourage local stem cells to migrate to the implant region.

DH I think that is one variable here as to how much or how well you provide suitable marrow when you do this surgery. These most uncommon stem cells will also go on to differentiate into bone marrow stem cells, so they could be very useful in many ways in this system.

GL So what do we actually do when we do the bone grafting procedures? With the bad bone, for instance, in the rheumatoid patient, we bring in fresh bone marrow from some other part of the body and reinforce the bone, giving mechanical stability. But, we probably also bring in many other things. What do you say, Professor Brånemark?

PIB I don't know. I believe that the marrow is the mother of bone. That is not scientific, but I believe in it. It worked very well in the rheumatoid patient, and when you see the bone quality of those cases, it is remarkable how well it works. So I believe that marrow spongious bone is important and the key in those cases.

Could I also suggest that we consider the circulation in the bone, particularly in the endosteal parts? Is it possible to preserve it or be very kind to it? Is it possible to improve the circulation in the bone; in these small bones and in the long bones? Has anyone considered circulation in the bone of the hip, for instance? In relation to the outcome of the procedures, that might be one important factor.

Björn Gunterberg As Professor Brånemark mentioned, smoking was a very negative factor. I think that is correct and relates exactly to the circulation.

PIB Thank you very much!

BG It is a factor in any type of surgery. You have bad results related to smoking.

GL When we are performing bone grafting procedures, do we bring in living bone, or do we bring in dead bone? Does the bone graft survive? Is it so that we introduce our screw into a piece of dead bone that first has to be revascularized before the osseointegration process starts? Or does it start right away? We have heard that you shouldn't exceed a gap over 40 μm between the implant and the living bone to get the osseointegration process to work. Now when you bring in graft bone, how does it work?

PIB Well, I might not agree with those dimensions, but since we have an open circulation of the marrow, it may be that there is a very early recirculation. We know for sure that in a skin graft there is an open circulation after 24 to 48 hours, and that might be the reason why this works very well.

GL I think we should switch over now to the clinical implications of the osseointegration principle. How about the tumor resection situation, Dr Gunterberg? What are the possible implications, and what are the future perspectives?

BG As I outlined yesterday, there are a few very imminent applications, for instance, the alternative to amputate or not to amputate. There you have a question to answer because we believe that maybe the amputee will have a higher quality of life. This needs individual assessment every time, of course, and if you can't predict the outcome with an osseointegrated prosthesis reliably, it is difficult to make this decision. But I think in the future, we may be able to really outline to the patient what quality of life he or she may expect with this technique. Some of the prosthetic replacements in tumor surgery may benefit from osseointegration.

Now the prostheses we use are supposed to osseointegrate, and I believe some of them do. But the results do not indicate that this always happens, so it is a little unpredictable. There are of course new areas of development in this perspective. I showed you pelvic resections, which are now fairly well reconstructed. I think much improvement can be developed here, and possibly, we can preserve joints in a more reliable way. I could also touch on problems of the spine, but I think this is not appropriate here.

WM I was just thinking about the role of osseointegration in tumor resections. In the resection arthrodoses, we use modular nails above and below the resection. It makes sense, thinking about what we have done with cementless stems, that we can shorten the nails quite a bit and achieve good initial stability. We could do resection arthrodoses using osseointegration as a method of fixing instead of using nails. I suspect you could probably apply this to many of the joints in the hand as has already been done, and to the wrist as well.

BG There are many methods of achieving this resection. One way is using long enough nails with bone graft plates or other methods. But if I may elaborate on that, I think that if you can have a spacer that is osseointegrated, then you don't really need to have an extensive incision. So maybe you can save a lot of trouble and surgical trauma with osseointegration.

PIB I remember now that many years ago Ed Chao suggested that kind of approach. I don't know what happened, but apparently we may have more information from long bones and extensive experience from the craniomaxillofacial region.

GL Moving into the limb prosthesis business: Dr Childress, what would your thoughts be on the indications for using the osseointegration approach?

Dudley Childress I only have the perspective of what I have seen and heard here. From what I hear, you basically are considering osseointegration for short, above-knee prostheses, which are very difficult to fit, or have some problems with conventional kinds of fitting. What I hear is the rationale for doing this now. But as I say, I do not actually have experience with the method.

BR I think the initial approach that was taken was somewhat conservative—carefully selecting patients who had documented problems with the standard socket prostheses. There is also the possibility of major failures where you would have some kind of recourse. However, within the evaluation program, we are trying to assess the functional level to which these patients can be rehabilitated, thereby possibly giving us insight into the indications and possibilities of expanding into other areas. But that also takes into account muscle function, motor control, and the afferent side of the osseoperception phenomenon. We are carefully evaluating each case now, and hopefully we will be able to define in the not-too-distant future the indications in more detail.

DC I think this is the way the endoprostheses started. At first, they were the only surgeons willing to do hip surgery for people who were 70 years old. As time went on, they found out how well it worked and that it was a logical procedure. The only problem is that in the hip replacement case, there were more patients. The numbers were larger, and so it would be desirable if the experience could be expanded beyond the borders of Sweden.

GL Dr Brånemark, what would we say are the main advantages of an osseointegrated limb prostheses as opposed to the conventional types?

PIB With osseointegration, the patients feel that the prostheses are part of their bodies.

WM I think in the hip prostheses patients, especially a young patient who is active, you can get a sense that "osseoperception" is even playing a role there. And if you get a young active patient, he or she will say: "I have forgotten that you have done surgery. This hip is part of me. I can do everything." Even though we tell them not to, they do everything. They play tennis and never sense that the prosthesis is not a part of their body. If you look at the radiograph, it is obviously an osseointegrated stem.

On the other hand, you may have a patient who says: "You know, I am doing pretty well, but I can feel something at the end of the day. I think I have muscle strain. I don't like going upstairs, particularly." You look at that radiograph and see an implant that is unstable or has fibrous ingrowth. So I think even in the hip patient, you can sense the role of osseointegration.

I want to ask you a question: How happy are you with the current state of the art standard BK prostheses? Are those patients very enthusiastic or interested in the techniques presented? Do you think that is something that you are satisfied with at the current level, or do you think that there is still room for significant improvement?

DC I don't think any one of us is ever satisfied at any level. Certainly, we are expecting much better prostheses than what we have. The bearing of weight by soft tissues that were not designed for that, I think, has always been considered to be the wrong approach. But we did not have alternatives. This is what osseointegration gives—a new alternative.

BG As I see it, the main problem with any amputee is not the prostheses; it is the contact area between skin and socket where you have the problems: volume, skin friction, and many others. That is why the patient takes a prosthesis off as often as possible. They use it for function, not for rest.

DC Not only is the interface troubled, but the whole tissue as well.

GL Do we understand the phenomenon of osseoperception? Do we know what it is? Is bone a conductor of something to be registered by exterior structures, or are there sensors in the bone?

PIB I don't know. There could be mechanoreceptors at nerve endings that could be the sensors. I think this is just a new theme. It is remarkable what amputees can do, so I think we should make use of it. Not only necessarily in major prostheses, but perhaps in the finger prostheses as well.

GL Whatever the sensors are, the end point of reception seems to be the brain, which is able to process whatever afferent information comes to it and make use of it. This plasticity is probably more pronounced in the younger than in the older patients.

DC I would be surprised if these signals projected to the brain. I would think they would project in the spinal cord. There may be perhaps some autonomic action there. But the numbers of receptors are tremendous within muscle, tendon, and skin, of course, compared with the receptors in ligaments and cartilage. I expect that to be the same even compared to bone, so I suspect that proprioception is due to other aspects in the muscular system.

The system is very sensitive to positioning, and so if very minor positions were changed, a signal might be generated. If you push on the bone, you may change the position ever so slightly. It doesn't have to be the bone that is responding. It is all these things that are attached on the bone that basically integrate all this information.

GL It seems there are many things to explore in the coming years.

PIB Could I ask Dr Osmond to comment on the role of the bone marrow?

DO The bone marrow has a rich innervation. Its function has long been a matter of some mystery. Nerve fibers come into the marrow from periosteal nerves and along the nutrient artery, distributed along with the artery. Some are nonmyelinated sympathetic nerves, which have a motor effect, encouraging release of particular blood cells from the marrow, and a constrictive action in the smaller arterial vessels.

There are also rich supplies of myelinated nerve fibers that appear to peter out as naked nerve fibers throughout the marrow. Their function remains unknown, but they are generally thought to be afferent fibers, perhaps contributing to the sort of discomfort that one feels during bone marrow aspiration, for example. Concerning the formation of a secondary medullary cavity around interpreted prostheses, one may wonder whether these nerves may not have some relevance in this situation.

LMB I think that the nervous system has an enormous capacity to use the secondary signals if the primary is not there. For example, after impairment of the system of vision, you use your ears or your tactile senses when you cannot use your eyes. I think that when you amputate a finger, the brain tries to substitute the tactile senses of that finger using any information source that is available, like vision or the proprioreceptors of the muscles, etc. You try to gather information, and as time goes by that kind of substitution becomes subconscious. You don't actually notice that you are doing it, and you believe that you actually are sensing with your fingers.

GL If you use one finger often, for instance, when you become blind and start reading in braille, that finger will expand its projection enormously in the sensory as well as in the motor cortex of the brain.

Moving to the endoprostheses, what would be the optimal joints in the future for application of this principle? You have to take into consideration that techniques we are using today are working well. What kind of joints will be optimal to have the osseointegration process take place? We have experiences now from the finger joints and from the ankle joint. What else is needed now?

ÅC At present, I think for the ordinary patient, the conventional prostheses for the hip and knee are functioning very well. But there are certain situations in which we lack the optimal implant. Here, I think the principles of osseointegration could be applied. One example: the proximal femur may have an abnormal anatomy as a result of congenital dislocation of the hip or other childhood diseases; therefore, one may be forced to use a custom-made prosthesis or find another solution.

There are also other difficult joints where we do not have good solutions at present. One example is the elbow, and another is the ankle; both are heavily loaded joints. There is no really good implant available for the time being, so for these joints, the principles of osseointegration may have a future role.

PIB With some of the ankle joints, osseointegrated fixtures may be the right solution. Part of the problem is our old thoughts about fixing the knee and the elbow due to the complexity of the motion. In the knee, there is a screw mechanism, and in the elbow, there is rotation. I think those add to the problems in the fixation. To adjust that, you might have to redefine the anatomy and try to eliminate some degrees of motion. I mean, in the knee and perhaps in the elbow, to make the motion just depend on pure flexion and extension, try to avoid the supination and pronation and make it flexion-extension. Maybe the rotation may lead to failure, but you have the ability to redefine the anatomy and control.

ÅC Especially in the elbow, you often have problems with very poor bone. Most of the patients that require elbow replacement suffer from rheumatic disease.

WM I guess I have a dissenting opinion. It is my bias that in any joint, in a high percentage of the cases, an implant that is not osseointegrated is not going to work. Regardless of the implant material or the surgical technique, a successful implant in active patients is osseointegrated or it won't be a successful implant.

I think, at least in the hip, we have two main problems: One is the revision case, in which we need a way to build or replace bone stock and ideally get osseointegration at the graft-implant interface. I think that we are going to need some help from biologists.

The second area is that the most reliable cementless stems are the ones that are fixed in the diaphysis. It happens to also be the worst situation in terms of mechanically induced stress shielding in the proximal femur. We again maybe need help from the biologist. We need to design systems that will give us a higher percentage of osseointegration in the metaphysis, and therefore avoid some of the mechanically induced bone loss.

GL Dr Brånemark, can we have a final remark from you on endoprosthesis perspectives?

PIB I think the important question is related to whether there is a need for improvements or not. I believe that true osseointegration with the right bone function is the best solution, but that has to be proven, and it will take another couple of decades.

What is surprising, as I see it, is that it has worked in very severe rheumatoid cases. We have a few ankle joints of severe rheumatoids where it also seems to be working without many problems. Maybe, when we learn more about how to modify the healing in the bone and add whatever we add to optimize bone regeneration, that might give us a routine procedure. At present, I agree that it is not working routinely. Then it is just a luxury, and that is not cost-effective. I think this is the most important question: Can we make this a routine, predictable procedure? I think we probably are there in finger joint rehabilitation.

GL Thank you. With this, I think we should close this general discussion, and it is my great pleasure to leave the final word to Professor Skalak, who is going to give some concluding remarks for this meeting.

Chapter 19

Concluding Remarks

Richard Skalak

Thank you. Perhaps you have noticed that I have been taking notes the whole time. I will try to summarize what I think I, as an observer, am hearing here. The interesting thing to me, looking back over 20 years or so, is that in this meeting I heard no challenge to the fact that osseointegration occurs. In fact, what I am hearing is claims that, historically, it has occurred often, and in many situations. So this is different from the early dental experience of osseointegration. Many people said it was not possible and that it would not last. So this is my first comment: that the concept of osseointegration seems well established, and that what we are talking about now is sort of an optimization and exploration of the possible extensions, benefits, and advantages. This is a second level of development and exploration.

It also seems to me, in listening to this meeting, that the key remaining clinical problems are the phenomena of stress shielding and wear debris. It is an unusual situation where an implant is placed and the bony structures to which it is attached begin to disappear while in use. I think we should try to design prostheses and procedures in which the stress shielding is not as severe a long-term problem. I don't have the solution on hand, but I think there could be better solutions.

The other major problem that has come out is the deleterious effects of debris production. I don't think the current design will be improved to the extent that debris will not be produced. So thinking about an engineering solution, it seems that the debris has to be contained in some kind of flexible envelope. The design of the prosthesis must be such that the debris cannot get into places where it can cause damage. We have a somewhat similar problem in designing earth dams, and we put in various membranes for a similar purpose. It is a matter of providing blocking membranes, but at the moment I don't have specific solutions.

Another thing that has come from this meeting is that the possibilities of using growth factors, genetic engineering, molecular biology, and cell biology really need to be considered. When I hear a detailed discussion on adhesion molecules of cells, mobility, bone marrow stem cells, hematopoietic reproduction, etc, it is clear that cellular and molecular biology need to be better explored.

The discussions in this meeting seem to be divided into two parts: First, there are discussions that proceed on a macroscopic basis: we say it resorbs, becomes inflamed, and so on. Second, there are discussions that say there was integrin there, and if you don't have the integrin, you don't get the adhesion effect. I think the molecular viewpoint needs to be brought in more fully.

Finally, I'd like to say that I think the field is ripe for continuing the process that Dr Chien mentioned at the beginning. The experience of osseointegration in dentistry should be useful for extension into the orthopedic field. This calls for the study of development of bones and of cartilage, but I think we should study the fibrous tissue generation as well. It appears that the behavior of skin at a transcutaneous prosthesis junction is much less well defined biologically or mechanically, compared to the status of integration to bone. I think it would be worthwhile to study, just as thoroughly as osseointegration has been studied, how to anchor skin in transcutaneous connections. It looks to me like a similar situation. In the dental case, anything that was placed in the way of an implant 30 years ago came out. Now it stays in place. I think the transcutaneous prosthesis can be equally successful.

Finally, a miscellaneous comment. In all of this work, I see that medical science knows a good deal about what I would call preventive medicine. A lot of good could come from this field if the knowledge that is accumulating here on how teeth, joints, and bones behave was somehow taught to the public to a greater extent. I'd like to close there and thank you all for the honor of giving you this brief summary and observations.